IEE Electrical Measurement Series 3
Series editors A.C. Lynch
                A.E. Bailey

Microwave measurement

Previous volumes in this series:

Microwave measurement

Edited by

A.E.Bailey

Peter Peregrinus Ltd
On behalf of The Institution of Electrical Engineers

Published by: Peter Peregrinus Ltd., London, UK.

© 1985: Peter Peregrinus Ltd.

Cover photograph courtesy of Peter Cochrane—British Telecom Research Laboratories.

The photograph shows a 6.1 m effect Cassegrain antenna used for slant path propagation measurements above 10 GHz.

ISBN 0 86341 048 0.

Printed in England by Short Run Press Ltd., Exeter

Contents

Preface

This book contains the lecture notes for the IEE Vacation School on Microwave Measurements to be held at Canterbury in September 1985.   The programme covers a wide range of topics and much of the material to be presented is based on recent research results.   It will be of value to those working in the field to have it brought together in permanent form.

Microwave measurements are necessarily made in distributed systems where the dimensions of the equipment may be many wavelengths.   We start therefore with a survey of the essential theory of transmission lines, waveguides and microwave circuits.   This is followed by chapters on the practical aspects of lines, matching and signal detection.   The next four chapters deal with measurements of the basic quantities power, attenuation, circuit parameters and noise, followed by descriptions of more specialised techniques for measurements in the time and frequency domains.   Free field measurements of antennas and power flux density are then dealt with.   The book concludes with chapters on more general topics: automation, the statistical treatment of errors and a brief look at the economics of measurement.

Our thanks are due to the authors who have taken the additional trouble needed to prepare their contributions in a form suitable for publication some time in advance of the Vacation School and to the Organising Committee who have done all the hard work of arranging a relevant and balanced programme.   The result of their efforts is a series of chapters which reveal an immediacy and expertise which could have been achieved in no other way.

List of contributors

S.Arnold,
MEL,
Manor Royal,
Crawley,
West Sussex,
RH10 2PZ.
(Chapter 20)

R.N.Clarke,
Division of Electrical Science,
National Physical Laboratory,
Teddington,
Middlesex,
TW11 OLW.
(Chapter 14)

Dr P.Cochrane,
British Telecom Research Laboratories,
Martlesham Heath,
Ipswich,
IP5 7RE.
(Chapters 11, 12 and 16).

Dr R.J.Collier,
Electronics Laboratories,
University of Kent,
Canterbury,
CT2 7NT.
(Chapters 1 and 4).

Professor A.L.Cullen,
Department of Electronic and Electrical Engineering,
University College,
Torrington Place,
London,
WC1E 7JE.
(Chapters 2 and 3).

C.H.Dix,
Crowgate,
Bishop Monkton,
North Yorkshire,
HG3 3QP.
(Chapter 22).

Dr A.E.Fantom,
Division of Electrical Science,
National Physical Laboratory,
Teddington,
Middlesex,
TW11 OLW.
(Chapter 19).

S.J.Gledhill,
Marconi Instruments Limited,
Longacres,
St Albans,
Herts,
AL4 OJN.
(Chapter 13).

E.J.Griffin,
Royal Signals and Radar Establishment,
St Andrews Road,
Malvern,
Worcs,
WR14 3PS.
(Chapters 5 and 6).

L.J.T.Hinton,
British Calibration Service,
National Physical Laboratory,
Teddington,
Middlesex,
TW11 OLW.
(Chapter 21).

Dr L.C.Oldfield,
Royal Signals and Radar Establishment,
St Andrews Road,
Malvern,
Worcs,
WR14 3PS.
(Chapter 7).

Dr R.D.Pollard,
Department of Electrical and Electronic Engineering,
The University,
Leeds,
LS2 9JT.
(Chapter 17).

M.W.Sinclair,
Royal Signals and Radar Establishment,
St Andrews Road,
Malvern,
Worcs,
WR14 3PS.
(Chapter 10).

F.L.Warner,
Royal Signals and Radar Establishment,
St Andrews Road,
Malvern,
Worcs,
WR14 3PS.
(Chapters 8 and 9 and Appendix A).

D.A.Williams,
Marconi Electronic Devices Limited,
Doddington Road,
Lincoln,
LN6 OLF.
(Chapter 15).

R.W.Yell,
Division of Electrical Science,
National Physical Laboratory,
Teddington,
Middlesex,
TW11 OLW.
(Chapter 18).

IEE Vacation School on Microwave Measurements
University of Kent, Canterbury
8 - 14 September 1985

ORGANISING COMMITTEE

Dr R.J.Collier, Chairman

Mr S.Arnold

Dr P.Cochrane

Mr R.W.Yell

Miss S.P.Morrison, Secretary.

Chapter 1

Transmission lines

R. J. Collier

1. INTRODUCTION

A transmission line is a device for guiding electromagnetic waves. The lines can take many forms and these will
be discussed in the lecture on 'Modern Lines'. The purpose
of this lecture is to go through the main properties of
transmission lines in preparation for the lectures that
follow in this Vacation School. Obviously one lecture
cannot hope to cover all the theory of transmission lines
and various references are given at the end of these notes
for those interested in pursuing aspects of this subject at
a greater depth.

These lecture notes will begin with definitions of the
main properties of transmission lines. First of all the
lines need dividing into two categories. There are those
lines in which it is much easier to describe them in terms
of the voltage and current measured at some points, e.g.
coaxial cable and twisted pair lines. In the second
category there are those lines in which the electric and
magnetic fields prove to be easier parameters to use e.g.
waveguides and optical fibres.

2. LINES DESCRIBED BY VOLTAGE AND CURRENT

2.1. These lines usually have an equivalent circuit containing distributed impedances. Typically, the lines consist of
two conductors, which together have a capacitance per metre
C and an inductance per metre L between them. The equivalent circuit for these lines is given in Fig.1.

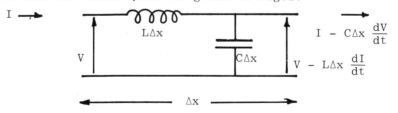

Fig.1.   The equivalent circuit of a short length
of transmission line

If a voltage, V, is applied at the input along with a current, I, then the output will differ from the input due to the voltage drop across the inductance and the current lost through the capacitance. Put in mathematical terms:

$$\frac{\partial V}{\partial x} = -L\,\frac{\partial I}{\partial t} \quad ; \quad \frac{\partial I}{\partial x} = -C\,\frac{\partial V}{\partial t}$$

These equations are called the telegraphist's equations. Differentiating these two equations gives:

$$\frac{\partial^2 V}{\partial x^2} = -L\frac{\partial^2 I}{\partial t \partial x} = L\,C\,\frac{\partial^2 V}{\partial t^2}$$

$$\frac{\partial^2 I}{\partial x^2} = -C\,\frac{\partial V}{\partial t \partial x} = L\,C\,\frac{\partial^2 I}{\partial t^2}$$

The equations are called the wave equations. If only sinusoidal solutions are considered for the moment, then the solutions to the wave equation are:

$$V = V_0\,\sin(\omega t - \beta x) + V_1\,\sin(\omega t + \beta x)$$

where these are the forward and reverse voltage waves. Also

$$I = I_0\,\sin(\omega t - \beta x) + I_1\,\sin(\omega t + \beta x)$$

are the forward and reverse current waves. Substituting these solutions into the wave equation gives

$$-\,\beta^2 V = -\,\omega^2 LC\ V$$
$$-\,\beta^2 I = -\,\omega^2 LC\ I$$

β is called the phase constant and is given by

$$\beta = \omega\sqrt{LC}$$

for lossless lines.

2.2 Phase Velocity

A wave travelling from an oscillator down a transmission line might look like that shown in Fig. 2.

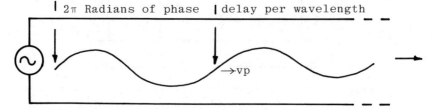

|2π Radians of phase |delay per wavelength

→vp

Fig.   2.

This wave will have a phase velocity v_p, where the phase velocity is the velocity of a point of constant phase on the waveform. Since the wave takes a finite time to arrive, it is delayed in phase as it goes along the line. The amount of phase delay is 2π radians per wavelength, λ. The phase delay per metre is β and is called the phase constant. So $\beta = 2\pi/\lambda$, multiplying top and bottom by f the frequency:-

$$\beta = \frac{2\pi f}{\lambda f} = \frac{\omega}{v_p}$$

Hence

$$v_p = \frac{\omega}{\beta} = \frac{\omega}{\omega\sqrt{LC}} = \frac{1}{\sqrt{LC}}$$

2.3   Characteristic Impedance

Both the current and voltage waves travel at the same velocity and the relationship between V and I is called the characteristic impedance. Using one of the telegraphist's equations

$$\frac{\partial V}{\partial x} = -L\frac{\partial I}{\partial t} \quad ; \quad V = V_o \sin(\omega t - \beta x) + V_1 \sin(\omega t + \beta x)$$

$$\frac{\partial V}{\partial x} = -\beta V_o \cos(\omega t - \beta x) + \beta V_1 \cos(\omega t + \beta x)$$

$$\frac{\partial I}{\partial t} = \frac{\beta}{L} V_o \cos(\omega t - \beta x) - \frac{\beta}{L}V_1 \cos(\omega t + \beta x)$$

$$I = \frac{\beta}{\omega L} V_o \sin(\omega t - \beta x) - \frac{\beta}{\omega L} V_1 \sin(\omega t + \beta x)$$

but $I = I_o \sin(\omega t - \beta x) + I_1 \sin(\omega t + \beta x)$

So for the forward wave, i.e. that travelling in the positive x direction,

$$\frac{V_o}{I_o} = \frac{V_o \omega L}{\beta V_o} = \frac{\omega L}{\omega \sqrt{LC}} = \sqrt{L/C} = Z_o$$

$$\frac{V_1}{L} = -\sqrt{L/C} \text{ for the reverse wave}$$

This ratio of voltage to current is called the Characteristic Impedance, Z_o, of the line and is positive for forward waves and negative for reverse waves.

2.4   Propagation Constant

Few lines approximate to the equivalent circuit given in Fig. 1. and this needs to be modified to include any resistive losses that may occur in the lines. Like C and L these will be distributed. It is normal to take a distributed resistance R in series with L to represent losses in the conductors and a distributed conductance G in parallel with C to represent losses in any dielectric between the conductors. In Fig. 3. is shown equivalent circuit of a transmission line with losses.

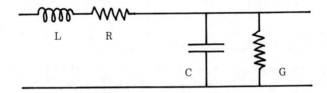

Fig. 3. <u>The equivalent circuit of a line with losses.</u>

Taking sinusoidal waves it can be seen that the telegraphist's equations are

$$\frac{\partial V}{\partial x} = - Z I \quad ; \quad \frac{\partial I}{\partial x} = - Y V$$

and the wave equations

$$\frac{\partial^2 V}{\partial x^2} = ZYV \quad ; \quad \frac{\partial^2 I}{\partial x^2} = ZYI$$

where $Z = R + j\omega L$
           $Y = G + j\omega C$

The solution to these equations is best described in exponential form, viz:

$$V = V_o \exp(j\omega t - \gamma x) + V_1 \exp(j\omega t + \gamma x)$$

$$I = I_o \exp(j\omega t - \gamma x) + I_1 \exp(j\omega t + \gamma x)$$

where γ is the propagation constant and is given by

$$\gamma = \alpha + j \; \beta$$

where α is the attenuation constant of the line. Substituting in the wave equation gives

$$\gamma^2 V = ZYV \quad ; \quad \gamma^2 I = ZYI$$

So
$$\gamma = \sqrt{ZY}$$

$$\gamma = \sqrt{(R+j\omega C)\ (G+j\omega C)}$$

Also using the first of the telegraphist's equations

$$\frac{\gamma V}{\partial x} = -\gamma V_O \exp(j\omega t - \gamma x) + \gamma V_1 \exp(j\omega t + \gamma x)$$

$$I = \frac{Y}{Z} V_O \exp(j\omega t - \gamma x) - \frac{Y}{Z} V_1 \exp(j\omega t + \gamma x)$$

$$= I_O \exp(j\omega t - \gamma x) + I_1 \exp(j\omega t + \gamma x)$$

So
$$Z_O = \frac{V_O}{I_O} = \frac{V_O Z}{V_O} = \sqrt{\frac{R+j\omega L}{G+j\omega C}}$$

For lines with small losses then the following approximations prove useful:

If $R \ll \omega L$; $G \ll \omega C$

$$\gamma = \sqrt{j\omega L(1 + \frac{R}{j\omega L}) \times j\omega C(1 + \frac{G}{j\omega C})}$$

$$\gamma = j\omega \sqrt{LC}(1 + \frac{R}{2j\omega L} - \ldots)(1 + \frac{G}{2j\omega C} - \ldots)$$

$$\gamma = j\omega \sqrt{LC} + \frac{R}{2Z_O} + \frac{GZ_O}{2} + \ldots$$

Hence for low loss lines
$$\gamma \cong \frac{R}{2Z_O} + \frac{GZ_O}{2} + j\omega \sqrt{LC}$$

$$\cong \alpha + j\beta$$

$$\alpha = \frac{R}{2Z_O} + \frac{GZ_O}{2} \quad \text{nepers/m}$$

$$\beta = \omega \sqrt{LC} \quad \text{(as before) radians/m}$$

$$Z_O = \sqrt{\frac{L}{C}} \quad \text{(as before) } \Omega$$

2.5 Standing Waves and VSWR

Standing waves occur when both a forward and a reverse wave are present on a line. Since both waves change phase in opposite directions there will be regular points where the wave is in phase and points where the opposite is true. When the waves are in phase their voltages add and there will be a maximum amplitude at this point on the line. The currents will conversely cancel at this point giving a minimum amplitude. This is because the characteristic impedance for a reverse wave is negative. Since all waves change phase at 2π radians per wavelength then the maxima are separated by half a wavelength, i.e. the forward wave delayed by 180° whilst the reverse wave is advanced by 180°. In Fig. 4. is shown a standing wave.

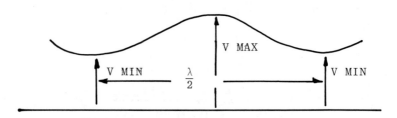

Fig. 4.   A standing wave

If $V = V_O \sin(\omega t - \beta x) + V_1 \sin(\omega t + \beta x)$

$$V_{MAX} = \left| V_O \right| + \left| V_1 \right|$$

$$V_{MIN} = \left| V_O \right| - \left| V_1 \right|$$

It is relatively easy to measure the values of V_{MAX} and V_{MIN} and the Voltage Standing Wave Ratio, S, (V.S.W.R.) is given by

$$S = \frac{V_{MAX}}{V_{MIN}}.$$

2.6 Reflection Coefficient

If a line whose characteristic impedance is Z_O is terminated in an impedance Z_L then an incident wave will be partly reflected. If the voltage at the termination is V_L given by

$$V_L = V_O \sin(\omega t - \beta x) + V_1 \sin(\omega t + \beta x)$$

and the current is I_L given by

$$I_L = I_o \sin(\omega t - \beta x) + I_1 \sin(\omega t + \beta x)$$

Then since $\dfrac{V_L}{I_L} = Z_L$

then:
$$Z_L = \frac{V_o + V_1}{I_o + I_1}$$

If the origin is taken at Z_L, i.e. $x = o$ at Z_L. Since

$$\frac{V_o}{I_o} = Z_o \text{ and } \frac{V_1}{I_1} = -Z_o,$$

then:
$$Z_L = \frac{V_o + V_1}{\dfrac{V_o}{Z_o} - \dfrac{V_1}{Z_o}}$$

If a reflection coefficient is defined as

$$\Gamma = \frac{V_1}{V_o}$$

then
$$Z_L = \frac{V_o(1+\Gamma)}{\dfrac{V_o}{Z_o}(1-\Gamma)}$$

or $Z_L - \Gamma Z_L = Z_o + \Gamma Z_o$

$$\Gamma = \frac{Z_L - Z_o}{Z_L + Z_o}$$

Also
$$S = \frac{V_{MAX}}{V_{MIN}} = \frac{|V_o| + |V_1|}{|V_o| - |V_1|} = \frac{|V_o|(1+|\Gamma|)}{|V_o|(1-|\Gamma|)}$$

$$S = \frac{1+|\Gamma|}{1-|\Gamma|}$$

$$|\Gamma| = \frac{S-1}{S+1}$$

2.7   Power Flow

When a line is terminated in its characteristic impedance Z_O, all the incident power is absorbed in the terminat-
ion.   If the incident wave had an amplitude $|V_O|$ then the
power in Z_O would be:

$$\frac{|V_O|^2}{2 \ Z_O}$$

Hence this must also be the power in the incident wave.   In
general the reflected power would be

$$\frac{|V_1|^2}{2 \ Z_O} = \frac{|\Gamma|^2 V_O|^2}{2 \ Z_O}$$

and $|\Gamma|^2 \ = \dfrac{\text{Reflected Power}}{\text{Incident Power}}$

and $\dfrac{\text{Transmitted Power}}{\text{Incident Power}} = |\tau|^2 = 1 - |\Gamma|^2$

3.   LINES DESCRIBED BY ELECTRIC AND MAGNETIC FIELDS

These lines do not have an equivalent circuit in the
same way as those in the previous section.   This is because
some of the lines have only one conductor, e.g. waveguides
and some have no conductors, e.g. optical fibres.   So it is
not possible to discuss voltage and current without also
defining the exact point at which they are measured.   Since
these lines guide their waves by internal reflection it is
easier to discuss the wave and fields and then relate them
to transmission line parameters.   First of all, a plane
wave must be described and then rectangular metallic waveguide will be discussed in detail to show how these principles
are applied.

3.1 Plane Electromagnetic Wave

A plane electromagnetic wave has no electric or magnetic
fields in the direction of propagation.   It is also called a
TEM wave (Transverse Electromagnetic Wave) as the fields lie
only in a direction perpendicular to the direction of propagation.

The electromagnetic wave equations are as follows:

$$\nabla^2 \overline{E} = \mu \, \epsilon \, \frac{\partial^2 \overline{E}}{\partial t^2}$$

$$\nabla^2 \overline{H} = \mu \, \epsilon \, \frac{\partial^2 \overline{H}}{\partial t^2}$$

μ is the permeability and ϵ is the permittivity of the medium in which the wave is propagating. In cartesian form these equations become

$$\frac{\partial^2 \overline{E}}{\partial x^2} + \frac{\partial^2 \overline{E}}{\partial y^2} + \frac{\partial^2 \overline{E}}{\partial z^2} = \mu \, \epsilon \, \frac{\partial^2 \overline{E}}{\partial t^2}$$

and similarly for $\overline{H}$.

Since the fields are plane, if the direction of propagation is in the z direction, the Electric field can be anywhere in the x,y plane. A solution would be

$$E = E_o \exp(j\omega t - \gamma z) + E_1 \exp(j\omega t + \gamma z)$$

where γ is the propagation constant and by substitution

$$\gamma^2 = -\omega^2 \, \mu\epsilon$$

$$\gamma = j\omega\sqrt{\mu\epsilon}$$

It is customary to use k rather than β for the phase constant. k is called the wave number and

$$\gamma = jk \qquad k = \omega\sqrt{\mu\epsilon}$$

$$(c.f. \quad \beta = \omega\sqrt{LC})$$

The free space values of μ and ϵ are

$$\mu_o = 4\pi.10^{-7} \ H/m$$

$$\epsilon_o - 8.854.10^{-12} \ F/m$$

So
$$k_o = \omega\sqrt{\mu_o \, \epsilon_o}$$

As in section 2.1

$$v_p = \frac{\omega}{k} = \frac{1}{\sqrt{\mu\epsilon}}$$

$$v_p \ \text{in free space} = \frac{1}{\sqrt{\mu_o \, \epsilon_o}}$$

$$= 2.99792458.10^8 \ m/s$$

$$\cong 3.10^8 \ m/s$$

If we choose the electric field to lie in the x direction then from Maxwells third equation

$$\nabla \times \overline{E} = - \frac{\partial \beta}{\partial t}$$

$$\nabla \times \overline{E} = \begin{vmatrix} \overline{a}_x & \overline{a}_y & \overline{a}_z \\ \frac{\partial}{\partial x} & \frac{\partial}{\partial y} & \frac{\partial}{\partial z} \\ E_X & 0 & 0 \end{vmatrix} = - \mu \frac{\partial H}{\partial t}$$

$$E_X = E_0 \exp(j\omega t - \gamma z) + E_1 \exp(j\omega t + \gamma z) \, \overline{a}_x$$

where $\overline{a}_x$, $\overline{a}_y$ and $\overline{a}_z$ are unit vectors in the cartesian directions. Hence,

$$\nabla \times \overline{E} = \frac{\partial E_X}{\partial z} \overline{a}_y$$

$$= -\gamma \, E_0 \exp(j\omega t - \gamma z) + \gamma E_1 \exp(j\omega t + \gamma z) \, \overline{a}_y$$

$$= -j\omega\mu \, H_0 \exp(j\omega t - \gamma z) - j\omega\mu \, H_1 \exp(j\omega t + \gamma z)\overline{a}_y$$

$\overline{H}$ must lie only in the y direction to be consistent. So $\overline{E}$ and $\overline{H}$ are orthogonal; and the wave impedance, η, is given by:

$$\eta = \frac{E_0}{H_0} = \frac{j\omega\mu}{\gamma} = \frac{j\omega\mu}{\omega\sqrt{\mu\epsilon}} = \sqrt{\frac{\mu}{\epsilon}}$$

$$\eta = \frac{E_1}{H_1} = -\sqrt{\frac{\mu}{\epsilon}} \quad \text{for reverse waves, where the "handedness" of the axes is reversed.}$$

In free space

$$\eta_0 = \sqrt{\frac{\mu_0}{\epsilon_0}} = 376.61 \ \Omega$$

3.2   Rectangular Waveguide

When a wave is incident on a short circuit the reflected wave is out of phase so that the total tangential electri field goes to zero at the surface. This is also true for obliquely incident waves. Fig. 5. shows a section of rectangular metallic waveguide in which a plane wave is moving in direction lying in the x z plane. The electric field is in the y direction.

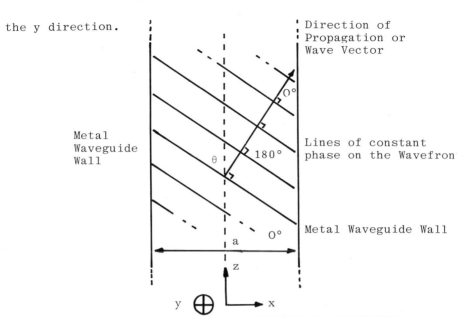

Direction of
Propagation or
Wave Vector

Metal
Waveguide
Wall

Lines of constant
phase on the Wavefron

Metal Waveguide Wall

Fig. 5.   A PLANE WAVE IN A RECTANGULAR WAVEGUIDE

The rate of change of phase in the direction of propagation is k_O, the wave number i.e. 2π radians of phase change per wavelength. As this wave is incident on the metal wall on the right of the diagram it will form another plane wave as shown in the Fig. 6. The angle of reflection will be equal to the angle of incidence according to the usual laws of reflection.

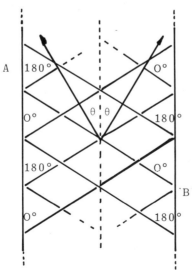

Fig. 6.   TWO PLANE WAVES IN A RECTANGULAR WAVEGUIDE

On further reflection this wave must 'rejoin' the original wave. The phase fronts of the two waves must fit together as shown so that they are always 180° out of phase at the two walls. Moving from A→B the phase of the original wave is constant at 0°. So in order for the waves to fit the phase shift of the reflected wave along AB must be 2mπ where m = 1,2,3...

The rate of change of phase along BA is the wave number, k_o, resolved in that direction, i.e.

$$k_o \sin 2\theta.$$

If the distance between the waveguide walls is a, then

$$AB = \frac{a}{\cos \theta}$$

So the phase shift is

$$k_o \sin 2\theta \times \frac{a}{\cos \theta} = \frac{2k_o a \sin\theta \cos\theta}{\cos \theta}$$

and this must be equal to 2mπ, so

$$k_o a \sin \theta = m\pi \qquad m = 1,2,3$$

3.3   Phase Velocity

The phase velocity in the z direction is obtained from the z component of the wave number i.e. v_p in z direction = $\frac{\omega}{k \text{ in z direction}}$

$$v_p = \frac{\omega}{k_o \cos \theta} = \frac{\omega}{k_o (1-\sin^2\theta)^{\frac{1}{2}}}$$

As

$$k_o = \omega \sqrt{\mu\epsilon} = \frac{2\pi}{\lambda_o}$$

$$v_p = \frac{\frac{1}{\sqrt{\mu\epsilon}}}{(1 - \sin^2\theta)^{\frac{1}{2}}}$$

Now

$$\sin \theta = \frac{m\pi}{k_o a} = \frac{m\pi}{2\pi a} \lambda_o = \frac{m\lambda_o}{2a}$$

$$v_p = \frac{\frac{1}{\sqrt{\mu\epsilon}}}{\left[1 = \left(\frac{m\lambda_o}{2a} \right)^2 \right]^{\frac{1}{2}}}$$

m is called a mode number and when a wave 'fits' this is called a waveguide mode.

3.4 Wave Impedance (TE Modes)

As the electric field is always in the y direction, these modes are called Transverse Electric modes (or TE modes) and the magnetic fields associated with the 2 plane waves are in the x, z plane. The wave impedance is usually taken as the ratio of the transverse electric and magnetic fields, i.e.

$$Z_{TE} = \frac{E_Y}{H_x} = \frac{E_o \; \exp \; j(\omega t - kz)}{H_o \; \cos \; \theta \; \exp \; j(\omega t - kz)}$$

where z is the direction of propagation.

Since

$$\frac{E_o}{H_o} = \sqrt{\frac{\mu}{\epsilon}}$$

Then the transverse impedance is Z_{TE} given by

$$Z_{TE} = \frac{\sqrt{\dfrac{\mu}{\epsilon}}}{\left[1 - \left(\dfrac{m\lambda_o}{2a} \right)^2 \right]^{\frac{1}{2}}}$$

3.5 Wave Impedance (TM Modes)

The above arguments also apply to the TM modes, i.e. transverse magnetic modes. In these modes the magnetic field is purely transverse. All the equations for the velocity still apply but the impedance is inverted as the transverse component of E is now required:

$$Z_{TM} = \frac{E_X}{H_Y} = \frac{E_o \; \cos \; \theta \; \exp \; j(\omega t - kz)}{H_o \; \exp \; j(\omega t - kz)}$$

$$= \frac{\sqrt{\dfrac{\mu}{\epsilon}}}{\left[1 - \left(\dfrac{m\lambda_o}{2a} \right)^2 \right]^{\frac{1}{2}}}$$

3.6 General Solution

In general, a wave must also fit between the other pair of walls and the condition for this is

$$k_o \, b \sin \phi = n\pi \qquad n = 1,2,3.$$

n can be zero if the electric field is perpendicular to these faces. Hence, the mode description

$$TE_{mn} \longleftarrow \text{no. of half sines in the y direction}$$
no. of half sines in the x direction

or TM_{mn}

Combining the above results, since the wave number can have 3 components

and
$$k = k_x \overline{a}_x + k_y \overline{a}_y + k_z \overline{a}_z$$

and
$$|k| = k_o = (k_x^2 + k_y^2 + k_z^2)^{\frac{1}{2}}$$

$$k_x = k_o \sin \theta$$
$$k_y = k_o \sin \phi$$

Then
$$k_o = ((\frac{m\pi}{a})^2 + (\frac{n\pi}{b})^2 + k_z^2)^{\frac{1}{2}}$$

$$k_z = (k_o^2 - (\frac{m\pi}{a})^2 - (\frac{n\pi}{b})^2)^{\frac{1}{2}}$$

$$k_z = k_o \left(1 - \left(\frac{m\lambda_o}{2a}\right)^2 - \left(\frac{n\lambda_o}{2b}\right)^2 \right)^{\frac{1}{2}}$$

For TE and TM modes:

$$v_p = \frac{\frac{1}{\sqrt{\mu\epsilon}}}{k}$$

$$Z_{TE} = \frac{\sqrt{\frac{\mu}{\epsilon}}}{k}$$

$$Z_{TM} = \sqrt{\frac{\mu}{\epsilon}}\, k$$

$$k = \frac{k_z}{k_o} = \left[1 - \frac{m\lambda_o}{2a}^2 - \frac{n\lambda_o}{2b}^2 \right]^{\frac{1}{2}}$$

3.7 Cut-off Frequency

Each mode has a cut-off frequency at which the wave number in the direction of propagation is zero, i.e. $k_z = 0$. For the TE modes only either m or n can be zero. So the cut-off diagram can be constructed as shown in Fig. 7.

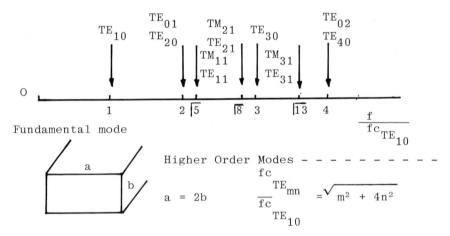

Fig. 7. RELATIVE CUT-OFF FREQUENCIES FOR RECTANGULAR WAVEGUIDES.

Above the cut-off frequency the phase velocity for each mode is initially infinite and then as frequency increases it approaches the velocity for plane waves in the medium inside the guide. Similarly, the wave impedance is initially infinite for TE modes and zero for TM modes but again approaches that of free space at higher frequencies.

Below the cut-off frequency, the modes do not propagate and since the wave number is imaginary have an exponential decay with no phase shift down the guide.

3.8 Group Velocity

Any modulated plane wave in free space has a group velocity equal to its phase velocity. Clearly, in a rectangular waveguide the waves are propagating obliquely to the direction of the guide. So the group velocity or the velocity of a modulating signal will be the component of the free space velocity in the z direction of the guide. That is

$$v_g \text{ (group velocity)} =$$

$$\frac{1}{\sqrt{\mu\epsilon}} \left[1 - \left(\frac{m\lambda_o}{2a}\right)^2 - \left(\frac{n\lambda_o}{2b}\right)^2 \right]^{\frac{1}{2}}$$

For rectangular waveguide

$$v_p \times v_g = \frac{1}{\sqrt{\mu\epsilon}}$$

4.   REFERENCES

1.   RAMO, WHINNERY and VAN DUZER: "Fields and Waves in Communications Electronics", Wiley.

2.   DAVIDSON, C.W.:   "Transmission Lines for Communications", Macmillan.

3.   JORDON, E.C.:   "Electromagnetic Waves and Radiating Systems", Prentice-Hall.

4.   CHIPMAN, R.A.:   "Transmission Lines", Schaum's Outline Series, McGraw-Hill.

Chapter 2

Scattering coefficients

A. L. Cullen

1. INTRODUCTION

The definition of a scattering coefficient is best introduced through the simpler idea of a reflection coefficient.

Referring to Fig.1, we consider a transmission line in which only one mode can propagate, terminated in a one-port device which we assume to be linear and passive.

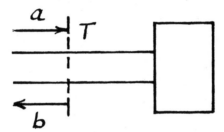

Fig.1. Reflection by one-port device

We select a terminal plane T in the transmission line; this may be (but is not necessarily) the physical junction between the line and the device, and we take this plane as the origin of a co-ordinate z measured along the axis of the line towards the device. Then the voltage at any point along the line, sufficiently far from the device for evanescent modes to be negligible, can be written

$$V = a\,e^{-j\beta z} + b\,e^{+j\beta z} \qquad (1)$$

If the characteristic impedance of the line is chosen as the unit of impedance in what follows, we have also

$$I = a\,e^{-j\beta z} - b\,e^{+j\beta z} \qquad (2)$$

We assume a time-factor $e^{j\omega t}$ which is suppressed in all our formulae. The term $a\,e^{-j\beta z}$ represents a wave travelling towards the device, whilst the
term $b\,e^{+j\beta z}$ represents a wave travelling away from the
device, and is thought of as <u>reflected</u> by the device. The
ratio of the reflected to the <u>incident</u> wave is called the
<u>reflection coefficient</u>, and we shall denote it by Γ .
Thus, we can write

$$\Gamma = \frac{b}{a} \tag{3}$$

as the reflection coefficient of the device at the terminal plane T.
    If the transmission line is a simple TEM line, vol-
tage and current have a simple and familiar significance,
and the complex impedance Z at T is simply

$$Z = \frac{V_0}{I_0} \tag{4}$$

where the zero suffices indicate that V and I are to be
evaluated at z = 0.
    Using (4) with (1) and (2) we find

$$Z = \frac{a + b}{a - b} \tag{5}$$

    Using (3) with (5), we find a simple relationship
between Z and T as follows

$$Z = \frac{1 + \Gamma}{1 - \Gamma} \tag{6}$$

Similarly the admittance $Y\ (=1/Z)$ at T is simply

$$Y = \frac{1 - \Gamma}{1 + \Gamma} \tag{7}$$

The ideas of impedance and admittance are often useful, even when the simple interpretation as a voltage-to-current ratio is not valid in its most obvious sense. It is always possible to define, purely formally, a voltage and a current from which the electric and magnetic fields in a line or waveguide of any kind can be derived when needed. We shall see later, however, that the impedance (admittance) concept cannot be generalised so completely as can the idea of reflection coefficient. In spite of this drawback, it can sometimes be used to great advantage in simplifying certain types of calculation as we shall also demonstrate later.

2. SCATTERING COEFFICIENTS

The one-port device of Fig.1 is completely character-ised by the reflection coefficient Γ if this is specified at the frequency or frequencies of interest.

The situation becomes more complicated, however, for devices with more than one port. Consider, for example, the 2-port device of Fig.2.

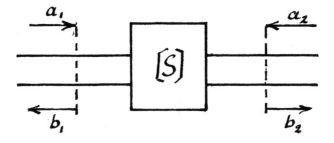

Fig.2. Scattering by 2-port device

If we measure the ratio b_1/a_1 at port T_1, the result will depend on conditions at port 2, and will no longer be determined uniquely by the characteristics of the 2-port device itself. In fact, the four quantities a_1, a_2, b_1, b_2 are linearly related since we are assuming linear systems throughout. It is particularly convenient to express the outward-going waves b_1 and b_2 in terms of the inward-going waves a_1 and a_2 thus:

$$b_1 = S_{11} a_1 + S_{12} a_2 \qquad\qquad (8)$$

$$b_2 = S_{21} a_1 + S_{22} a_2 \qquad\qquad (9)$$

The quantities S_{11} , S_{12} etc are called <u>scattering coefficients</u>.

Equations (8) and (9) can be written as a single matrix equation as follows:

$$[b] = [S][a] \tag{10}$$

or

$$\begin{bmatrix} b_1 \\ b_2 \end{bmatrix} = \begin{bmatrix} S_{11} & S_{12} \\ S_{21} & S_{22} \end{bmatrix} \begin{bmatrix} a_1 \\ a_2 \end{bmatrix} \tag{11}$$

The 2-port junction is completely characterised if four scattering coefficients, which in general are complex numbers, are specified at the frequency or frequencies of interest.

It is sometimes more convenient to work in terms of voltages and currents. We define

$$V_i = a_i + b_i$$

$$I_i = a_i - b_i \tag{12}$$

as the voltage and current at the i'th port, where i = 1 or 2 in the present case. We can then relate the V's to the I's by a pair of equations thus

$$V_1 = Z_{11} I_1 + Z_{12} I_2$$

$$V_2 = Z_{21} I_1 + Z_{22} I_2 \tag{13}$$

or by a single matrix equation

$$\begin{bmatrix} V_1 \\ V_2 \end{bmatrix} = \begin{bmatrix} Z_{11} & Z_{12} \\ Z_{21} & Z_{22} \end{bmatrix} \begin{bmatrix} I_1 \\ I_2 \end{bmatrix} \tag{14}$$

Alternatively we can write

$$[V] = [Z][I] \qquad (15)$$

There is, however, a drawback in using the impedance matrix in general theoretical work; it may not exist. Fig.3 gives a simple example. The voltages V_1 and V_2 cannot be expressed solely in terms of the currents I_1 and I_2 in the form of equations (13), and so for this simplest of all 2-ports, the impedance matrix representation cannot be used.

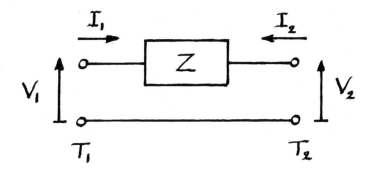

Fig.3. A simple 2-port device

The scattering matrix representation is <u>always</u> possible, however, and for example, the scattering coefficients for the series impedance of Fig.3 are

$$S_{11} = S_{22} = \frac{Z}{2 + Z}$$

$$S_{12} = S_{21} = \frac{2}{2 + Z} \qquad (16)$$

3.   RECIPROCITY, SYMMETRY AND LOSSLESSNESS

We can now conveniently introduce three properties which are important not only for 2-port junctions, but for the multiport junctions we shall be considering later.

Those properties are:

(i)   reciprocity
(ii)   symmetry
(iii) losslessness

In principle, a multiport junction may possess any or none of these properties. For example, the two-port junction we have examined, and which is described by equations (16), possesses the property of reciprocity, since $S_{12} = S_{21}$ and symmetry, since $S_{11} = S_{22}$. This last property follows at once by inspection. The junction may possess the property of losslessness, and will do so if

$$Z = \pm jX,$$ a pure reactance.

We note that in this case

$$|S_{11}|^2 = \frac{X^2}{4 + X^2}$$

$$|S_{12}|^2 = \frac{4}{4 + X^2}$$

(17)

so that

$$|S_{11}|^2 + |S_{12}|^2 = 1$$

(18)

It follows from (18), if $a_2 = 0$, that

$$|b_1|^2 + |b_2|^2 = |a_1|^2$$

(19)

The physical interpretation of this equation is that the power reflected from port 1 plus the power emerging from port 2 is equal to the power incident on port 1. Since $a_2 = 0$, there is no power incident on port 2 and so power is conserved by the junction, and it is indeed lossless.

This rather obvious property of the scattering coefficients is complemented by a less obvious one.
If we write

$$S_{ij} = |S_{ij}| e^{j\phi_{ij}}$$

(20)

it is easily shown from equation (16) that

$$\phi_{12} = \frac{\phi_{11} + \phi_{22}}{2} \; \stackrel{+}{-} \; \frac{\pi}{2} \qquad (21)$$

This equation is in fact true for any reciprocal loss-free 2-port junction, though we have derived it for a very special case.

At this point we shall move from the particular to the general, and discuss the properties of reciprocity, symmetry and losslessness for an N-port junction. We begin with losslessness.

First, by an obvious extension of the previous 2-port discussion, we write the matrix equation for an N-port junction thus:

$$\begin{bmatrix} b_1 \\ b_2 \\ \cdot \\ \cdot \\ \cdot \\ b_N \end{bmatrix} = \begin{bmatrix} S_{11} & S_{12} & \cdot & \cdot & \cdot & \cdot & S_{1N} \\ S_{21} & S_{22} & & & & & \\ \cdot & & & & & & \\ \cdot & & & & & & \\ \cdot & & & & & & \\ S_{N1} & \cdot & \cdot & \cdot & \cdot & \cdot & S_{NN} \end{bmatrix} \begin{bmatrix} a_1 \\ a_2 \\ \cdot \\ \cdot \\ \cdot \\ a_N \end{bmatrix} \qquad (22)$$

Note that this equation can be written in compact notation, and is then identical with (11).

If the junction is loss-free, the total power flowing out of the junction minus the total power flowing in must be zero. Thus

$$|b_1|^2 + |b_2|^2 + \cdots + |b_N|^2 = |a_1|^2 + |a_2|^2 + \cdots + |a_N|^2$$

or

$$\sum_{i=1}^{N} |b_i|^2 - \sum_{i=1}^{N} |a_i|^2 = 0 \qquad (23)$$

Now $|a_i|^2 = a_i a_i^*$, and using the extended matrix notation each sum in (23) can be written as the product of a row matrix and a column matrix, thus

$$\sum_{i=1}^{N} |b_i|^2 = \{b_1, b_2, \cdots b_N\} \begin{bmatrix} b_1^* \\ b_2^* \\ \cdot \\ \cdot \\ \cdot \\ b_N^* \end{bmatrix}$$

(24)

The column matrix on the right hand side of (24) is simply the conjugate of (22), and the row matrix on the right hand side of (24) is simply the transpose (ie rows and columns interchanged) of the column matrix on the left hand side of (22). In the compact matrix notation, therefore, equation (23) can be written as follows:

$$\widetilde{[s][a]}[s^*][a^*] - \{a\}[a^*] = 0 \qquad (25)$$

the tilde denoting the transpose. The curly brackets indicate a row matrix. Further progress requires a theorem in matrix algebra which states that the transpose of the product of two matrices is equal to the product of the transposes taken in inverse order. Thus

$$\widetilde{[a]}\widetilde{[s]}[s^*][a^*] - \{a\}[a^*] = 0$$

or

$$\{a\}\widetilde{[s]}[s^*][a^*] - \{a\}[a^*] = 0$$

(26)

Introducing the unit matrix

$$[I] = \begin{bmatrix} 1 & 0 & \cdot & \cdot & \cdot & 0 \\ 0 & 1 & \cdot & \cdot & \cdot & 0 \\ \cdot & & & & & \\ \cdot & & & & & \\ \cdot & & & & & \\ 0 & 0 & \cdot & \cdot & \cdot & 1 \end{bmatrix}$$

(27)

equation (26) can be written

$$\{a\}\left([\widetilde{s}][s^*] - [I]\right) = 0 \tag{28}$$

Since this must hold for all possible excitations of the junction - that is, for all possible values of the N elements of $[a]$ it follows that

$$[\widetilde{s}][s^*] = [I] \tag{29}$$

and the scattering matrix is said to be <u>unitary</u>.

Equation (29) is a key equation in the theory of loss-less waveguide junctions, and we shall return to it. But before leaving it, we illustrate its application by deriving equations (18) and (21) directly from (29) without making any assumptions about the 2-port other than losslessness.

Specifically

$$\begin{bmatrix} S_{11} & S_{21} \\ S_{12} & S_{22} \end{bmatrix} \begin{bmatrix} S_{11}^* & S_{12}^* \\ S_{21}^* & S_{22}^* \end{bmatrix} = \begin{bmatrix} 1 & 0 \\ 0 & 1 \end{bmatrix} \tag{30}$$

Forming each element of the unit matrix in turn by row into column multiplication of the 2 x 2 matrices on the left hand side, we find

$$S_{11}S_{11}^* + S_{12}S_{21}^* = 1 \tag{31}$$

$$S_{12}S_{12}^* + S_{22}S_{22}^* = 1 \tag{32}$$

$$S_{11}S_{12}^* + S_{21}S_{22}^* = 0 \tag{33}$$

$$S_{12}S_{11}^* + S_{22}S_{21}^* = 0 \tag{34}$$

We see at once that (31) is identical with (18). (32) is the corresponding equation when the junction of Fig.2 is excited from the right hand side and $a_1 = 0$. Equations (33) and (34) are complex conjugates of one another and only one, say, (33), need be considered.

Using the notation introduced in (20) we find from (33)

$$|S_{11}||S_{12}|e^{j(\phi_{11} - \phi_{12})} + |S_{21}||S_{22}|e^{j(\phi_{21} - \phi_{22})} = 0$$

(35)

If the junction is reciprocal,

$$|S_{21}| = |S_{12}|$$

(36)

and

$$\phi_{21} = \phi_{12}$$

(37)

It follows from (36), (31) and (32) that

$$|S_{22}| = |S_{11}|$$

(38)

Using (36), (37) and (38) in (35) we find

$$e^{j(\phi_{11} - \phi_{12})} + e^{j(\phi_{12} - \phi_{22})} = 0$$

or

$$e^{j\left(\frac{\phi_{11} - \phi_{22}}{2}\right)}\left[e^{j\left(\frac{\phi_{11} + \phi_{22}}{2} - \phi_{12}\right)} + e^{-j\left(\frac{\phi_{11} + \phi_{22}}{2} - \phi_{12}\right)}\right] = 0$$

or

$$\cos\left(\frac{\phi_{11} + \phi_{22}}{2} - \phi_{12}\right) = 0$$

so that

$$\phi_{12} = \frac{\phi_{11} + \phi_{22}}{2} \pm \frac{\pi}{2} \qquad (39)$$

and we notice that this is identical with (21). We have therefore verified the statement made just below (21) that this equation is valid for any loss-free reciprocal 2-port whatever its structure may be.

We have already used the property of reciprocity for a 2-port junction, namely

$$S_{12} = S_{21}$$

The general form of reciprocity relation for an N-port junction is

$$S_{ij} = S_{ji} \qquad (40)$$

This result can be proved from the Lorentz reciprocity formula

$$\int_S \left(\vec{E}' \times \vec{H}'' \right) \cdot d\vec{S} = \int_S \left(\vec{E}'' \times \vec{H}' \right) \cdot d\vec{S} \qquad (41)$$

in which $\left(E', H' \right)$ and $\left(\vec{E}'', \vec{H}'' \right)$ are two different electromagnetic fields on a closed surface containing only materials whose permittivity, permeability and conductivity can be represented as symmetric tensors. (This excludes magnetised ferrites, for example).

If the surface S is chosen to pass through the terminal planes T_1, T_2, T_3, ...T_N of the N-port junction, (41) can be replaced by

$$V_1' I_1'' + V_2' I_2'' + ----- + V_N' I_N''$$

$$= V_1'' I_1' + V_2'' I_2' + ----- + V_N'' I_N' \qquad (42)$$

Using the incident-wave and reflected-wave representation of the voltages and currents, we find

$$(a_i' a_i'' - a_i' b_i'' - a_i'' b_i' - b_i' b_i'')$$

$$+ \quad - \ - \ - \ - \ -$$

$$+(a_N' a_N'' - a_N' b_N'' + a_N'' b_N' - b_N' b_N'')$$

$$= (a_i'' a_i' - a_i'' b_i' + a_i' b_i'' - b_i'' b_i')$$

$$+ \quad - \ - \ - \ - \ -$$

$$+ (a_N'' a_N' - a_N'' b_N' + a_N' b_N'' - b_N'' b_N')$$

which reduces to

$$a_i'' b_i' + a_2'' b_2' + \ - \ - \ - \ - \ + a_N'' b_N'$$

$$= a_i' b_i'' + a_2' b_2'' + \ - \ - \ - \ - \ + a_N' b_N'' \tag{43}$$

We are, of course, at liberty to choose the excitation of the various ports to suit our convenience.

We assume that all ports except the i th and the j th are terminated by matched loads so that $a_1 = 0$, $a_2 = 0$, etc, but $a_i \neq 0$, $a_j \neq 0$.

Then (43) reduces to

$$a_i'' b_i' + a_j'' b_j' = a_i' b_i'' + a_j' b_j'' \tag{44}$$

Next we express the b's in terms of the a's thus

$$
\left.
\begin{aligned}
b'_i &= S_{ii}\, a'_i + S_{ij}\, a'_j \\[1mm]
b'_j &= S_{ji}\, a'_i + S_{jj}\, a'_j \\[1mm]
b''_i &= S_{ii}\, a''_i + S_{ij}\, a''_j \\[1mm]
b''_j &= S_{ji}\, a''_i + S_{jj}\, a''_j
\end{aligned}
\right\}
\tag{45}
$$

Putting (45) into (44) and simplifying yields

$$
S_{ij}\left(a''_i a'_j - a'_i a''_j\right) = S_{ji}\left(a''_i a'_j - a'_i a''_j\right)
\tag{46}
$$

Since the incident waves a''_i etc are quite arbitrary, the only possible solution of (46) is

$$
S_{ij} = S_{ji}
$$

which is equation (40) and was to be proved.

The only remaining property is that of symmetry. We have already considered one simple example, a symmetrical 2-port junction. When there are more than two ports, more complicated symmetries are possible, and for the most general treatment it is necessary to employ group theory.

For most practical applications the symmetry relationships are so simple that group theory is not necessary, and we shall avoid its use here.

We consider two examples. The reciprocal symmetrical waveguide T-junction of Fig.4(a) is represented by a 3 x 3 matrix in which

$$
S_{11} = S_{22}
\tag{47}
$$

$$
S_{13} = \pm S_{23}
\tag{48}
$$

S_{12} and S_{21} must also be specified.

The upper sign applies in (48) if figure 4(a) represents an H-plane junction, and the lower sign applies for an E-plane junction.

The reciprocal symmetrical H-plane 4-port junction of Fig. 4(b) is represented by a 4 x 4 matrix in which the following equations hold

$$S_{11} = S_{22} = S_{33} = S_{44} \tag{49}$$

$$S_{12} = S_{23} = S_{34} = S_{41} \tag{50}$$

$$S_{13} = S_{24} \tag{51}$$

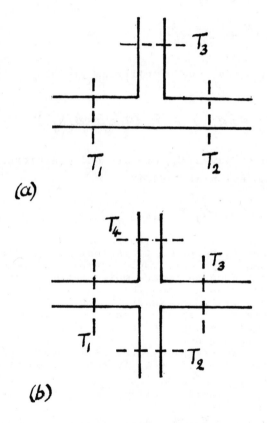

(a)

(b)

Fig.4   Symmetrical junctions

It is interesting to note that only <u>three</u> complex numbers are needed to specify the characteristics of the 4-port junction, although a completely arbitrary recip-rocal 4-port junction would require the specification of <u>ten</u> complex numbers to characterise it completely. This illustrates how powerful a factor symmetry can be in reducing the number of independent parameters needed to describe the behaviour of a waveguide or transmission line junction.

There are actually two different ways in which symmetry may be useful in microwave measurements.

Firstly, if the junction is accurately made, it can serve as a useful check on the accuracy of the measuring apparatus, through the use of a set of equations such as (49), (50) and (51).

Secondly, if the measuring apparatus is sufficiently accurate, the same set of equations can be used as a check on the accuracy of the construction of the junction.

When the impedance or admittance matrix representation is used, there are equivalent results for the special cases of reciprocity, symmetry and losslessness.

The reciprocity result follows very simply from (42) if we take two ports at a time, the remaining ports being open-circuited. Suppose we take the m th and n th ports. Suppose for the single-primed excitation the n th port is open circuited, so that

$$I_n' = 0 \tag{52}$$

Suppose that for the double-primed excitation the m th port is open-circuited, so that

$$I_m'' = 0 \tag{53}$$

The only non-zero currents are then I_m' and I_n'' and (42) reduces to

$$V_n' I_n'' = V_m'' I_m'$$

Using (13) we find at once

$$Z_{mn} = Z_{nm} \tag{54}$$

which is formally identical with (40).

The symmetry conditions for the impedance matrix are the same as for the scattering matrix, and, for example, the results (47) to (51) are true if every s is replaced by a Z without changing the suffices.

The condition for losslessness is

$$\sum_{n=1}^{N} re\, V_n I_n^* = 0 \tag{53}$$

or

$$\sum_{n=1}^{N} re\left(Z_{n1} I_1 + \cdots + Z_{nN} I_N\right) I_n^* = 0 \tag{54}$$

Suppose once again we consider a special excitation, in which I_m and I_n are the only non-zero currents. In other words, we may imagine all ports except the m th and the n th to be open-circuited. Then (54) reduces to

$$re \left(Z_{mm} I_m + Z_{mn} I_n \right) I_m^*$$

$$+ re \left(Z_{nn} I_n + Z_{nm} I_m \right) I_n^* = 0$$

(55)

Since I_m and I_n are idependent, (55) is equivalent to the three separate equations below.

$$re \; Z_{mm} I_m I_m^* = 0 \tag{56}$$

$$re \; Z_{nn} I_n I_n^* = 0 \tag{57}$$

$$re \left(Z_{mn} I_n I_m^* + Z_{nm} I_m I_n^* \right) = 0 \tag{58}$$

From (56) and (57) it follows at once that

$$re \; Z_{mm} = re \; Z_{nn} = 0$$

That is, all the "self-impedances" are purely reactive, as we should expectintuitively, and we can write

$$Z_{mm} = j X_{mm} \tag{59}$$

For the mutual impedances, the result is not quite as simple.
Let us write

$$\left. \begin{array}{l} I_m I_n^* = c + jd \\[2mm] I_n I_m^* = c - jd \end{array} \right\} \tag{60}$$

so that

Let

$$Z_{mn} = R_{mn} + j X_{mn}$$

$$Z_{nm} = R_{nm} + j X_{nm}$$

(61)

Substituting (60) and (61) into (58) leads to

$$R_{nm}\, c + R_{mn}\, c - X_{nm}\, d + X_{mn}\, d$$

Since c and d are independent, this leads to

$$\left. \begin{array}{l} R_{nm} = - R_{mn} \\ X_{nm} = X_{mn} \end{array} \right\}$$

(62)

or

$$Z_{nm} = - Z_{mn}^{*}$$

(63)

In the special case of a <u>reciprocal</u> lossless junction, (54) must hold simultaneously with (63), and the only possible solution then is

$$R_{nm} = R_{mn} = 0$$

$$Z_{nm} = Z_{mm} = j X_{mn}$$

(64)

Thus, for a lossless reciprocal junction, all the impedance coefficients, self and mutual, are purely imaginary. This is a much simpler result than the corresponding result for the scattering matrix, in which nothing can be said about the individual scattering coefficients; all that can be done is to make statements about certain combinations of the scattering coefficients according to the unitary matrix condition. There are a number of occasions in which the greater simplicity of the impedance matrix for a lossless reciprocal junction can be exploited usefully in microwave network theory.

4.   SPECIAL CASES

4.1. There are a number of interesting properties which arise when we consider the constraints which the foregoing general theory imposes on specific N-port devices. We have already seen one example; the relationship (39) between the phases of scattering coefficients S_{12}, S_{11} and S_{22} of a lossless reciprocal 2-port device.

For devices with 3 or more ports, some much more interesting results emerge. We shall deal with some of these in the rest of this section.

Before doing so, however, we define two specific classes of N-port waveguide junction whose properties are of great practical importance.

(i)   Completely matched N-port junction

A completely matched N-port junction has the property that each port presents a matched load to its input wave-guide if all the remaining ports are terminated in perfectly matched loads.

(ii)   Circulator

A circulator is a completely matched N-port junction having the property that a wave incident on port 1 emerges from port 2 only (with a suitable numbering scheme), and that a wave incident on port 2 emerges from port 3 only and so on. A wave incident on port N emerges from port 1, and that completes the sequence. It is assumed in the above that all ports except the input port are terminated by perfectly matched loads. Clearly, a circulator is a non-reciprocal device, and ideally it is lossless.

4.2. 3-port junctions

The asymmetrical 3-port junction of Fig.5 is supposed to be lossless and reciprocal, but is otherwise perfectly general; the three input guides could all be different, for example, one microstrip, one coaxial line, one waveguide.

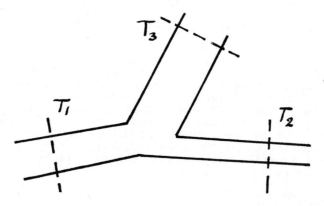

Fig.5 Asymmetrical 3-port junction

We assume that they are coupled electromagnetically, so that a wave incident on any one port will give rise to waves emerging from the other two ports.

The decoupling theorem

We now prove the decoupling theorem, which is true for any such 3-port junction.
It is convenient to work with voltages and currents and an impedance matrix in this case.
The sign conventions are so chosen that if V_n and I_n are positive and real, power flows towards the junction at the n th port.
Bearing in mind that we have assumed losslessness and reciprocity, the following equations apply.

$$
\left.
\begin{aligned}
V_1 &= j X_{11} I_1 + j X_{12} I_2 + j X_{31} I_3 \\
V_2 &= j X_{12} I_1 + j X_{22} I_2 + j X_{23} I_3 \\
V_3 &= j X_{31} I_1 + j X_{23} I_2 + j X_{33} I_3
\end{aligned}
\right\}
\quad (65)
$$

(We have chosen to retain cyclic order in the suffices.)
We wish to show that if port 3 is terminated by a reactance such that

$$
\frac{V_3}{-I_3} = j X_3 \quad (66)
$$

then port 2 can be completely decoupled from port 1 if X_3 is correctly chosen. (Note that we must write $-I_3$ for the current flowing into the reactance because of the sign convention we have adopted.) For simplicity, we choose to have port 2 open-circuited.
Put $I_2 = 0$ in (65) and use (66) to eliminate V_3. Thus, the last equation of (65) gives

$$
j X_3 I_3 - j X_{33} I_3 = j X_{31} I_1
$$

or
$$
I_3 = -\left(\frac{X_{31}}{X_{33} + X_3} \right) I_1 \quad (67)
$$

Putting (67) into the second equation of (65) gives

$$
V_2 = \left[j X_{12} - j X_{23} \left(\frac{X_{31}}{X_{33} + X_3} \right) \right] I_1 \quad (68)
$$

It follows at once that we can make V_2 zero if the square-bracketed factor in (68) vanishes, and this will be the case if

$$X_3 = \frac{X_{23} X_{31}}{X_{12}} - X_{33} \tag{69}$$

A more practical version of the decoupling theorem states that <u>a correctly placed short-circuit in one arm of the junction will decouple the other two arms</u>.

The input reactance of a length d of guide connected to port 3, (and assumed to be of unit characteristic impedance) is

$$j \tan \beta_3 d$$

Thus, the decoupling position for a sliding short in arm 3 is d_3 where

$$\tan \beta_3 d_3 = \frac{X_{23} X_{31}}{X_{12}} - X_{33} \tag{70}$$

Because of the use of cyclic order of the suffices, the corresponding decoupling positions d_1 and d_2 for short-circuits in arms 1 and 2 can be written down at once by permuting the suffices in (70), thus

$$\left. \begin{array}{l} X_1 = \tan \beta_1 d_1 = \dfrac{X_{31} X_{12}}{X_{23}} - X_{11} \\[4mm] X_2 = \tan \beta_2 d_2 = \dfrac{X_{12} X_{23}}{X_{31}} - X_{22} \end{array} \right\} \tag{71}$$

Since no power can enter port 1 when port 3 is terminated by X_3 as given by (69), the input impedance to port 1 must be purely reactive. The value of this reactance can be found from the first equation of (65), putting $I_2 = 0$ and using (67). Thus

$$j X_1^i = \frac{V_1}{I_1} = j \left(X_{11} - \frac{X_{31} X_{12}}{X_{23}} \right) \tag{72}$$

Comparison of (72) with (71) shows that X_1^i is just the negative of the reactance X_1 required to decouple port 2 from port 3.

Now suppose that new reference planes are chosen to coincide with the characteristic planes defined by (70) and (71). That is

$$T_1 P_1 = d_1 \quad , \quad T_2 P_2 = d_2 \quad , \quad T_3 P_3 = d_3$$

Then it follows from (72) that $X_1^i = 0$ when $X_3 = 0$,

i.e. nodes of electric field will appear at the characteristic planes of arm 1 when a short-circuit is placed at one of the characteristic planes of arm 3. The concept of characteristic planes was useful experimentally in the days before the network analyser became generally available and before really wide bandwidths were required, so that slotted-line and sliding-short techniques were quite practacable. This is only rarely the case today, but the characteristic plane concept remains a useful theoretical concept, as we shall now demonstrate.

At the characteristic planes, a new set of equations like (65) must hold, but with new coefficients, say x_{11}, x_{12}, etc. For this special choice of terminal plane locations, $X_n^i = 0$ so that the new reactance coefficients must be related according to (70) and (71), as follows:

$$\left.\begin{array}{l} x_{11} = \dfrac{x_{31}\, x_{12}}{x_{23}} \\[3mm] x_{22} = \dfrac{x_{12}\, x_{23}}{x_{31}} \\[3mm] x_{33} = \dfrac{x_{23}\, x_{31}}{x_{12}} \end{array}\right\} \tag{73}$$

or, more conveniently

$$x_{11}\, x_{22} = x_{12}^{\,2} \quad etc.$$

whence

$$\left.\begin{array}{l} x_{12} = \sqrt{x_{11}\, x_{22}} \\[3mm] x_{23} = \sqrt{x_{22}\, x_{33}} \\[3mm] x_{31} = \sqrt{x_{33}\, x_{11}} \end{array}\right\} \tag{74}$$

Thus, at the characteristic planes, we have

$$V_1 = j\,x_{11}\,I_1 + j\sqrt{x_{11}x_{22}}\,I_2 + j\sqrt{x_{33}x_{11}}\,I_3$$

$$V_2 = j\sqrt{x_{11}x_{22}}\,I_1 + j\,x_{22}\,I_2 + j\sqrt{x_{22}x_{33}}\,I_3 \qquad (75)$$

$$V_3 = j\sqrt{x_{33}x_{11}}\,I_1 + j\sqrt{x_{22}x_{33}}\,I_3 + j\,x_{33}\,I_3$$

This at once leads to a simple transformer equivalent circuit for the 3-port junction, as shown in Fig.6.

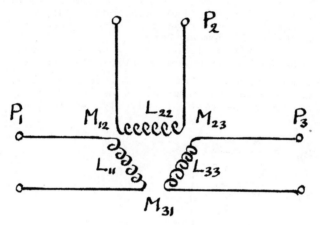

Fig,6   Transformer equivalent of 3-port

Inspection shows that

$$M_{12} = \sqrt{L_{11}L_{22}} \qquad (76)$$

so that the coupling coefficient between coils 1 and 2 must be unity.   Equivalence then requires that

$$\left. \begin{aligned} \omega L_{11} &= x_{11} \\ \omega L_{22} &= x_{22} \\ \omega L_{33} &= x_{33} \end{aligned} \right\} \qquad (77)$$

Matching theorem - I

This matching theorem applies to a symmetrical
T-junction of the kind shown in Fig.4(a). We take arms 1
and 2 to be symmetrical with respect to arm 3. The theorem
may now be stated as follows.
If port 2 is terminated by a perfectly matched load,
then port 1 can be made to present a matched load to its
input waveguide if a short-circuit is appropriately
positioned in arm 3.
This proof is easily obtained with the help of equations (75). First we deduce from the symmetry of ports 1
and 2 that $x_{11} = x_{22}$, and so write for this special
symmetry

$$V_1 = jx_{11}I_1 + jx_{11}I_2 + j\sqrt{x_{33}x_{11}}\,I_3$$
$$V_2 = jx_{11}I_1 + jx_{11}I_2 + j\sqrt{x_{33}x_{11}}\,I_3$$
$$V_3 = j\sqrt{x_{33}x_{11}}\,I_1 + j\sqrt{x_{33}x_{11}}\,I_2 + jx_{33}I_3 \qquad (78)$$

We see at once by inspection that

$$V_1 = V_2 \qquad (79)$$

We are given that port 2 is matched, so that

$$\frac{V_2}{-I_2} = 1 \qquad (80)$$

We require that port 1 should be matched looking
towards the junction so that

$$\frac{V_1}{I_1} = 1 \qquad (81)$$

From the last three equations we deduce that

$$I_2 = -I_1 \qquad (82)$$

It then follows from (78) that

$$\frac{V_3}{-I_3} = -jx_{33} \qquad (83)$$

and this is the reactance to be connected to port 3 to
achieve a match looking into port 1 when port 2 is
terminated by a matched load.

The appropriate position d_3' for the short-circuit in arm 3 is given by

$$\tan\beta_3\, d_3' = -x_{33} \qquad (84)$$

Note that d_3' is measured from a characteristic plane P_3, in the direction away from the junction.

Thus, the theorem is proved.

Matching theorem - II

This theorem can be stated as follows.

It is impossible to make a completely-matched lossless reciprocal 3-port junction.

The proof is most readily done from the scattering matrix approach using the concept of characteristic planes.

In general, for any arbitrary choice of terminal planes we have

$$\left.\begin{aligned}
b_1 &= S_{11}\,a_1 + S_{12}\,a_2 + S_{31}\,a_3 \\
b_2 &= S_{12}\,a_1 + S_{22}\,a_2 + S_{23}\,a_3 \\
b_3 &= S_{31}\,a_1 + S_{23}\,a_2 + S_{33}\,a_3
\end{aligned}\right\} \qquad (85)$$

We now deduce the special relationships between the S_{ij} when the terminal planes are chosen to coincide with the characteristic planes.

From the earlier discussion we know that a short-circuit at P_3 $(b_3 = -a_3)$ will completely decouple port 2 $(b_2 = a_2 = 0)$ from port 1, to which the source is assumed to be connected. We also know that there will be a voltage node $(b_1 = -a_1)$ at the characteristic plane P_1 in arm 1.

Putting these results into (85) yields the following equations

$$-a_1\,(1 + S_{11}) = S_{31}\,a_3$$

$$0 = S_{12}\,a_1 + S_{23}\,a_3$$

$$-a_3\,(1 + S_{33}) = S_{31}\,a_1$$

from which we find

$$1 + S_{11} = \frac{S_{31}\,S_{12}}{S_{23}}$$

$$1 + S_{22} = \frac{S_{12}\,S_{23}}{S_{31}}$$

$$1 + S_{33} = \frac{S_{23}\,S_{31}}{S_{12}} \qquad (86)$$

For the junction to be completely matched, we must have

$$S'_{11} = S_{22} = S_{33} = 0 \qquad (87)$$

Putting (87) into (86) yields

$$\frac{S_{31} S_{12}}{S_{23}} = 1 \quad , \quad \frac{S_{12} S_{23}}{S_{31}} = 1$$

Multiplying these two equations yields

$$\left.\begin{array}{l} S'^2_{12} = 1 \\[6pt] S^2_{23} = 1 \\[6pt] S^2_{31} = 1 \end{array}\right\} \qquad (88)$$

But conservation of power (or the unitary matrix theorem) can be used to obtain a conflicting requirement, thus

$$|a_1|^2 = |b_2|^2 + |b_3|^2$$

or

$$|a_1|^2 = \left(|S_{12}|^2 + |S_{31}|^2\right)|a_1|^2$$

so that

$$|S_{12}|^2 + |S_{31}|^2 = 1 \qquad (89)$$

Since (88) and (89) cannot be satisfied simultaneously the theorem is proved.

Matching Theorem - III

So far we have been assuming reciprocity in our 3-port junction. We next remove this constraint, and find that a completely-matched 3-port junction is now a possibility. The theorem we shall prove is as follows.

Any completely matched lossless 3-port junction is a circulator.

We begin by writing down the scattering matrix for a perfectly matched 3-port.

$$[S] = \begin{bmatrix} 0 & S_{12} & S_{13} \\ S_{21} & 0 & S_{23} \\ S_{31} & S_{32} & 0 \end{bmatrix} \qquad (90)$$

We now introduce the condition of losslessness which states that $[S]$ must be a unitary matrix. The product of the transpose of $[S]$ with the complex conjugate of $[S]$ must therefore produce the unit matrix. Carrying out the multiplication and identifying corresponding elements in the product matrix and the unit matrix yields the following equations:-

$$S_{31}S_{32}^* = S_{21}S_{23}^* = S_{12}S_{13}^* = 0 \qquad (91)$$

$$|S_{21}|^2 + |S_{31}|^2 = 1 \qquad (92)$$

$$|S_{12}|^2 + |S_{32}|^2 = 1 \qquad (93)$$

$$|S_{13}|^2 + |S_{23}|^2 = 1 \qquad (94)$$

Examination shows that there are only two possibilities. Either

$$S_{12} = S_{23} = S_{31} = 0 \; ; \qquad |S_{21}| = |S_{32}| = |S_{13}| = 1 \qquad (95)$$

or

$$S_{21} = S_{32} = S_{13} = 0 \; , \qquad |S_{12}| = |S_{23}| = |S_{31}| = 1 \qquad (96)$$

Both possibilities represent circulators, the difference between them being simply in the direction of the circulators, as shown in Fig.7(a) and (b).

As Fig.7(c) shows, two 3-port circulators can be combined to form one 4-port circulator, and it is clear that the addition of a 3-port circulator to any of these four ports will produce a 5-port circulator, and so on. Thus, without the need for formal proof, we see that the circulators of any arbitrary number of ports may be constructed.

Similarly, we can demonstrate the possibility of an isolator without the need for formal proof. If a matched load is placed on port 3 of the 3-port circulator of Fig. 7(a), a wave entering port 2 will be absorbed in the matched load, and no wave will emerge from port 1.

The resulting scattering matrix is

$$[S] = \begin{bmatrix} 0 & 0 \\ e^{-j\phi} & 0 \end{bmatrix} \qquad (97)$$

for the ideal lossless case.

In practice, of course, isolators are made as single units, as are 4-port circulators; we are only concerned here with theoretical possibilities as opposed to practical design.

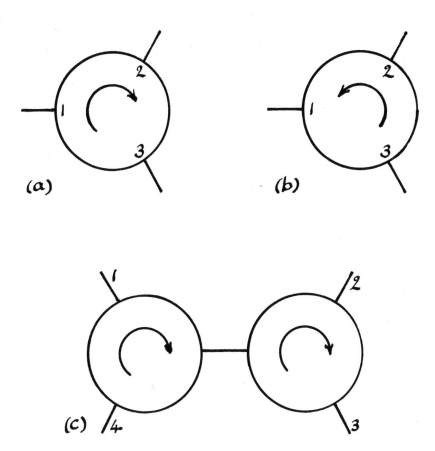

Fig.7   Circulators

4.3.   4-port junctions

In this section we prove two important theorems relating to loss-free reciprocal 4-port junctions, and we also derive some additional results.

Matching theorem - IV

This theorem may be stated thus:
If in a 4-port lossless reciprocal junction there is no coupling between two ports, each of which looks matched

when the other three ports are terminated by matched loads, then the remaining two ports are also decoupled and matched. Let ports 3 and 4 be decoupled and matched, so that

$$S_{33} = S_{34} = S_{44} = 0 \tag{98}$$

$$[S] = \begin{bmatrix} S_{11} & S_{12} & S_{13} & S_{14} \\ S_{12} & S_{22} & S_{23} & S_{24} \\ S_{13} & S_{23} & 0 & 0 \\ S_{14} & S_{24} & 0 & 0 \end{bmatrix}$$

The unitary matrix principle leads to the following results for the diagonal elements of the unit matrix.

$$|S_{11}|^2 + |S_{12}|^2 + |S_{13}|^2 + |S_{14}|^2 = 1 \tag{99}$$

$$|S_{12}|^2 + |S_{22}|^2 + |S_{23}|^2 + |S_{24}|^2 = 1 \tag{100}$$

$$|S_{13}|^2 + |S_{23}|^2 = 1 \tag{101}$$

$$|S_{14}|^2 + |S_{24}|^2 = 1 \tag{102}$$

Adding (99) and (100), and adding (101) and (102) gives:

$$|S_{11}|^2 + 2|S_{12}|^2 + |S_{22}|^2 + |S_{13}|^2 + |S_{23}|^2 + |S_{14}|^2 + |S_{24}|^2 = 2$$

$$|S_{13}|^2 + |S_{23}|^2 + |S_{14}|^2 + |S_{24}|^2 = 2$$

Comparing these two equations, we see at once that

$$S_{11} = S_{12} = S_{22} = 0 \tag{103}$$

and the theorem is proved.

A 4-port junction satisfying (98) and (103) is called a directional coupler. The nomenclature is obviously appropriate when the configuration of the ports is as shown in Fig. 8.

When (98) and (103) are satisfied, a wave incident on
the junction from the left, at either port 1 or 2, will be
wholly transmitted to the right emerging in a certain ratio
from ports 3 and 4.

Conversely, a wave incident from the right on port 3 or
port 4, will be wholly transmitted to the left, emerging
from ports 1 and 2.

The theorem we have just proved is self-evident if
applied to a junction having symmetry about a central plane
as at AA' in Fig.8, so that (103) follows from (98) by
nothing more compicated than a symmetry argument.

It is, however, quite striking when applied to a 4-port
junction such as the magic-tee junction of Fig.9.

It is obvious by symmetry that ports 3 and 4 of the
magic-tee are decoupled; it is equally obvious that for this
very reason ports 3 and 4 can be matched independently by
tuning screws when hte remaining ports are terminated by
matched loads. It is far from obvious, and indeed at first
sight unlikely, that this will result in ports 1 and 2
being decoupled and that it will also result in port 1, for
example, looking matched when ports 2, 3 and 4 are terminated in matched loads.

By similar methods another matching theorem for 4-port
junctions can be proved, though we shall simply state it
here without proof.

Matching theorem - V

Any completely-matched lossless reciprocal 4-port is a
directional coupler

To complete this section on scattering coefficients, we
shall prove a useful result about the phases of the scattering coefficients of a symmetrical directional coupler.

We consider a directional coupler having symmetry about
the planes AA' and BB' of Fig.8. The scattering matrix can
be written.

$$[S] = \begin{bmatrix} 0 & 0 & \alpha & \beta \\ 0 & 0 & \beta & \alpha \\ \alpha & \beta & 0 & 0 \\ \beta & \alpha & 0 & 0 \end{bmatrix}$$

(104)

Applying the unitary principle we find

$$|\alpha|^2 + |\beta|^2 = 1 \tag{105}$$

$$\alpha\beta^* + \alpha^*\beta = 0 \tag{106}$$

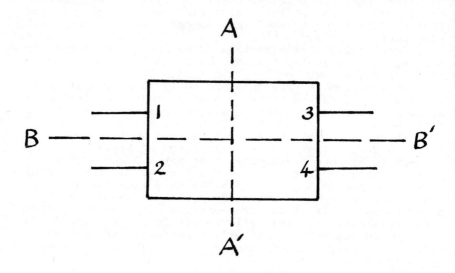

Fig.8   4-port junction

Equation (105) is simply a statement of conservation of energy for the junction. Equation (106) is more interesting, and leads to a very important practical result. We first put

$$\alpha = |\alpha| e^{j\phi_\alpha}$$
$$\beta = |\beta| e^{j\phi_\beta} \qquad (107)$$

Putting (107) into (106) gives

$$2|\alpha||\beta| \cos(\phi_\alpha - \phi_\beta) = 0 \qquad (108)$$

Thus, so long as $|\alpha|$ and $|\beta|$ are both non-zero, we must have

$$\phi_\alpha - \phi_\beta = \pm \frac{\pi}{2} \qquad (109)$$

This states, for example, that if a wave is incident on port 1 of a symmetrical ideal directional coupler, shown schematically in Fig.8, the waves emerging from ports 3 and 4 will differ in phase by 90°. (The sign ambiguity in (109) is not usually important; it depends on the sign conventions

adopted as well as on the nature of the coupling mechanism.)
This 90⁰ phase difference is, moreover, independent of
how the power is shared between ports 3 and 4; it depends
only on the symmetry of the junction and on the assumption
that ports 1 and 2 are perfectly decoupled and perfectly
matched. Thus, it is a property which frequently holds over
a very wide frequency range, even though the power sharing
between ports 3 and 4 may vary widely.

Chapter 3

Circuit analysis

A. L. Cullen

1.   DETERMINATION OF SCATTERING COEFFICIENTS

A 2-port junction such as that in Fig.2 can be represented by a 2 x 2 scattering matrix,

$$[S] = \begin{bmatrix} S_{11} & S_{12} \\ \\ S_{21} & S_{22} \end{bmatrix} \qquad (1)$$

associated with the equations

$$b_1 = S_{11} a_1 + S_{12} a_2$$

$$b_2 = S_{21} a_1 + S_{22} a_2$$

which we have already introduced in **chapter 2,** but which
are reproduced here for convenience.
To determine the scattering coefficients experimentally
we can adopt the following definitions:

$$\left. \begin{aligned} S_{11} &= \frac{b_1}{a_1} \\ S_{21} &= \frac{b_2}{a_1} \end{aligned} \right\} \quad \text{with port 2 perfectly matched}$$

$$\left. \begin{aligned} S_{12} &= \frac{b_1}{a_2} \\ S_{22} &= \frac{b_2}{a_2} \end{aligned} \right\} \quad \text{with port 1 perfectly matched}$$

These definitions are particularly convenient when a network analyser is available, and in this case, the scattering matrix representation is certainly the simplest.

A corresponding simplicity with the impedance (admittance) matrix representation is only possible if the voltage and currents can be measured with the relevant port open (short)-circuited. This is not a convenient condition for wide-band microwave measurements, and the popularity of the scattering matrix over the impedance (or admittance) matrix for measurements, as opposed to certain theoretical calculations, stems largely from this fact.

It is a simple matter to extend the measurement scheme outlined above to the measurement of the scattering coefficients of a general N-port network. All that is necessary is to take two ports at a time, matching the remaining ports, and then to apply the above procedure as if a 2-port junction were being measured. By repeating the procedure as necessary, all the scattering coefficients can be found. Symmetry and reciprocity may reduce the number of scattering coefficients to be measured, but in practice it is wise to ignore these possibilities since the redundancy thus provided is useful in checking the accuracy of the measurements. In general, a 2N-port junction may be measured as N^2 independent 2-port junctions.

2. THE WAVE TRANSMISSION MATRIX

When an N-port junction is considered in isolation, it is convenient to divide the waves in the guides connected to the N ports into two groups, those going towards the junction, denoted by a_i, and those going away from the junction, denoted by b_i.

When the same guide connects two junctions, this grouping is ambiguous until the junction under consideration is specified. No great difficulty arises, of course, so long as one is systematic and careful in carrying out the calculations.

There is, however, a type of calculation that occurs so commonly that it is worth while to adopt a different convention, especially suitable for the purpose, when dealing with it. This is the situation in which one has a chain of 2-port devices connected in tandem. We then divide the waves into a group travelling from left to right, denoted by A_i, and a group travelling from right to left, denoted by B_i. In each case primes may be used to distinguish the waves associated with the different junctions.

For the 2-port junction of Fig.9, we write

$$\begin{bmatrix} A_1 \\ B_1 \end{bmatrix} = \begin{bmatrix} a & b \\ c & d \end{bmatrix} \begin{bmatrix} A_2 \\ B_2 \end{bmatrix}$$

For the left-hand 2-port junction of Fig. 10,

$$\begin{bmatrix} A_1' \\ B_1' \end{bmatrix} = \begin{bmatrix} a' & b' \\ c' & d' \end{bmatrix} \begin{bmatrix} A_2' \\ B_2' \end{bmatrix} \qquad (3)$$

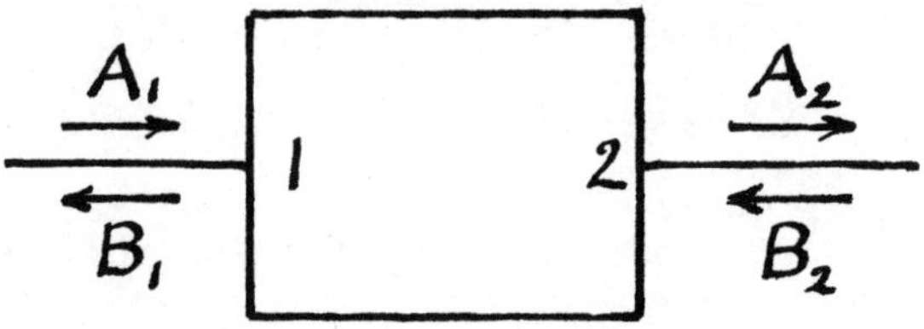

Fig.9   2-port junction

Inspection of Fig. 10 shows that

$$\begin{bmatrix} A_2' \\ B_2' \end{bmatrix} = \begin{bmatrix} A_1 \\ B_1 \end{bmatrix} \qquad (4)$$

Substituting in (3) from (2) and (4) gives

$$\begin{bmatrix} A_1' \\ B_1' \end{bmatrix} = \begin{bmatrix} a' & b' \\ c' & d' \end{bmatrix} \begin{bmatrix} a & b \\ c & d \end{bmatrix} \begin{bmatrix} A_2 \\ B_2 \end{bmatrix} \qquad (5)$$

Thus, the resultant wave transmission matrix of two cascaded 2-port junctions is the product of the individual transmission matrices. Note that the left-to-right ordering of the 2-port devices must be preserved in the multiplication of their associated matrices.
Symbolically,

$$\begin{bmatrix} W_r \end{bmatrix} = \begin{bmatrix} W' \end{bmatrix} \begin{bmatrix} W \end{bmatrix} \qquad (6)$$

where W_r is the resultant matrix.
It follows from (6) and (5) that

$$\begin{bmatrix} W_r \end{bmatrix} = \begin{bmatrix} (a'a + b'c) & (a'b + b'd) \\ (c'a + d'c) & (c'b + d'd) \end{bmatrix} \qquad (7)$$

It is useful to be able to calculate the wave-transmission matrix of (2) from the corresponding scattering matrix。

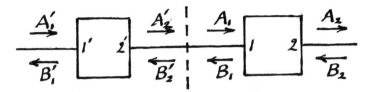

Fig。10   Cascaded 2-ports

With the convention of Fig。9, the scattering equations can be written

$$a = \frac{j}{S_{21}}$$

$$b = -\frac{S_{22}}{S_{21}}$$

$$c = \frac{S_{11}}{S_{21}}$$

$$d = \frac{S_{12}S_{21} - S_{11}S_{22}}{S_{21}}$$

(8)

We can show from (8) that

$$ad - bc = \frac{S_{12}}{S_{21}}$$

(9)

Thus, for a reciprocal 2-port device, for which

$$S_{21} = S_{12}, \qquad ad - bc = 1$$

(10)

and so <u>the determinant of the wave transmission matrix of a reciprocal 2-port device is unity</u>。
If the 2-port is lossless,

$$A_1 A_1^* - B_1 B_1^* = A_2 A_2^* - B_2 B_2^*$$

(11)

Applying the same type of arguments that we have used for the scattering matrix, we find

$$|a|^2 - |c|^2 = 1$$

$$|d|^2 - |b|^2 = 1$$

$$re\,(ab^* - cd^*) = 0$$

(12)

For a lossless reciprocal 2-port device we also have $|S_{11}| = |S_{22}|$ since $S_{12} = S_{21}$; the result follows directly from the conservation of energy. Using (8) this implies that $|b| = |c|$ and from (12) we see that this in turn implies that $|d| = |a|$

Thus

$$|b| = |c|$$

$$|d| = |a|$$

(13a)

These results, together with the last equation of (12) leads to the result (in an obvious notation) that:

$$\phi_a \pm \phi_d = \phi_b \pm \phi_c$$

(13b)

Now using (10) together with (13a) and (13b) we find

$$|a|^2 e^{j(\phi_a + \phi_d)} - |b|^2 e^{j(\phi_a + \phi_d)} = 1$$

and the only way this can be reconciled with the second equation of (12) is if

$$\phi_a + \phi_d = 0$$

Thus we find, finally

$$\phi_a + \phi_d = \phi_b + \phi_c = 0$$

(14)

These results can be used as a check on numerical results for a 'lossless' device whether obtained theoretically or experimentally.

3. WAVEGUIDE NETWORKS

In a microwave system such as a radar antenna network, components such as bends, twists, tapers, etc., may be cascaded together with lengths of waveguide between them. The wave transmission matrix provides a convenient means of handling such a situation. It is only necessary to represent the waveguide in transmission matrix form, and this can be done by inspection.

For a lossless guide, of length L_1, the equations can be written

$$A_1 = A_2 \, e^{+j\beta L_1}$$

$$B_1 = B_2 \, e^{-j\beta L_1}$$

and so the wave transmission matrix is

$$W_1 = \begin{bmatrix} e^{+j\beta L_1} & 0 \\ 0 & e^{-j\beta L_1} \end{bmatrix} \tag{15}$$

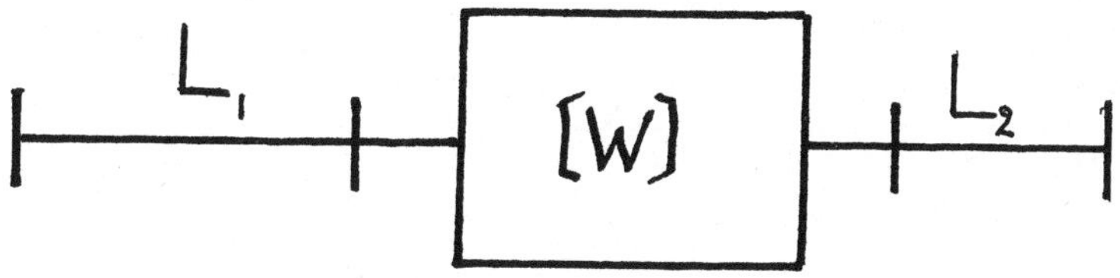

Fig.11 Simple microwave network

We may now derive the transmission matrix for the simple network of Fig.11 thus:-

$$[W_r] = [W_2][W][W_i] \qquad (16)$$

where $[W_2]$ is the wave transmission matrix of a length L_2 of guide. We carry out the multiplication to find:

$$[W_r] = \begin{bmatrix} a\,e^{j\beta(L_1+L_2)} & b\,e^{j\beta(L_1-L_2)} \\ c\,e^{-j\beta(L_1-L_2)} & d\,e^{-j\beta(L_1+L_2)} \end{bmatrix} \qquad (17)$$

We note that if $L_1 = L_2$, the off-diagonal terms are unaffected, since the change in phase due to a wave A travelling from left to right through a length L is exactly opposite to the change in phase of a wave B travelling from right to left through the same length of guide.

4.   TRANSFORMATION OF REFLECTION COEFFICIENT

An advantage of reflection coefficient as opposed to impedance (or admittance) in theoretical work is that its transformation through a length of guide is so simple.
Using (15) we can write

$$\left.\begin{array}{l} A_1 = A_2\,e^{j\beta L} \\ B_1 = B_2\,e^{-j\beta L} \end{array}\right\} \qquad (18)$$

So if we define

$$\left.\begin{array}{l} \Gamma_1 = \dfrac{B_1}{A_1} \\[2ex] \Gamma_2 = \dfrac{B_2}{A_2} \end{array}\right\} \qquad (19)$$

we get

$$\Gamma_1 = \Gamma_2\,e^{-j2\beta L} \qquad (20)$$

The transformation of reflection coefficient by an arbitrary passive linear device is less simple, but equally important, and it has some interesting general properties which we shall now examine. From (2) we get

$$A_1 = a\,A_2 + b\,B_2$$

$$B_1 = c\,A_2 + d\,B_2$$

whence

$$\frac{B_1}{A_1} = \frac{c\,A_2 + d\,B_2}{a\,A_2 + b\,B_2}$$

or using (19) again

$$\Gamma_1 = \frac{c + d\,\Gamma_2}{a + b\,\Gamma_2} \tag{21}$$

Such a transformation is called a bilinear trans-formation, for obvious reasons. Bilinear transformations of special kinds are frequently encountered in microwave network theory – the transformation of impedance along a line, of impedance to reflection coefficient to impedance (or admittance) are a few examples – and so they are worthy of some study.

Mathematically speaking the bilinear transformation of (21) is the most general functional relationship between Γ_1 and Γ_2 for which Γ_1 is uniquely defined when Γ_2 is given and Γ_2 is uniquely defined when Γ_1 is given.

It is convenient to simplify (21) by dividing through top and bottom by a. We can then introduce new complex constants A, B, C and write the transformation as follows:

$$\Gamma_1 = \frac{A\,\Gamma_2 + B}{C\,\Gamma_2 + 1} \tag{22}$$

Since the form of the wave transmission matrix is unchanged by multiplication by other transmission matrices, (corresponding physically to cascade connection of a number of 2-port components to form a composite transmission system) it follows that the transmission of reflection

coefficient through any linear passive transmission system
will always have the bilinear form of (22).
Note the following connections:

$$
\left.
\begin{aligned}
A &= \frac{d}{a} \\
B &= \frac{c}{a} \\
C &= \frac{b}{a}
\end{aligned}
\right\}
\tag{23}
$$

As a specific example, the matrix $\left[W_r\right]$ of (17) has the
following elements:

$$
\left.
\begin{aligned}
a_r &= a\, e^{j\beta(L_1 + L_2)} \\
b_r &= b\, e^{j\beta(L_1 - L_2)} \\
c_r &= c\, e^{-j\beta(L_1 - L_2)} \\
d_r &= d\, e^{-j\beta(L_1 + L_2)}
\end{aligned}
\right\}
\tag{24}
$$

Substituting in (23) the bilinear transformation
constants become:

$$
\left.
\begin{aligned}
A &= \frac{d}{a}\, e^{-j2\beta(L_1 + L_2)} \\
B &= \frac{c}{a}\, e^{-j2\beta L_1} \\
C &= \frac{b}{a}\, e^{-j2\beta L_2}
\end{aligned}
\right\}
\tag{25}
$$

In this particular case, the modification of the
bilinear transformation may be more readily obtained by
transforming the reflection coefficients Γ_1 and Γ_2 through
the appropriate lengths of waveguide L_1 and L_2 respectively.
However, in general, the matrix multiplication rule is
always available and can be used when no short cut is
possible.

5. PROPERTIES OF THE BILINEAR TRANSFORMATION

We conclude this section by proving the important result that a bilinear transformation transforms circles into circles. To avoid suffices we take two complex variables w and z, and we write

$$w = \frac{Az + B}{Cz + 1} \tag{26}$$

Let us rearrange (26) in the following way

$$w = \frac{\frac{A}{C}(Cz + 1) + \left(B - \frac{A}{C}\right)}{Cz + 1}$$

$$w = \frac{A}{C} + \frac{\left(B - \frac{A}{C}\right)}{Cz + 1}$$

or

$$\frac{w - \frac{A}{C}}{B - \frac{A}{C}} = \frac{1}{Cz + 1} \tag{27}$$

We now introduce the new variables W and Z defined by

$$W = \frac{w - \frac{A}{C}}{B - \frac{A}{C}}$$

$$Z = Cz + 1 \tag{28}$$

so that

$$W = \frac{1}{Z} \tag{29}$$

We assume that Z describes a circle of radius R and centre at S in the complex Z-plane. The complex distance between a point Z on the circle and its centre is clearly Z - S, and we must have

$$|Z - S| = R$$

or

$$(Z - S)(Z^* - S^*) = R^2$$

which leads to

$$ZZ^* - SZ^* - S^*Z - R^2 = O \qquad (30)$$

The general form of the equation is

$$ZZ^* + PZ + P^*Z^* + QQ^* = O \qquad (31)$$

If we encounter an equation of the form of (31) we know that it represents a circle in the complex plane with centre S and radius R given by

$$S = -P^*$$

$$R = \sqrt{(PP^* - QQ^*)} \qquad (32)$$

We now return to (29). Z describes a circle and therefore satisfies (31). What is the locus of W? Substituting (29) in (31) gives

$$\frac{1}{WW^*} + \frac{P}{W} + \frac{P^*}{W^*} + QQ^* = O$$

or

$$WW^* + \left(\frac{P^*}{QQ^*}\right)W + \left(\frac{P}{QQ^*}\right)W^* + \frac{1}{QQ^*} = O \qquad (33)$$

This is of the same form as (31) and so the proof is completed.

One of the most important practical results of this transformation is that the locus of the reflection coefficient at the input port of a 2-port (or of a cascaded system of 2-ports) terminated by a movable short-circuit, is a circle. This follows at once when we recognise that the locus of the reflection coefficient at the output port is a circle of unit radius centred on the origin in the complex reflection-coefficient plane, and that this reflection coefficient undergoes a bilinear transformation from the output port to the input port.

It is easily shown that the radius of the circle described by the input reflection coefficient is

$$\frac{|S_{12}|^2}{1 - |S_{11}|^2}$$

if the device is reciprocal and if the short-circuited guide is connected to port 1. This is the <u>transmission efficiency</u> of the network (defined as the power reaching a matched load connected to port 2 when unit power is incident on port 1 from a matched source). The <u>transmission efficiency in the opposite direction is not necessarily the same, even for reciprocal devices</u> since it is given by

$$\frac{|S_{12}|^2}{1 - |S_{22}|^2}$$

and $|S_{11}|$ and $|S_{22}|$ are not necessarily equal for 2-port devices having losses.

6. SIGNAL FLOW GRAPHS

The method of signal flow graphs enables calculations of the properties of complicated microwave networks to be carried out simply and systematically. It is of particular value in dealing with networks connected in cascade. In essence, it consists of representing complex wave amplitudes by points in a diagram and scattering parameters of the networks by directed lines connecting these points so that, for example, s_{21} of a two-port is represented by a line joining the input wave amplitude a_1 to the output b_2. By applying certain rules it is then possible to simplify the diagrams of complicated circuits and deduce their overall properties.

A brief account of the method of signal flow graphs and its uses is given in Appendix A, page 453.

Chapter 4

Modern lines

R. J. Collier

1. INTRODUCTION

In recent years the number of new transmission lines
that have become available for use in microwave circuits has
greatly increased. The designer may choose a particular
transmission line for one part of a circuit because its
properties match those required. Indeed, a circuit may be
a combination of many different forms of line. Unfortunately, direct measurements of many of these circuits is often
impossible. This is because most microwave measurement
techniques use coaxial lines at lower microwave frequencies
(~ up to 20 GHz) and various rectangular waveguides for the
higher frequencies. So every measurement must be made via
a transition which converts the electromagnetic wave from a
mode on, say, coaxial line to a mode on another line. These
transitrons are far from perfect and the techniques used to
measure 'beyond' the transition will be described in later
chapters. This chapter will attempt to summarise some of
the properties of modern lines and discuss the problems of
transitions between them. The chapter on transmission lines
covered rectangular waveguide in some detail so that will
not be repeated here.

Most microwave circuits contain active devices which
have to be matched to the characteristic impedance of the
input and output lines. In many applications a wideband
circuit is required and, since the impedance of the device
and the line vary with frequency, this can prove to be a
major design problem. A transmission line may be selected
which 'best fits' the active device.

In some matching networks and filters, the phase
constant of the line is a critical parameter. Again, as
this is not a linear function of frequency, considerable
design problems arise. Indeed, special dispersive sections
of line are often used to compensate for frequency variation.

In this age of micro-electronics, small transmission
lines may be used to reduce the volume and to make them
compatible with small devices. Although this reduces the
problem of higher order modes the losses may dramatically
increase due to the compression of the currents. At higher
frequencies, the skin effect losses and those resulting from
the surface roughness of metallic conductors can be
prohibitively large.

Finally, some lines are used either because of their good coupling to adjacent lines, or their ability to form radiative elements of a linear phase array, or their low losses giving high 'Q' resonators, or their ease of manufacture, or their compatibility with solid state devices, or their low cost! The lines described in this chapter will be those using mainly metallic guiding structures - coaxial, rectangular waveguide, microstrip and fin line, and those using mainly dielectric guiding structures - Image guide, Dielectric Guide and H-Guide.

2.   METALLIC WAVEGUIDES

Metallic waveguides use two distinct methods of guiding waves. The first uses a pair of balanced lines (i.e. where the currents induced are equal and opposite) which are arranged to be sufficiently close so that their far fields and, in particular, their radiation fields, all cancel. This leaves fields just in the vicinity of the lines and also with rapidly decaying fields outside. All of these structures support a mode which will propagate down to 0Hz and higher order modes only occur when one of the dimensions transverse to the direction of propagation is approaching a wavelength. If the line has only one dielectric and the attenuation is negligible, then the line will be almost non-dispersive i.e. the phase velocity and the Characteristic impedance will be independent of frequency. For coaxial lines the fields outside cancel completely for the fundamental mode. If a metallic sheet is placed between the lines in a balanced system a single line induces currents in the sheet which are equivalent to a balanced pair. Thus, microstrip and its many forms use this principle. In the cases where mixed dielectrics are used the modes will be slightly dispersive. At very low frequencies the waves will spread into all the dielectrics and have a low effective dielectric constant. This always gives a higher velocity and, in some cases, a lower characteristic impedance. At higher frequencies the wave will propagate only in the region of highest dielectric constant. This will make the effective dielectric constant equal to the dielectric constant for this region. In this case the wave will have a lower velocity and usually a higher characteristic impedance.

2.1   CO-AXIAL LINES

Co-axial lines are normally used with one mode propagating, the transverse electromagnetic mode, (T.E.M. mode), and the following equations apply for its design:

$$L = \frac{\mu}{2\pi} \ \log_e \left\{ \frac{b}{a} \right\} \qquad\qquad C = 2\pi\varepsilon \left(\log_e \left\{ \frac{b}{a} \right\} \right)^{-1}$$

$$Z_o = \sqrt{\frac{L}{C}} \; = \; \sqrt{\frac{\mu}{\epsilon}} \; \log_e \left\{ \frac{b}{a} \right\} \qquad v_p \; = \; \frac{1}{\sqrt{\mu\epsilon}}$$

At high frequencies the skin effect loss is:

$$R \; = \; \frac{R_S}{2\pi} \left(\frac{1}{b} + \frac{1}{a} \right)$$

The dielectric loss term is:

$$G \; = \; 2\pi\sigma \left(\log_e \left\{ \frac{b}{a} \right\} \right)^{-1}$$

$$\lambda \; = \; \frac{R}{2Z_o} \; + \; \tfrac{1}{2} G Z_o \quad \text{nepers} \quad m^{-1}$$

where a, and b are the inner and outer radius as shown in
the diagram below. μ is the permeability of the medium
between the conductors, normally $\mu = \mu_o$ or $4\pi.10^{-7}$ Hm^{-1}.
ϵ is the permittivity of the medium between the conductors
and $\epsilon = \epsilon_R \epsilon_o$. ϵ_R is the relative dielectric constant and
ϵ_o is 8.854 p F m^{-1}. R_S is the skin resistance of the
conductors. σ is the conductivity of the dielectric. λ is
the attenuation constant.

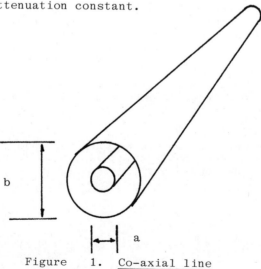

b

a

Figure   1.   Co-axial line

Co-axial lines can easily be made to have characteristic
impedances in the range 20Ω to 100Ω and their dispersion
characteristics are good. A detailed account of those lines
is given in Chipman (1) and Baden Fuller (2). It can be
seen from the equations for resistance that larger values of
a and b reduce the series resistance. However, as the cable
increases in diameter higher order modes may be able to
propagate. As a general guide, when the free-space wave-
length is less than $2\pi\, b\sqrt{\epsilon_R}$ then the higher order modes can
be present. So most cables are used with a diameter small
enough to avoid higher order modes and yet large enough to
give acceptable attenuation. The exact theory is given in
Baden-Fuller (2). The limit for most co-axial cables is
around 25 GHz with b = 3mm. Most measurement techniques use
co-axial parts up to this frequency. Since co-axial cables
are much smaller than rectangular waveguides at the lower
microwave frequencies, they are used as input and output
ports and for connections between sub-assemblies. Transitions between co-axial lines and most other lines exist.
Particularly common are those to waveguide and microstrip;
the latter transition being more complex as it possesses
radiation losses and has poor repeatability.

2.2 RECTANGULAR WAVEGUIDE

Along with co-axial cable, rectangular waveguide is the
most widely used microwave transmission line for input and
output ports on many devices particularly above 25 GHz.
Since rectangular waveguide is limited to fixed frequency
bands most of the waveguide devices are similarly restricted.
For measurement systems an entire series of interchangeable
waveguides is required to cover the entire microwave range.
For example, consider X-band guide. The cut off frequency
for the lowest order mode, the TE_{10} mode, is 6.557 GHz.
However the guide is unuseable above this frequency as both
the phase velocity and the characteristic impedance are
changing rapidly with frequency. So it is normal practice to
use the guide well above the cut off frequency where those
parameters are not so frequency dependent i.e. at 8.2 GHz,
and are a factor of 1.66 away from their free-space values.
The next higher order modes arrive at 13.114 GHz (twice the
cut-off frequency for the TE_{10} mode). However, in order to
avoid these modes it is normal to use the guide up to 12.4
GHz, where the parameters are a factor of 1.18 away from
their free-space values. This means that a waveguide bandwidth is much less than an octave being 8.2 GHz to 12.4 GHz
at X-Band. Waveguides are dispersive, particularly at the
lower frequencies. At low microwave frequencies, waveguides
are able to handle more power than most other guides due
mainly to their size. Figure   2 gives the properties of 3
guides to illustrate the points made.

Band Letter	Cut off Frequency GHz	Frequency Range GHz	Power Rating MW	Atten- uation dB m^{-1}	Dimensions "a"	"b"
S	2.078	2.60-3.95	2.5	0.033	2.84	1.34
X	6.557	8.20-12.4	0.2	0.164	0.9	0.4
Q	21.081	26.5 -40.0	0.022	0.82	0.28	0.14

Figure   2. Waveguide Properties (3)

2.3 RIDGED WAVEGUIDE

A technique for increasing the bandwidth of rectangular waveguides is to use a central ridge - see figure   3 - which lowers the cut-off frequency of the lowest order mode. This can make the bandwidth increase by a factor of as high as three. The wave is concentrated within the region of the ridge and the guide has better dispersion properties, but larger losses than the ordinary rectangular guide.

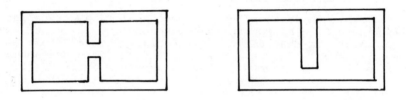

Figure   3. Ridged Waveguides

It may help understanding this concept by considering the cut-off frequency using the transverse impedance method. For ordinary rectangular waveguides at cut-off there is a plane wave moving in just the transverse direction and forming a standing wave. The impedance at the walls is zero for the transverse electric fields. Using a transmission line analogy, the impedance a distance a away from the walls is

$$Z_{1N} = j\eta_o \tan (K_o a)$$

where η_o is the intrinsic impedance of free-space and K_o is the free-space wave number. By finding the values of K_o for $Z_{1N} = 0$ the various cut-off wavelengths, λ_c, can be found.

$$Z_{1N} = O \text{ if } \lambda_c = \frac{2a}{n} \qquad n = 1, 2, 3.$$

For a ridged guide, the ridge can be represented as a capacitance, C, mounted halfway across the guide.

The expression for Z_{1N} is now

$$Z_\mu = j\eta_o \left\{ \frac{2 - \eta_o \, \omega \, C \tan \theta}{\eta_o \omega c - 2\cot 2\theta} \right\}$$

where $\theta = \frac{1}{2}K_o a$. Now the cut-off frequencies are found by setting $Z_{1N} = O$ as before. So $Z_{1N} = O$ if $\eta_o \, w \, C \tan \theta = 2$

$$\text{or } \lambda_c = \frac{a}{n} \qquad n = 1, 2, 3.$$

The first of these conditions gives a solution for a mode cut-off at a much lower frequency than the rectangular guide - see figure **4.** The second condition gives the cut-off frequencies for the TE_{20}, TE_{40} etc. modes which are unaffected by the ridge as they have a zero of electric fields at the centre of the guide.

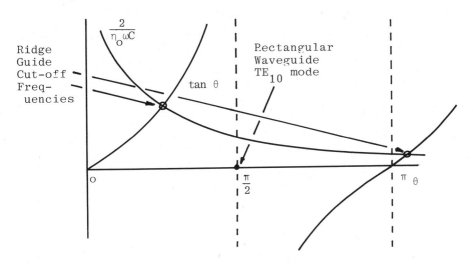

Figure 4.   Ridge Waveguide Cut-Off Frequencies

A reasonable coverage of the properties of these guides can be found in the Microwave Engineer's Handbook (3).

2.4 FIŃ LINE

A modern line which uses several of the properties of ridged waveguide is called Fin Line (see Figure   5). The ridge is made by inserting a printed circuit into the rectangular guide. The printed circuit can be made very accurately using photolithographic techniques. Since the fields are strongest between the 'fins', the dimensions and surface roughness of the rest of the guide is not critical. Various papers by (4), (5), (6) cover the properties of Fin Line and the transitions between it and rectangular waveguide. The use of the printed circuit board means that solid state components can be mounted directly onto the board before

insertion into the guide and successful fin line circuits
have been made up to 100 GHz.

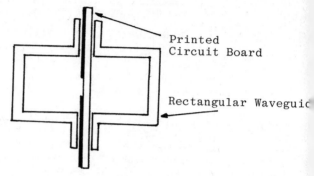

Printed
Circuit Board

Rectangular Waveguide

Figure   5. Fin Line

The characteristic impedance is typically 10 to 400
ohms depending on the configuration. The phase velocity
is just below the free space velocity, due to the dielectric,
and does show some dispersion. The phase velocity falls
slightly with increasing frequency.

2.5 MICROSTRIP

Microstrip transmission lines are used in many circuit
applications over most microwave frequencies. Figure 6
shows the cross-section of a microstrip line. This is a
microwave form of printed circuit and can be produced by
conventional photolithographical techniques with accuracies
approaching ± 0.5 μm.

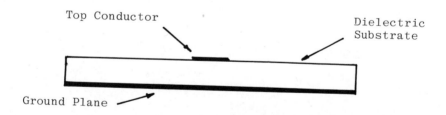

Top Conductor

Dielectric
Substrate

Ground Plane

Figure   6. Microstrip Cross-section

Since the waves travelling down the microstrip move
partly in the dielectric substrate and partly in the air
above the substrate, the velocity of the waves is given by:

$$\frac{3.10^8 \text{ ms}^{-1}}{\sqrt{e_R}} < \nu_p < 3.10^8 \text{ ms}^{-1}$$

where ε_R is the relative dielectric constant of the substrate. Thus, the waves move faster than a plane wave in the dielectric substrate but slower than a plane wave in air. This is usually described by an effective dielectric constant, ε_R effective, which is given by

$$\nu_p = \frac{3.10^8 \text{ ms}^{-1}}{\sqrt{\varepsilon_R \text{ effective}}}$$

and so

$$1 < \varepsilon_R \text{ effective} < \varepsilon_R$$

Dispersion in microstrip is much less than waveguide. As the frequency increases ε_R effective $\to \varepsilon_R$ and Z_0 also increases with frequency . A 50Ω line at 0Hz might become a 54Ω line at X Band and the phase velocity falls by 10% over the same frequency range. The width of the top conductor, w, and the thickness of the dielectric, h, determine the values of ν_p and Z_0. For a 50Ω line at X Band typical values might be $\varepsilon_R = 9.7$ (Alumina) $w = h = 0.6$mm, $\nu_p = 1.15.10^8$ ms^{-1}, ε_R effective $= 6.76$, $\alpha = 3$dBm^{-1}.
The attenuation of microstrip is much greater than that for rectangular waveguide but as the circuits are often only a few wavelengths long, this is not a critical factor. A good summary of papers on microstrip is given in references (3), (7), (8) and (9). The range of characteristic impedance available in microstrip is around 5Ω to 150Ω and so is ideal for both solid state devices and filter construction. Solid state devices can be mounted directly onto the microstrip circuit without encapsulation. In cases where a large number of circuits is required, microstrip is an ideal form for monolithic technology. Various forms of microstrip exist e.g. Slot line (similar to fin line), inverted microstrip, stripline, triplate and coplanar line and these will be found in the literature - see references (11) and (12).

3.   DIELECTRIC WAVEGUIDES

One of the problems which arise at high microwave frequencies and continue through to optical frequencies is the increase losses in metallic structures. These can be summarised as follows:

(1)   The Skin effect increases losses proportional to $f^{\frac{1}{2}}$ and above 35 GHz even more due to the anomalous skin effect (10).

(2)   The reduction in the size of the transmission line, to avoid higher order, increases the losses proportional to f.

(3)   The surface or edge roughness, particularly in regions of high current density will also increase the losses, particularly above 100 GHz.

So any metallic waveguide has increasing losses at higher frequencies even if the losses are expressed as dB per wavelength. Although dielectric structures do have losses, their loss tangents can be very low at all frequencies. At optical frequencies the very high purity of optical fibres can give attenuations less than 3 dB Km^{-1}. So it is not surprising that the answer to increased losses in metallic waveguides is to use non-metallic waveguides (i.e. dielectric waveguides). Since the basic principle of all dielectric waveguides is that of total internal reflection, the surface of the guides has to be smooth to avoid radiation losses. Also the medium surrounding the guide has also to be free of losses and be uniform in the near region where the fields are still significant in amplitude. At optical frequencies the guide may be only 3μm in diameter so it is relatively easy to surround it with another dielectric medium which will totally control the fields. A typical outside diameter might be 150μm. At microwave frequencies a dielectric guide might be 2mm square and would need surrounding by 100mm diameter dielectric. It is the practice to use air as the surrounding medium, in most cases.

3.1. RECTANGULAR DIELECTRIC WAVEGUIDES AND IMAGE GUIDES

The cross-section of these guides is shown in figure 7. The modes which propagate in these guides are similar to those found in rectangular waveguides except that there is, in addition, a slow wave in the surrounding medium. So the wave in the dielectric cannot go faster than the velocity of a plane wave in the surrounding medium. Using the concept of effective dielectric constant the velocity, v_p, lies between the following:

$$\frac{3.\ 10^8\ ms^{-1}}{\sqrt{\epsilon_2}} \quad > \quad \frac{3.\ 10^8\ ms^{-1}}{\sqrt{\epsilon_{Reffective}}} = v_p \quad > \quad \frac{3.\ 10^8\ ms^{-1}}{\sqrt{\epsilon_1}}$$

$$\epsilon_2 \quad < \quad \epsilon_{Reffective} \quad < \quad \epsilon_1$$

Where ϵ_1 is the relative dielectric constant of the waveguide and ϵ_2 is that of the surrounding medium.
So the waves in the waveguide are fast waves like those in rectangular metallic guides but they are never infinite. The dispersion is therefore much less than rectangular waveguides, with:

$$\epsilon_{Reffective} \quad = \quad \epsilon_2 \qquad \text{at the cut off frequency}$$

$$\epsilon_{Reffective} \quad = \quad \epsilon_1 \qquad \text{at very high frequencies}$$

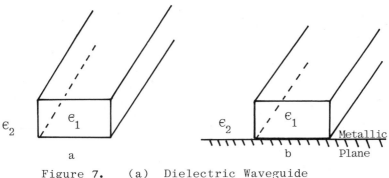

Figure 7. (a) Dielectric Waveguide
            (b) Image Guide

References (13), (14) and (15) give some of the
properties of these waveguides. The usuable bandwidth is
greater than that of rectangular waveguide. Near to cut-off
the energy in the external slow wave is very high and the
fields are very extensive. So the guides are usually used
well above their cut-off frequencies. At higher frequencies
the next mode begins to propagate and this marks the limit
for monomode propagation.
    Dielectric Guides surrounded by air have the lowest
losses. However, it is sometimes convenient to have a ground
plane which eliminates some of the higher order modes and
increases the monomode bandwidth. This image guide form is
similar to dielectric guides although not equivalent and has
certainly been used for many microwave circuits. Several
components like directional couplers, isolators, filters etc.
have been designed and this technology will increase at the
higher microwave frequencies.

4. TRANSITIONS

    One of the obstacles to measurements on modern circuits
is the need for transitions. Indeed, in some cases, this
obstacle may determine whether or not it is used. For microstrip, the transition most commonly used is to coaxial cable.
For a transition to be satisfactory, the mode on one line
should be 'smoothly' transformed to the mode on the next line.
Part of the problem is to make sure that the cross-section of
the mode is small enough for a microstrip circuit and this
often involves miniature 3mm coaxial cables. Another part
of the problem is to make sure that the transverse field
patterns are similar and finally, to match any changes in
dielectric between the coaxial cable and the microstrip
substrate. It is never wise to assume a 50Ω cable will
connect to another 50Ω cable without checking some of the

above criteria.
  The most common transition, or launcher, to dielectric
guides is to use a horn. The dielectric is usually extended
into the waveguide where a long taper section ensures that
the wave becomes internally trapped. The purpose of the
horn is to launch the external slow wave which is essential
to make up the complete dielectric mode, particularly at
frequencies near cut-off.
  Transitions, in general, will cause mismatches and
losses. The mismatches can sometimes give rise to multiple
reflections which will cause ripples on swept frequency
displays. Some transitions also radiate and this can also
interfere with both adjacent circuits and the radiation
pattern of associated aerials. Considerable measurement
time can be spent on de-embedding the transition in order
to measure a circuit and the topic should be approached
with caution.

5.   REFERENCES

1.   Chipman, R. A.:   "Transmission lines", Schaums Outline
       Series, Mc Graw-Hill.
2.   Baden-Fuller, A. J.:   "Microwaves", Pergamon Press.
3.   "Microwave Engineers Handbook", Vols 1,2.   Artech House.
4.   Saad, A. M. K. and Begemann, G.:   "Electrical
       Performance of Fin Lines of various Configurations"
       IEE MOA, January 1977, Vol 2, pp 81-81.
5.   Meier, P. J.:   "Millimeter Integrated Circuits suspended
       in the E-plane of Rectangular Waveguide", IEEE
       MTT - 26, No. 10, 1978, p 727-733.
6.   Bates, R. N. and Coleman, M. D. :   "A Study of
       Integrated Fin Line as a transmission medium for
       Millimetre Waves", Philips Technical Note, No. 175%
7.   Wolff, I. :   "Microstrip Bibliography", H. Wolff,
       Aachen, Germany.
8.   "Advances in Microwaves", Vol. 8, 1974, Academic Press.
9.   Edwards, T. C. :   "Foundations for Microstrip Circuit
       design", Wiley.
10. Tischer, F. J. :   "Excess Conduction losses at Millimetr
       Wavelengths", IEEE Mtt - 24, No. 11, 1976,
       p 853 - 858.
11. Gunston, M. A. R. :   "Microwave Transmission Line
       Impedance Data", Van Nostrand, Reinhold.
12. Hilberg, W. :   "Electrical Characteristics of
       Transmission Lines", Artech House.
13. Unger, H.G. :   "Planar Optical Waveguides and Fibres",
       Oxford.
14. Kapany, N. S. and Burke, J. J. :   "Optical Waveguides",
       Academic Press.
15. Owyang, G. H. :   "Foundations of Optical Waveguides",
       Arnold.

Chapter 5

Reflections and matching

E. J. Griffin

1. INTRODUCTION

RF measurements of the performance of equipment, in terms of output power, gain, impedance, effective noise temperature and similar quantities, can generally be made only at the input or output waveguide flanges or coaxial connectors. Reflections of signals arising from inequality of impedances at these connections can cause uncertainty in the results of measurement: the significance of this 'mismatch uncertainty' will be treated in the lectures on specific measurement techniques. One way to reduce the uncertainty is to minimise the reflections with the aid of adjustable impedance transformers (tuners): the use and limitations of these is the main topic of this chapter.

2. TERMINOLOGY

Much of the literature on matching has only tenuous connection with the practical needs of the metrologist. This section sets our topic in the context of this literature.

2.1. Reflection is expressed quantitatively as the voltage reflection coefficient (VRC) relative to the characteristic impedance Z_o of the mode transmitted by the transmission line or waveguide. For a device of complex impedance Z_L, the VRC relative to Z_o is expressed precisely as

$$\Gamma = (Z_L - Z_o)/(Z_L + Z_o) \quad \text{(Chap. 1, section 2.6)}$$

The word 'match' is used in various ways in the literature. For clarity, we define

a) an impedance match when $Z_L = Z_o = R_o + jX_o$, when there is no reflection,

b) a conjugate match when $Z_L = Z_o^* = R_o - jX_o$, when there is maximum power transfer.

If the transmission system is lossless, $X_o = 0$ and the two conditions are identical: we shall use the term 'resistance match' to describe this. Because practical transmission lines and waveguides usually have only small losses, the approximation is usually made that impedance and resistance match conditions are the same.

2.2.   This approximation is valid for most work but the
losses in the standard of Z_O (e.g. a precision coaxial line)
may be significant at vhf and must then be accounted for in
assigning a value to Γ to refer it to the nominal R_O of the
standard line: otherwise the results of independent measurements of Γ can only be compared with reference to the same
physical standard.   These losses in impedance standards
arise from the variation with frequency of the current distribution (1) in the conductor(s).   Whether or not they are
significant for particular measurements of Γ may be judged
from the calculated theoretical variation of $|Z_O|$ of a
nominal 50 ohm air-spaced 14 mm coaxial line:

Frequency in GHz:-	0·1	1·0	10·0
$|Z_O|$ for brass conductors in ohms	50·531	50·110	50·031
$|Z_O|$ for silver conductors in ohms	50·090	50·028	50·008

Surface finish imperfections cause the surface resistivity
and internal inductance to exceed these figures calculated
from bulk match resistivity values (2) so that for highest
accuracy measurements of Γ below about 2 GHz, the frequency
dependence of Z_O must be measured (3, 4, 5).

2.3.   Conjugate matching to achieve maximum power transfer
from a source is treated later but a warning may be appropriate here about manufacturers' specification of 'minimum
power output' from a source.   This sometimes means the
minimum power delivered by the source to a load which is the
conjugate match to the (unstated) impedance of the source,
rather than that delivered to a load matched to the impedance
of the appropriate transmission line.   The need to instal
and manipulate a tuner to extract the specified minimum power
at each frequency may be intolerable in a wideband measurement system.

2.4.   For the rest of this chapter, unless otherwise stated,
the term 'matched' means matched to the characteristic
impedance Z_O of a transmitted mode as an approximation to
the nominal characteristic impedance R_O in a lossless line.

2.5.   The literature on matching sometimes discusses "broad-
band" and "narrowband" matching, loosely meaning bandwidths
of 5-10% and <1% respectively.   A matching network is
essentially a filter which ideally achieves zero reflection
at frequencies within the passband.   Even an ideal filter
can provide zero reflection only at discrete frequencies, so
the word 'bandwidth' is meaningful only if the frequency
range is accompanied by a statement of the acceptable VRC.
Generally, 'narrowband' in this context means tuned to
provide a match at a single frequency, the bandwidth over
which the VRC is acceptable being more less fortuitous.

2.6.   Classical filter theory can be used to design matching
networks up to about 100 MHz using air-cored coils and miniature capacitors as lumped components.   At higher frequencies
printed circuit components can be used (6) and filter theory
has been developed for matching networks using distributed

components at microwave frequencies (7). When the equipment
reflection coefficients that can be achieved by these means
cause unacceptably high mismatch uncertainties we may use
tuners. These are essentially series, shunt or cascade
connections of adjustable reactive (or susceptive) elements
producing a match at the measurement frequency.

3. TUNER ELEMENTS

3.1. Impedance can be transformed by a length of transmission line or waveguide; if its length can be varied by a
sliding short circuit, an adjustable reactance can be realised. In this section we examine these and other 'building
blocks' for tuners.

3.2. A length l of lossless line transmitting to an impedance Z_L a mode of characteristic impedance R_0 has an input
impedance:

$$Z_{in} = R_0(Z_L + jR_0\tan\beta l)/(R_0 + jZ_L\tan\beta l) \qquad (1)$$

where $\beta = 2\pi/\lambda$, λ being the wavelength of the mode and l
being positive in the direction away from the load toward
the source. The use of the Smith chart to represent this
transformation of complex impedance is treated in the appendix at the end of this chapter. We shall use it later to
illustrate the effect of positioning variable susceptances
along the line. Meanwhile we note that if $Z_L = 0$, then
from equation (1) the input impedance is

$$Z_{sc} = jR_0\tan\beta l \qquad (2)$$

so that an adjustable reactance can be realised by varying
the position of a sliding short-circuit along a line.

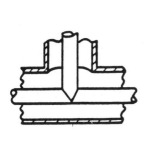

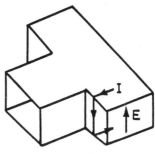

FIG. I a.   FIG. I b.

COAXIAL LINE
T - JUNCTION

H - PLANE JUNCTION

3.3. Shunt connection of this short-circuited line requires
a coaxial-line T-junction, fig 1a, or its waveguide counterpart, an H-plane junction, fig 1b, in which the junction is
formed in the narrow wall of the guide. (An analogy between

the two is that the voltage between the two conductors of the
coaxial T is uniform at low frequencies and, correspondingly,
the E field is continuous throughout the H-plane junction).
In both, maximum power transfer occurs when the signals at
the two ports of the main arm are in phase.   In neither is
the calculation of the susceptance arising from electromagnetic field distortion at the discontinuities simple (8,9)
so that neither is very amenable to calculation for design,
although both are satisfactory for use with a short-circuited
stub in a tuner.   Commercial coaxial-line stubs have up to
100 mm of adjustment with rod actuation of the short circuit
and up to 750 mm when it is moved by pins through slots in
the outer conductor to achieve a reduction in the overall
length; lines with helical inner conductors are also used to
attain about a 7:1 reduction in size, so that they may be
used down to about 40 MHz.   In waveguide, the shunt stub is
used either with a cascade connected guide or in an E-H
tuner (see section 6 below).

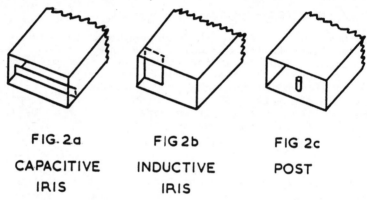

FIG. 2a   FIG 2b   FIG 2c

CAPACITIVE   INDUCTIVE   POST

IRIS   IRIS

3.4.   Susceptance can also be introduced into waveguide by
obstacles such as posts or irises, fig 2.   An abrupt change
of guide cross-section causes distortion of the electromagnetic field which results in scattering of energy both
sides of the obstacle (the limiting case being reflection
from a short circuit).   This scattering stimulates evanescent TE or TM modes whose amplitude decays exponentially with
distance because the dimensions of the waveguide preclude
their unattenuated propagation.   The energy storage associated with these modes gives rise to the susceptance.   With
TM modes the susceptance is capacitive and with TE inductive.   This implies that if the edges of the obstacle are
parallel to the E field it is inductive and conversely, as
may be seen from fig 2.   This is an approximation true only
if the obstacle has zero thickness for only then does it
behave as a pure shunt susceptance; finite thickness introduces series reactance (10).   The literature contains tabulated values relating to a variety of shapes of iris, step
and post (11,12) to serve as a guide for experiment.   The
aperture of an iris is not generally adjustable for tuning
because of imperfect contact between the diaphragm and waveguide walls.   Posts and capacitative irises cause degradat-

ion of the power rating of the waveguide because of decreased spacing in the E field direction and the high field intensity at sharp edges, so neither is best suited for tuners for high power systems.

3.5. The simplest adjust-able post is a screw pene-trating through the broad face of a waveguide or the outer conductor of a coax-ial line. In waveguide, the susceptance of a thin post is positive (capacit-ative) for small insertion but increases to infinity as the penetration is increased to about $\lambda_g/4$. Further insertion beyond this resonance causes the susceptance to be negative (inductive) and to decrease with increasing insertion.

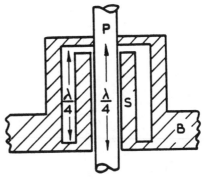

FIG.3.

HALF-WAVE FOLDED CHOKE

In coaxial line, increasing probe insertion causes only increasing capacitance. At higher microwave frequencies, power can be coupled out of the guide along the imperfectly contacting thread of a screw acting as a post and the posit-ion of the rf contact along the thread is often erratic. These troubles can be minimised by the incorporation of a choke such as the half-wave folded choke of fig 3. This comprises two cascaded $\lambda/4$ coaxial line sections, the two conductors of the first section being the probe P and sleeve S while those of the second section are the sleeve S and the body B. We see from equation (2) that a short-circuited quarter-wavelength line presents an open circuit at its input, so the short-circuited second section formed by S and B presents an open circuit at the junction with the first section and thus a short-circuit at the waveguide wall, so obviating leakage. The indeterminate impedance of the screw thread is in series with the open circuit at the junction, thus obviating also any effects arising from erratic contact of the thread. This kind of choke can be made effective over a waveguide bandwidth by making the ratio of the char-acteristic impedance of the second section to that of the first as large as possible.

3.6. Chokes are also often used in the remaining 'building block' we consider for tuners: the sliding short-circuit. Even a plane integral with the waveguide walls and perpen-dicular to them is slightly dissipative so that the VRC presented by a practical short circuit is finite and so, therefore, is the limiting reactance presented by a short-circuited stub. If a sliding plunger is fitted with spring contacts rubbing against the waveguide walls then the contact resistance of these will contribute to the imperfect perfor-mance of the short-circuit and a compromise is necessary

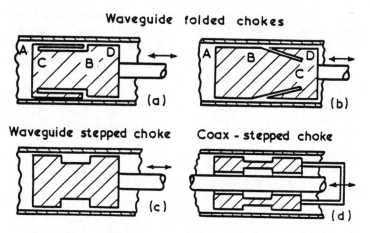

FIG.4 CHOKED SHORT CIRCUITS

between providing sufficient force for good contact and
avoiding mechanical wear.

3.7. In waveguide, the need for the plunger to make physical
contact with the walls can be avoided by using a folded choke
construction of the kind illustrated in fig 4a, which shows
a cross-section parallel to the narrow wall of the guide.
This half-wavelength choke is imilar in operation to that
described in section 3.5, the short-circuit at C being transformed by the two cascaded $\lambda_g/4$ sections (CB and BA) to
present a short-circuit at the front face A.   Whatever the
impedance of the sliding face BD, it is in series with the
open circuit at B so its properties have less relevance.   A
variation of this is shown at fig 4b, where the short-circuited portion of the choke is cut at an angle of about 15°
to ease manufacture and to provide less of a hiatus in the
electromagnetic field at B and hence a better approximation
to an open circuit there.   With hard anodised faces at BD,
this gives a repeatable value of $|\Gamma| \nless 0 \cdot 985$ over a waveguide
bandwidth and can be used at least up to Q band.   Another
type of choked plunger is shown in fig 4c: alternate sections
of $\lambda_g/4$ length of low, high and low characteristic impedance
guide are formed between the non-contacting sides of the
plunger and the waveguide wall.   If the impedances are Z_L
and Z_H, the impedance Z_B at the rear face of the plunger is
transformed to $Z_L^4/(Z_H^2 Z_B)$ at the front face (cf section 7);
a low impedance is thus presented at the short-circuit plane.
With five sections, rather than the three illustrated, values
of $|\Gamma| \nless 0 \cdot 990$ have been reliably obtained up to at least Q
band.   Similar plungers of circular cross-section may be
used in rectangular guide or in a circular extension to rectangular guide (employing the TE_{11} mode).   With lossy material
behind the plunger to absorb leakage, this last has been used
to provide $|\Gamma| \nless 0 \cdot 990$ at millimetre wavelengths.   The ratio of
Z_H to Z_L is at least 4: the higher this ratio, the less the
variation of $|\Gamma|$ with frequency.   Other constructions are used

commercially, some having slots in the front face of the
plunger (the purpose of which has not apparently been repor-
ted) and yielding values of |Γ| as high as 0·995.

3.8. For coaxial-line applications in which the bandwidth
is restricted to about the same value as for waveguide (i.e.
about 40%), a similar choke arrangement can be used, fig 4d.
The inner and outer cylindrical surfaces of the plunger form
with the line conductors chokes comprising three cascaded
$\lambda/4$ sections similar to fig 4c. Variations of the $\lambda/4$ line
have been used in which the contacts are made at a current
node (12). For general purpose coaxial line sliding short-
circuits, the frequency range is too great for $\lambda/4$ chokes
so spring contacts alone are often used. Probably the best
form of contact fingering is that used in the coaxial-line
cavities of some VHF oscillators in which the fingers are
held in torsion by the edges which are in contact with the
plunger and line conductor; however, this does not appear to
be used in commercial tuners.

4. TUNERS USING SHUNT SUSCEPTANCE

4.1. In this section we consider the properties of tuners
formed of variable susceptances spaced along a line.

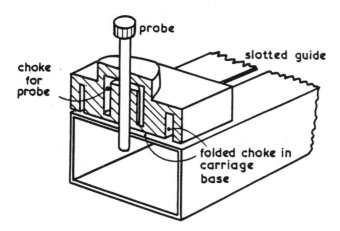

FIG. 5. SLIDING - PROBE TUNER

4.2. A simple adjustable impedance transformer is the
sliding probe tuner, fig 5, in which both the insertion of a
probe through a slot along the broad face of a waveguide and
its position along the slot can be adjusted. The insertion
provides an adjustable susceptance and the position along
the $\lambda_g/2$ long slot at the lowest frequency adjustment of
phase and magnitude of Γ . The probe and carriage are both
provided with chokes. RF electrical continuity is provided
by a half-wave choke formed by a $\lambda/4$ radial-line section
between the carriage and the guide and a further $\lambda/4$ deep
annular recess in the base of the carriage. Lossy material
is sometimes used along the slot to reduce leakage further.

Sliding probe tuners are available in most waveguide sizes
and for coaxial line (actually often constructed in slab
line) they are available for use over the range 1 - 12·4 GHz.

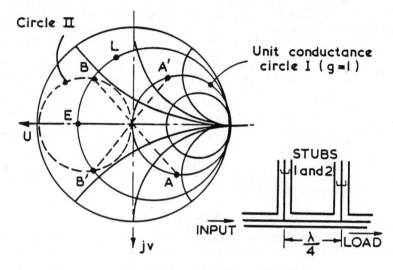

FIG. 6. TWO λ/4 SPACED STUBS

4.3.   A combination of two or more short-circuited stubs
spaced along the line can be used as a tuner.   The range of
loads that can be matched depends on the spacing and number
of stubs.   For example, consider two shunt stubs separated
by a λ/4 length of line, fig.6.   The action of each stub is
to add positive or negative susceptance (add or 'subtract'
susceptance) and, viewed in the complex plane of Γ (the Smith
chart, see Appendix), the aim is to transform the admittance
L presented by the load in the plane of stub 1 to the centre
of the chart, representing Γ = 0.   By adjusting stub 1,
susceptance is subtracted to transform L to a point B (or B')
on circle II (the image of the unit conductance circle I in
the jv axis).   B and B' are respectively transformed by the
λ/4 length of line to points A and A' on circle I so that
sufficient susceptance may be added or subtracted by adjustment of stub 2 to bring the transformed load point to Γ = 0
at the centre.   It follows that any load presenting an
admittance in the plane of stub 1 lying within the unit conductance circle cannot be matched by this arrangement. However
if the distance of one stub from the nearest connector of
the tuner differs from the corresponding distance of the
other stub by λ/4, the tuner can be physically reversed to
interchange the roles of circles I and II, enabling the
admittance of any passive load to be matched with one or
other connection.   This is true only for the frequency for
which the stub spacing is λ/4 and more stubs would be necessary to enable adjustment to a match for any load at
single frequencies over a range.

4.4. These two $\lambda/4$ spaced stubs illustrate the practical difficulty of tuning for a match. If a display of both the phase and magnitude of Γ at the tuner input is available then it is relatively simple to adjust to a match. But such a display of phase is rare and usually only the magnitude of Γ (or the equivalent VSWR) is known so that the most useful tuner would be one in which successive adjustments cause not only a reduction of $|\Gamma|$ but convergence towards $|\Gamma| = 0$. This is evidently not the case for the two-stub tuner because, from fig 6, adjustment of stub 1 to achieve minimum $|\Gamma|$ would transform the load point admittance L to E, which would be transformed by the $\lambda/4$ section of line to the plane of stub 2 as a point within the g = 1 circle, which renders matching by adjustment of stub 2 impossible. The most useful method is usually to adjust one stub a small amount for each attempt at matching by variation over the full range of the other stub.

4.5. Two advantages may be gained by spacing the stubs $3\lambda/8$ apart. The first is that reasoning similar to that above shows that only admittances within the g = 2 circle cannot be matched, thus increasing the range. The second is that the susceptance that needs to be introduced by the stubs for some near-match conditions is less than for $\lambda/4$ spacing, so avoiding difficulty in adjusting stubs near their resonant lengths.   It must be emphasised that such tuners cannot match every possible passive load over their frequency range. Typically, coaxial-line two-stub tuners require four models to cover 0·04 to 13·0 GHz in overlapping (approximately decade) ranges. Each model can match loads for which $|\Gamma| > 0·75$ in its frequency range; this performance is achieved by providing more than two positions along the line for connecting the stubs.   Micrometer adjustment of short-circuit position is often provided to facilitate both fine tuning and resetting to starting indications for tuning known loads: these features are of importance for precise measurement work.

4.6. The addition to two $\lambda/4$ spaced stubs of a third spaced $\lambda/4$ nearer the load allows a susceptance of -1 to be added in the plane of its connection, so reversing the roles of circles I and II of fig 6 and permitting any admittance to be matched. This is useful in enabling selection of the tuner with stub length and spacing appropriate to particular tasks; it does not help in the practical adjustment of tuners, for the reasons given above.

4.7. Although three-stub tuners may be used for high-power waveguide work, the most usual waveguide tuner is the three-screw tuner which is cheap to construct. In most commercial instruments the screw spacing is $3\lambda_g/8$ at the midband frequency: this implies a variation of spacing from 0·3 to 0·6 λ_g over the octave band.   This means that two tuners are needed to enable any passive load to be matched at any frequency in the band.   However, it has been shown that one tuner using three screws spaced at $\lambda_g/6$ at the lowest frequency should suffice (13). Because screws can be at near-resonant insertion depth, the use of chokes is desirable: the

better instruments use these and micrometer adjustment. With
these precautions, there is little to choose in ease of adjustment between three-screw and sliding-probe waveguide
tuners.   Two advantages of good sliding-probe tuners, however,
are the constancy of dissipative attenuation with probe
insertion (14) and the ease of separate adjustment of probe
position and insertion to predetermined settings by means of
stepper motors for a computer-controlled measurement system
(15).   Tuners with more than three screws are used, sometimes
allowing pre-setting for specific frequencies, sometimes all
with variable adjustment.   It was probably these last that
provoked the comment, over a quarter of a century ago (12):
'Complicated rules have been developed for tuning with multiple fixed position susceptances.   These rules serve chief-
ly to indicate that other methods of tuning should be used
wherever possible'.   The incorporation of chokes and good
mechanical drive arrangements makes this less valid today.

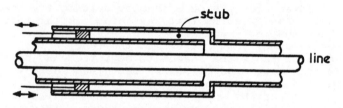

FIG.7a. COAX. SERIES STUB

FIG.7b. ALTERNATIVE COAX. SERIES
CONNECTION

5.   TUNERS USING SERIES REACTANCE

5.1.   Substitution of the words 'series' and 'impedance' for
'shunt' and 'admittance' throughout section 4 suggests that
series connected stubs could be used to provide reactance,
rather than susceptance tuning.   Two possible forms of series
connection in coaxial line are illustrated in fig 7: they are
more complex than shunt connections.   Apart from use in combination with a shunt stub (section 6), adjustable series
coaxial-line stubs offer no advantages and are not available
commercially.

5.2.   In waveguide, the connection of a series arm is formed
by an E-plane junction in the broad face of the waveguide,
as shown in figure 8.   Just as the series connection in a

coaxial line diverts
current into the series
arm, so the longitudinal
current in the broad
wall of a waveguide is
disturbed by the E-plane
junction.   In both dev-
ices antiphase supply to
the main arm junction
planes results in maximum
power transfer to the
series arm. Commercial
E-plane junctions are
available but a short-
circuited series arm as
a variable tuning element
is most often used in as
E-H plane tuner (sect.6).

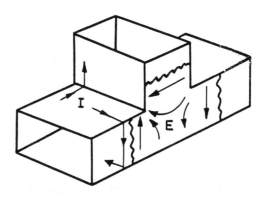

FIG. 8.   E - PLANE   JUNCTION

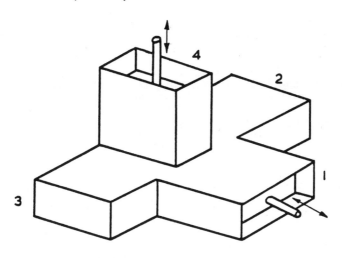

FIG. 9.   E - H   TUNER

6.   SERIES-SHUNT COMBINATIONS

6.1.   In waveguide, a combination of short-circuited series
and shunt stubs forms the E-H plane tuner, fig 9.   Because
each arm can provide an infinitely variable admittance, the
tuner is theoretically ideal.   In practice, the range of
$|\Gamma|$ that can be matched depends on losses at near-resonant
lengths and the quality of the short circuits.   Specified
limiting values of $|\Gamma| \not> 0 \cdot 9$ at frequencies below $12 \cdot 4$ GHz.
$|\Gamma| \not> 0 \cdot 8$ over $12 \cdot 4 - 50$ GHz, and $|\Gamma| \not> 0 \cdot 7$ above 50 GHz being
typical.   A counterpart in coaxial line is the combination
of series and shunt stubs shown in fig 10; this is usable
from about $0 \cdot 3$ to $6 \cdot 0$ GHz.

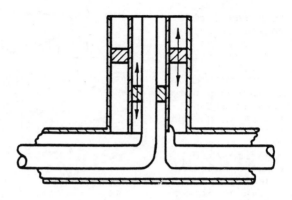

FIG.IO. SHUNT - SERIES COAX. TUNER

6.2. The use of an E-H plane tuner for an automatic tuner for microwave heating applications has been proposed (17). This might suggest that adjustment always converges for a symmetric structure. In practice the minimum $|\Gamma|$ attained is sometimes slightly greater than zero. The best procedure is to adjust the short circuits sequentially so that each is moved to a position just past minimum $|\Gamma|$, using the same direction of movement for each. This ensures that the loci of the reflections produced by the two arms on a polar plot of Γ intersect at the origin. In general there is more than one combination of settings to obtain a match; some may be more frequency-dependent than others.

7. TUNERS USING CASCADED LINES

7.1. The impedance transformation property of the lines used in the tuners so far considered is that achieved by phase shift, since the R_O of the tuner arms has been that of the measurement system. A line can transform both phase and magnitude, for if the mode characteristic impedance is R_T, then from equation (1) the input impedance is:

$$Z = R_T (Z_L + jR_T t)/(R_T + jZ_L t) \tag{3}$$

where Z_L is the impedance terminating the line and $t = \tan \beta \, 1$. Adjusting the length of the line to $\lambda/4$ makes $t = \infty$, so then

$$Z = R_T^2/Z_L \tag{4}$$

To match Z_L and R_O by a $\lambda/4$ line of characteristic impedance R_T, then, we need $R_T = R_O Z_L$, which is feasible only if $jX_L = 0$, i.e. if Z_L is real. Thus a quarter wavelength line in general transforms an impedance Z_L to R_T^2/Z_L, and specifically a resistance R_L to R_T^2/R_L.

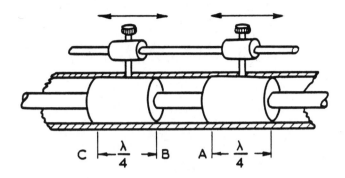

C $\vdash \dfrac{\lambda}{4} \rightarrow$ B   A$\vdash \dfrac{\lambda}{4} \dashv$

FIG.II. DOUBLE SLUG TUNER

7.2.   A tuner using $\lambda/4$ cascaded lines is the double slug tuner comprising two low impedance lines each $\lambda/4$ at the mid frequency of the useful range and separated from each other, as in fig 11.   The reduction of impedance below the R_O of the line is achieved by two cylindrical slugs of dielectric material of relative permittivity $\mathcal{E}_R \rangle 1$; the position of each along the line may be adjusted.   It can be shown that, provided both slugs can be moved over a distance $\lambda/2$ so that the phase of the signal reflected at C can be changed by 2π, the maximum VSWR that can be matched is $\mathcal{E}_R{}^2$, corresponding to $|\Gamma| = (\mathcal{E}_R{}^2 - 1)/(\mathcal{E}_R{}^2 + 1)$.   Typical commercial tuners of this kind for 50 ohm coaxial line cover frequency ranges of 0·3 to 1·7 GHz and 1·0 to 5·0 GHz, the maximum VRC that can be matched throughout being specified as 0·33 for dielectric slugs and 0·82 for slugs incorporating metal sleeves.   An analogous tuner has also been used for waveguide applications (12) and the principle of alternate high and low impedance sections is employed on non-contacting short circuits, as we saw in section 3.7.

7.3.   Another 'building block' is the variable length cascaded line component - a line-stretcher in coaxial line and a phase shifter in waveguide: clearly if the R_O of a line-stretcher is the same as that of the line in which it is used it merely changes the phase of the signal.   In practice there is an unavoidable change of cross-section which causes the VRC to vary by up to 0·2 over the commercially available range of adjustment of 300 mm ($\lambda/2$ at 0·5 GHz) or 600 mm for a folded 'trombone' section.   The latter has been used in waveguide but more convenient phase-shifters, such as the kind in which a longitudinal dielectric vane, parallel to the E-field and movable either across the guide or into the centre plane, guillotine fashion, can provide π phase change throughout a waveguide frequency band.

7.4.   Another application of a cascaded line section is in tuning a two-port containing a mainly susceptive admittance (e.g. a reverse biased diode).   A short-circuit stub terminating the two-port is in effect a cascade connection.

8.   SOURCE MATCHING

8.1.   A source is often required to deliver maximum power to a load via a transmission line.   This requires tuners at both ends of the line, so that the load impedance may be transformed to that of the line, which is in turn transformed to a conjugate match to the source impedance.   In practice the two tuners have to be adjusted alternately because adjustment of one causes a change of impedance transformed by the line at the other tuner.   In measurement applications, the load presented to the source may be subject to drastic change (e.g. by a near matched load being replaced by a short-circuit during a calibration process) but consequent changes of frequency and power must be avoided.   Partial isolation of the source from the load can be performed by an attenuator (pad) of 10 or 20 dB or by a ferrite isolator (typically providing 1 dB forward insertion loss and 20 dB isolation); the reflection coefficient at the output of these is usually smaller than that of the source alone.   For some applications, the source VRC should be as small as possible (e.g. to avoid the mismatch uncertainty in comparing watt-meters by substitution) so that the requirements for constant frequency and output are similar to those for a signal generator.

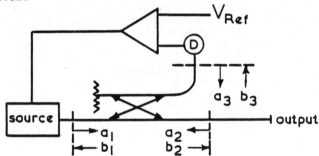

FIG. 12. STABILISATION OF SOURCE IMPEDANCE

8.2.   A matched source can be arranged by using a directional coupler to sample the power traversing the line, as in fig 12. The difference between the output of a detector receiving the sampled power and a stable reference voltage is used to control the power output from the source (16).   A ferrite attenuator can be used both for control and to provide some isolation to prevent frequency-pulling of the source by load variations.   If the coupler has infinite directivity, no power reflected from the load reaches the detector at port 3 and the power output from port 2 is constant irrespective of variations of load impedance, simulating a matched source at port 2.   In practice, the effective source VRC due to imperfect directivity can be derived from the scattering equations of the directional coupler:

$$b_1 = S_{11}a_1 + S_{12}a_2 + S_{13}a_3$$

$$b_2 = S_{21}a_1 + S_{22}a_2 + S_{23}a_3$$

$$b_3 = S_{31}a_1 + S_{32}a_2 + S_{33}a_3$$

where a_n and b_n are the amplitudes of the incident and reflected waves at port n (see chapter 2). If the VRC of the detector is Γ_d then $a_3 = \Gamma_d b_3$ and substituting for a_3 in the scattering equations and eliminating a_1 and b_1 gives:

$$b_2 = \left[S_{21}/S_{31} + \Gamma_d(S_{23}-S_{21}S_{33}/S_{31})\right]b_3 + (S_{22}-S_{32}S_{21}/S_{31})a_2$$

A source connected to an arbitrary line can be represented by:

$$a_1 = b_g + b_1\Gamma_g \tag{5}$$

where a_1 and b_1 are the amplitudes of the waves incident on and reflected from the line, b_g is the amplitude of the wave that would be supplied to a non-reflecting load, and Γ_g is the VRC of the source. If the emergent and incident waves b_2 and a_2 at port 2 of the coupler are redesignated A_1 and B_1 respectively, then:

$$A_1 = \left[S_{21}/S_{31} + \Gamma_d(S_{23}-S_{21}S_{33}/S_{31})\right]b_3 + (S_{22}-S_{32}S_{21}/S_{31})B_1 \tag{6}$$

It is evident that, with b_3 being maintained constant by the feedback loop, the first term on the right of (6) is, by comparison with equation (5), equivalent to b_g for the equivalent generator. Similarly, the second term, $(S_{22} - S_{32}S_{21}/S_{31}) = \Gamma_g$, the VRC of the equivalent generator. This term is solely a property of the directional coupler. With typical values of main guide VSWR of 1·05 ($|S_{22}| \fallingdotseq 0·025$) and a directivity of 40 dB ($|S_{32}/S_{31}| \fallingdotseq 0·01$) then, since $S_{21} < 1$, the worst possible phase combination gives $\Gamma_g = 0·035$, so that the equivalent source VSWR = 1·07. Before the equivalent reflection coefficient at port 2 can be measured, it is necessary to connect a signal source to this port and a variable load at port 1, this load being adjusted to obtain zero output from port 3. Under these conditions, the input VRC at port 2 is $(S_{22}-S_{32}S_{21}/S_{31}) = \Gamma_g$. Finally, if a tuner is connected at port 2, it may be adjusted to reduce the reflection at the output to zero.

8.3. For matching a high power source to a terminated line, a short slot hybrid (a form of 3dB directional coupler) can be used with two adjustable short circuits as in fig 13a to avoid the voltage breakdown liability of some other tuners; the device is sometimes gas pressurised to aid this. Adjustment of stubs can introduce any value of reflection into a matched line and, in theory, match any load. In practice, limitations are imposed by imperfections in the hybrid.

8.4. Another use of a directional coupler and short-circuited stubs for matching a high-power source is one in which weaker coupling than 3 dB is used and in which the stubs are connected to the secondary arms of the coupler,

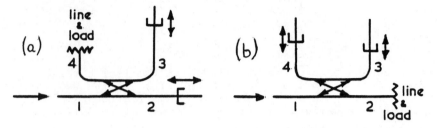

FIG.13. HIGH POWER TUNERS

as in fig 13b.   The smallest VSWR that can be attained by
this arrangement is about 1·11 using a 10 dB coupler or 1·01
for a 20 dB coupler.   Again, in practice the minimum values
may be somewhat greater than these because of imperfections
in the coupler and losses in the stubs near resonant lengths.

9.   CONCLUSIONS

These notes have presented material on the construction
and use of tuners for matching, in preparation for lectures
on rf measurement methods which often employ tuners to
minimise mismatch uncertainty.   We conclude with a summary
of the factors to be taken into account when choosing tuners
for such work:

(a) For most rf measurement work the power levels do not
exceed a few watts, so that it is only for high peak powers
that we need choose tuners that do not derate the main transmission system by avoiding high voltage gradients associated
with metal probes or irises.   Suitable tuners for such work
employ stubs, dielectric slugs or probes, E-H plane tuners,
or combinations of tuners with directional couplers.

(b) As the operating frequency rises, so leakage and erratic
tuning (arising from uncertainty of contact position) give
rise to greater problems; the use of tuners employing choked
probes and slots reduces both.

(c) Given a tuner without any accompanying specification of
the maximum voltage reflection coefficient that can be matched, some guidance can be obtained from the dimensions, probe
spacing, etc; in any case, the tuner can be tested when
terminated in a matched load.

(d) The insertion loss in a tuner is dependent on the load
impedance but about 1 dB should be allowed in planning a
measurement system.   The dissipative attenuation will be
small except when probe insertion or stub length approaches
resonance (unless the tuner leaks, of course, when the losses
may give the illusion of a broadband match having been att-
ined); this component should remain approximately constant
with constant probe insertion in a sliding probe tuner.

(e) For applications in which the frequency is often changed,
the facility of resetting to predetermined indications is

valuable. For this it is the quality of engineering in respect of fineness of control and indication, lack of backlash, and use of chokes to avoid erratic contact effects and leakage that is more important than the type of tuner.

(f) Similar considerations apply to ease of adjustment, except that a reduction of the number of variables is often worthwhile. The order of increasing difficulty of adjustment for VRC below about 0·005 (VSWR less than 1·01) is: sliding probe tuner, E-H plane tuner, multiple screw and stub tuners. Finally, it should be noted that for precise matching to minimise errors the choice should not be governed only by its initial cost. As for most equipment, the cost of time spent in operating and adjustment is likely to be high for poorly engineered tuners so that for frequent use expedients such as three screws inserted into a length of waveguide may well prove more costly than a choked, vernier and micrometer indicating, sliding probe tuner.

REFERENCES

1. McLachlan, N.W., 'Bessel functions for engineers', Oxford University Press, 1934, chapter 9.

2. Nelson, R.E. and Coryell, M.R., 'Electrical parameters of precision air-dielectric transmission lines', Nat Bur Stand (US) Monograph 96, 1966.

3. Spinney, R.E., 1963, 'The measurement of small reflection coefficients of components with coaxial connectors', Proc IEE part C, 110, 396.

4. Harris, I.A. and Spinney, R.E., 1960, 'The realisation of high-frequency impedance standards using air-spaced coaxial line', Trans IEEE, IM-9, 258-268.

5. Sanderson, A.E., 'Effect of surface roughness on propagation of the TEM mode', Advances in Microwaves 7, Academic Press, 1971, 2-56.

6. Aitcheson, C.L., 1971, 'Lumped-circuit elements at microwave frequencies', Trans IEEE, MTT-19, 928-937.

7. Matthei, G.L., Young, L. and Jones, E.M.T., 'Microwave filters, impedance-matching networks and coupling structures', McGraw-Hill, 1964.

8. Frank, N.H. and Chu, L.J., MIT Radiation Laboratory Reports nos 43-6 and 43-7, 1942.

9. Allanson, J.T., Cooper, R. and Cowling, T.G., 1946, 'The theory and experimental behaviour of right-angled junctions in rectangular section waveguides', Jnl IEE, 93, part III, 177.

10. Huxley, L.G.H., 'A survey of the principles and practice of waveguides', Cambridge University Press, 1947.

11. Moreno, T., 'Microwave transmission design data', Dover, 1958.

12. Ragan, G.L., 'Microwave transmission circuits', McGraw-Hill, 1948.

13.   Griffin, E.J., 1976, 'Design of 3-screw tuners',
      Electr Letters, **12**, 657.

14.   Sinclair, M.W., Sharpe-Neal, P.H.W. and Lappage, R.,
      'A range of improved radiometers for the UK noise
      calibration service', IEEE, CPEM Digest 1978.

15.   Abbott, D.A., Shurmer, H.V. and Temple, G.J., 'Automatic
      characterisation of 2-port components and devices in
      microstrip', IEE Digest 1979/22.

16.   Engen, G.F., 1958, 'Amplitude stabilization of a
      microwave signal source', Trans IEEE, **MTT-6**, 202-206.

17.   Boerner, W.B. and Chaudhuri, S.K., 1974, 'Design of
      automated microwave tuners', Arch für Elek und Uber-
      tragungsterhar, **28**, 358-362.

APPENDIX: THE SMITH CHART

A.1. Matching transformers, whether adjustable as tuners or
not, are two-port networks which, when terminated by a load
of non-zero voltage reflection coefficient, present a VRC of
zero at the input. Since the VRC is in general a complex
quantity $\Gamma = u + jv$, it is helpful to represent it on axes
u and v.   If we add to such a plot in the complex plane the
loci described by load impedance having constant resistance
and reactance, we have the familiar Smith chart. A derivation of this is presented here because the chart is helpful
in visualising the action of tuners: an appreciation of its
derivation often renders an accurate chart unnecessary. It
also demonstrates that the maximum Γ that can be introduced
by a tuner into a line is -1 times the maximum VRC that can
be matched with it; this is of obvious value in testing the
range of use of a tuner.

A.2.   The VRC of an impedance Z_L relative to the characteristic impedance R_0 of a transmitted mode on a lossless line
is given by: $$\Gamma_L = (Z_L - R_0)/(Z_L + R_0) \tag{1}$$
Normalising relative to R_0, by dividing throughout by R_0,
and putting $z_L = Z_L/R_0$,   we find

$$z_L = (1+\Gamma_L)/(1-\Gamma_L) \tag{2}$$

If Z_L terminates a length l of the transmission line (l being
positive in the direction away from the load towards the
source) and the phase constant is $\beta = 2\pi/\lambda$, then the input
impedance of the line is

$$Z = R_0(Z_L + jR_0\tan\beta l)/(R_0 + jZ_L\tan\beta l) \tag{3}$$

whence, normalising and substituting from (2), and writing
$t = \tan\beta l$, we find:

$$\Gamma = \Gamma_L(1-jt)/(1+jt) = \Gamma_L e^{-2j\beta l} \tag{4}$$

Hence, as we travel along the line AWAY from the load, the
phase angle of Γ DECREASES.   If we represent Γ as u+jv, we
can plot Γ_L and Γ on u and jv axes as shown in fig A.1.

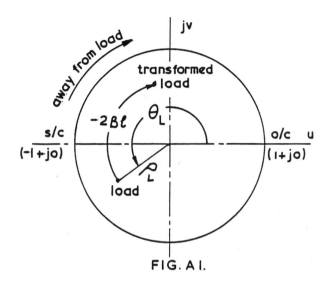

FIG. A I.

On this chart we can identify the values of Γ presented by various loads: for a short-circuit, $z=0$, $\Gamma = -1$; for an open circuit, $z=\infty$, $\Gamma = +1$, for a matched load, $z=1$, $\Gamma = 0$. For any passive load for which $r \geqslant 0$, it can be seen that $|\Gamma| \lesssim 1$.

A. 3. Any passive complex impedance normalised to R_0 can.be plotted, for from equation (2)

$$z = r + jx = (1 + \Gamma)/(1 - \Gamma).$$

Putting $\Gamma = u + jv$, separating real and imaginary parts, and eliminating x gives the locus for the constant resistance r as:

$$\left[u - r/(1+r)\right]^2 + v^2 = 1/(1+r)^2 \qquad (5)$$

which represents a circle of radius $1/(1+r)$, centred at $u = r/(1+r)$, $v = 0$.

Eliminating r from the same two equations yields

$$(u-1)^2 + (v-1/x)^2 = 1/x^2 \qquad (6)$$

so that the locus defined by constant reactance x is a circle of radius $1/x$, centred at $u = 1$, $v = 1/x$.

Adding these families of circles for various values of r and x gives the familiar Smith chart of fig A2, on which a normalised impedance can be plotted as the intersection of the two appropriate circles. We note from equation (5) that the unit resistance circle (for $r = 1$) is of radius $\frac{1}{2}$ and is centred at $u = \frac{1}{2}$, $v = 0$, so that it passes through the origin, as we should expect.

A. 4. A passive admittance $y = 1/z = g + jb$ can also be plotted, for, from equation (2),

$$g + jb = (1 - \Gamma)/(1 + \Gamma)$$

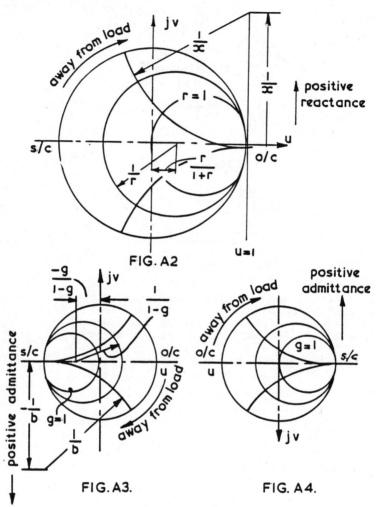

FIG. A2

FIG. A3.   FIG. A4.

and by the same process as before we find, eliminating b:

$$\left[u + g/(1+g)\right]^2 + v^2 = 1/(1+g)^2$$

so that the locus for constant conductance g is a circle of radius $1/(1+g)$ centred at $u = -g/(1+g)$ and $v = 0$.

Similarly, eliminating g, we find:

$$(u+1)^2 + (v+1/b)^2 = 1/b^2$$

so that the locus for constant susceptance b is a circle of radius $1/b$ centred at $u = -1$ and $v = -1/b$.

The admittance version of the Smith chart is thus as shown in fig A3, or in its more familiar form in fig A4, rotated through π. We note from this that the direction of the jv axis is reversed for admittance calculations.

Chapter 6

Detectors and detection for measurement

E. J. Griffin

1. INTRODUCTION

Previous chapters have defined concepts used for describing microwave circuits, such as transmission and reflection coefficients, and have treated methods of manipulating
these quantities to analyse circuits. Measurement standards
providing physical approximations to these concepts can be
realised in practice by circuit elements whose properties at
RF are either known or can be assumed, for example, a length
of waveguide or a short or open circuit. Comparison of
other elements with such standards requires the conversion
of RF voltage or power into DC or AF signals which can be
measured in terms of references whose properties are known
at these lower frequencies. This lecture provides an
introduction to some of the detectors and detection methods
employed for microwave measurement. It does not deal with
the detection of low-level signals in noise.

2. RF VOLTAGE

Previous lectures have shown that for TE modes in waveguide or the TEM mode in transmission line a voltage can be
defined proportional to the electric field intensity in the
transverse plane. Generally there will be reflection,
causing waves to be transmitted both from and to the RF
source. We can represent these by complex time-varying
voltages a and b respectively. Then the RF voltage ampli-
tude in the notional reference plane is $V = |a+b|$ and the
power flux from the source across the plane is

$$P = G_o(|a|^2 - |b|^2)$$

where $G_o = 1/Z_o$ (Z_o being the real characteristic impedance
calculated for a mode transmitted by a lossless waveguide or
transmission line). Most measurements require the determination of ratios of RF power or RF voltage amplitude but
the phase difference between two RF voltages is sometimes
required. Measurement of RF signal level, as such, needs
either the RF voltage amplitude or the RF power in a reference plane to be determined. Power measurement is discussed in a later chapter: here we consider what measurement of
RF voltage amplitude involves.

If the voltage reflection coefficient (VRC) of a detector is Γ and the power absorbed by the detector P, then the RF voltage V presented to the detector is given by

$$V^2 = P|1 + \Gamma|^2/(G_o(1 - |\Gamma|^2))$$

so that V may be found from a measurement of P.   Strictly the concept of voltage as a potential difference applies at RF only to a transmission line and in practice voltage amplitude can be measured only on transmission lines at frequencies up to about 1 GHz where the dimensions of a detector are small compared with a wavelength.   The physical phenomena exploited for such voltage measurements are the same as those used for determining voltage ratios in transmission lines or waveguides: only the constructional details of the mounts differ.

3.   DIODE DETECTORS

3.1.   General

The microwave detector most commonly used is the semi-conductor diode.   Cartridge-mounted point-contact diodes are used and more recently Schottky barrier diodes, mounted on thin film circuits in coaxial line components or on fin line circuits in waveguide, have become available.   These detectors contain an RF matching structure and a DC or video output circuit.   Mixers can also be used for heterodyne detection and homodyne detection: in the latter case the local oscillator frequency equals the RF so that in the absence of modulation only a DC output is obtained.   In this lecture we emphasise the aspects of performance which are of significance for measurements, rather than the physics of the p-n junction or its constructional details.

3.2.   Theory

Since there is no internal energy source, the current through a semiconductor diode is zero with zero applied voltage.   With reverse bias voltage less than that needed for Zener or avalanche breakdown, the reverse current is asymptotic to a constant saturation current I_s of a few microamps.   With forward voltage bias, the diode conducts and the current-voltage characteristic (neglecting the diode spreading resistance) approximates to:

$$i = I_s (\exp(v/D) - 1) \tag{1}$$

where D = nkT/q,
   i is the forward current due to the voltage v,
   q is the electronic charge, $1 \cdot 602 \times 10^{-19}$ coulomb,
   T is the temperature $\doteq$ 300 K for room temperature,
   k is Boltzmann's constant, $1 \cdot 3804 \times 10^{-23}$ J/K,
   n is a dimensionless constant.

n is sometimes called the "ideality factor".   Its value depends on the manufacturing process and varies from batch to batch.   It ranges from about $1 \cdot 2$ for low barrier height Schottky diodes to nearly 2 for point-contact silicon diodes.

Inserting numerical values with n = 2 gives D = 50 mV, so
the transfer characteristic of a diode detector depends on
whether the amplitude of the RF signal is greater or less
than about 50 mV. The form of the current-voltage
characteristic described by equation (1) is shown in fig 1(a).

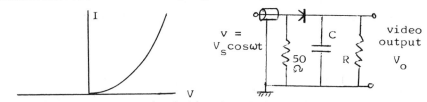

Fig 1(a) Diode characteristic. Fig 1(b) Series diode detector

Diodes are used in either series or shunt connection and
fig 1(b) shows the idealised circuit of a series detector.
The input 50 ohm resistor provides a match to the transmission line and a filter formed by capacitor C and video
load resistor R completes the circuit. For high level
signals ($\gg$ 50 mV) the diode conducts only to the extent
that the time constant CR allows the capacitor to charge.
Thus the circuit functions as a peak voltage detector
providing an output of nearly V_s, the peak value of the
applied RF signal $v = V_s \cos \omega t$.

For signals of less than 50 mV amplitude, CR will still
be large compared with the period of the RF cycle so that
the voltage across the diode will be $v = V_s \cos \omega t - V_o$,
where V_o is the voltage across the capacitor. Substituting
in (1) gives an instantaneous current through the diode of

$$i = I_s (\exp(-V_o/D) \exp((V_s/D)\cos \omega t) - 1)$$

But the DC output is R times the mean of i over the RF cycle
so that, writing Θ for ωt, we have

$$V_o = \frac{I_s R}{2 \pi} \int_0^{2\pi} (\exp(-V_o/D) \exp((V_s/D)\cos\Theta) - 1)\, d\Theta$$

$$\text{or}\quad V_o \exp(V_o/D) = \frac{I_s R}{2 \pi} \int_0^{2\pi} (\exp((V_s/D)\cos\Theta) - \exp(V_o/D))\, d\Theta$$

$$(2)$$

By series expansion of both sides and term by term
integration, it can be shown that for small signals

$$V_o = I_s R \left[\frac{D}{4(D+I_s R)} \left(\frac{V_s}{D} \right)^2 + \text{higher even powers of } V_s/D \right]$$

To the extent that the fourth and higher even powers of V_s/D
can be neglected, $V_o \propto V_s^2$. This analysis is a simplification
since an exact solution of (2) should involve the spreading
resistance, barrier capacitance and reactive terms to
represent the diode. But the conclusion that a diode
detector provides an output linearly proportional to large
RF voltages and approximately proportional to the square of

the amplitude of small RF voltages remains valid.

3.3.   Noise

Both point contact and Schottky barrier detectors exhibit
1/f noise (flicker noise) as well as a spectrum of white
noise which may be represented as an equivalent current
source $i_n = (4kTB/R_v)^{\frac{1}{2}}$
where R_v is the video load resistance and B the bandwidth.
The effect of noise is often determined experimentally by
using an oscilloscope to observe when the detected signal
is perceptibly above the noise and then to measure the RF
input power.   This is usually expressed in dB below 1 mW
and is termed the Tangential Sensitivity (TS) because the
characteristic of fig 1(a) is approximately tangential to
the voltage axis for small voltages.   Values of TS of
−50 to −55 dBm in a 10 MHz bandwidth are typical.   In a
detailed analysis of low level detection for a number of
types of diode, Cowley and Sorensen (1) introduced the
concept of Noise Equivalent Power (NEP), defined as the RF
power needed to produce a signal−noise ratio of unity for
a 1 Hz bandwidth and related NEP to TS by TS = $2 \cdot 5$ NEP $B^{\frac{1}{2}}$,
where B is the video bandwidth.   The usefulness of either
of these criteria depends not only on the bandwidth of
observation but also on the lowest frequency, because of
1/f noise.

3.4.   Practice

The noise produced by a Schottky diode is about 15 dB
below that of a point-contact diode (1).   Both types are
used in commercially available detectors for use with 50
ohm coaxial line and rectangular waveguide.   The coaxial
line detectors are broadband detectors covering frequencies
from about 10 MHz to an upper limit, determined by the
characteristics of the diode and the connector employed,
ranging from 12 to $26 \cdot 5$ GHz.   The short-term (i.e. a few
seconds) power limit is about 100 mW and the low level
sensitivity ranges from $0 \cdot 2$ to $0 \cdot 4$ mV/µW; both the output
level and the VSWR of these detectors vary with frequency.
For coaxial detectors the output variation ranges from
$\pm 0 \cdot 2$ dB/octave at VHF to a fall of 3 dB or more over the
18−$26 \cdot 5$ GHz band; the maximum VSWR is about 2.   Waveguide
detectors exhibit higher VSWR, with maxima ranging from
about $1 \cdot 5$ in X-band ($8 \cdot 2$ to $12 \cdot 4$ GHz) to 3 in Q-band (18−26
GHz) and at higher frequencies; the variation in output
similarly increases from $\pm 0 \cdot 3$ dB in X-band to ± 2 dB over
Q-band.

The value of load resistor affects the range of square
law operation.   Detectors optimising this are available
with a specified RF power level to provide not more than
some arbitrarily defined deviation from it (e.g. $0 \cdot 1$ dB (2),
$0 \cdot 2$ dB (3) or $0 \cdot 3$ dB (1)).   Reference (3) reports an
experimental comparison at 94 GHz of different diodes

showing that a departure from square law of $0 \cdot 2$ dB is experienced at levels ranging from -16 dBm to -10 dBm (25 to 100 μW). This is consistent with experience reported in the 1 to 8 GHz band of coaxial detectors (4) in which calibration was carried out against a thermistor detector to enable the detectors to be used for "square law" detection up to 250 μw. In both references (3) and (4) the change of magnitude of voltage reflection coefficient with power level is quantified showing about 10% for 10 dB at 94 GHz and 1% for the same power change in the 1 to 8 GHz range. Changes of output with temperature ranging from $-0 \cdot 15\%$ to $+2 \cdot 5\%$ per degree C have been reported (3, 4, 5) together with a law of temperature dependence for one type of detector which involves modification of equation (1) (5). The susceptibility of diode detectors to harmonics of the RF signal is well known (6) and a voltage doubler detector employing two diodes to reduce the effect of even harmonics has been described (7); in many applications this difficulty can be avoided by filtering the RF source. At frequencies up to about 1 GHz, the effects of non-linearity of a linear peak voltage detector have been nullified by opposing the DC output from an RF detector against a second diode supplied with a lower frequency (8).

3.5. Heterodyne and homodyne detection - theory

The sensitivity of detection can be increased by first translating the signal frequency and amplifying before final detection, thus minimising 1/f noise, and then restricting the bandwidth of the observed signal. This process requires either heterodyne or homodyne detection and in this section we treat these two methods in a qualitative way.

A simple explanation of heterodyne detection stems from considering the detector of figure 1(b). If this is operated at a level of about 1 mW or more, the CR filter biases the diode into linear operation. For heterodyne detection, this level is provided by a stable local oscillator (LO) locked to a frequency offset by a fixed amount from that of the RF signal source. If we now call the local oscillator voltage V_0 and its angular frequency ω_0, which for definiteness we may take to be less than the signal frequency ω_S, then an output signal of amplitude V_S, and at a frequency $\omega_S - \omega_0$ would be obtained by AC-coupling a video output from the circuit shown. This is an over-simplification, for at microwave frequencies an equivalent circuit for the diode must be invoked containing a voltage-dependent barrier capacitance and a current-dependent resistance, as well as constant reactive and resistive components (1) (5). Thus, full treatment of detection involving frequency changing, which is an exploitation of the non-linear time-dependent behaviour of a diode, involves consideration of both the generation of harmonics of the LO frequency and variation of V_0 by the superimposed RF signal V_s. At DC a conductance g_0 can be defined for the diode from the slope $\partial i/\partial v$ of the current-voltage characteristic of figure 1(a). Generation of harmonics of an assumed constant LO frequency can be

taken into account by defining in an analogous way a reactance at each angular frequency $n\omega_0$, where n is an integer or
zero, as:

$$\frac{\partial i}{\partial V_0} = \sum_{n=-\infty}^{n=+\infty} (g_n + jn\,\omega_0 C_n)\, e^{jn\,\omega_0 t}$$

The summation extends from $-\infty$ to $+\infty$ because we are
considering a continuous LO signal and ignoring transient
effects.   If the RF signal is less than, say $V_0/30$, i.e. is
30 dB or more below V_0, then it may be treated as a small
perturbation of V_0, so that $\delta V_0 = |V_s|\exp(j\omega_s t)$.   Substit-
ution of this in the expression above defining the reactance
at each frequency gives:

$$\delta i = |V_s| \sum_{n=-\infty}^{n=+\infty} (g_n + jn\,\omega_0 C_n)\, e^{j(n\omega_0 + \omega_s)t}$$

$$+ j|V_s|\omega_s \sum_{n=-\infty}^{n=+\infty} C_n\, e^{j(n\omega_0 + \omega_s)t}$$

Writing $f_0 = \omega_0/2\pi$ and $f_s = \omega_s/2\pi$, both assumed constant,
and substituting successive values of n in this expression
shows qualitatively that a voltage would be developed across
a video load through which δi flows having frequency components represented diagrammatically in figure 2: this does not
indicate relative amplitudes:

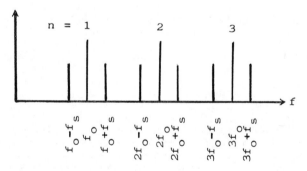

Figure 2

The amplitude of each component of this spectrum, including
the DC term, is proportional to V_s , so that the use of
frequency selective filters enables one component (usually
the lower sideband at f_0-f_s) to be selected using a narrow
bandwidth.

   This method of detection has the advantage that a fixed
intermediate frequency (IF) equal to f_0-f_s can be employed

for broadband measurements on components if the LO is locked to, but offset by the IF from, a variable-frequency RF source. This enables phase and magnitude references whose properties are known at the IF to be employed. This method has the disadvantage of requiring both an RF source and an LO; any variation of phase-locking between them causes a variation of IF signal arising from the use of highly selective filters in the IF amplifier. An alternative method uses one RF source whose output is divided into two portions (e.g. by a power divider), one forming a constant high-level reference signal (corresponding to the LO) and the other an amplitude-modulated varying test signal (corresponding to the RF signal V_s). This form of detection (in which $f_o = f_s$) has been termed synchrodyne (9) or homodyne detection. Its equivalence to commutative or phase sensitive detection has been derived in the literature (10). This reference provides a comparative theoretical survey of the effects of noise in coherent detection methods (i.e. heterodyne and homodyne) and demonstrates their advantage over incoherent (i.e. square law) detection in this respect.

3.6. Coherent detection - practice

Evidence of the sensitivity obtained with coherent detection is to be found in every superheterodyne receiver, of course. A report of an experimental investigation of the sensitivity of homodyne detection for microwave measurements including the effects of choice of modulation frequency, RF reference power and externally applied DC bias on the sensitivity and range of detectors using point-contact and Schottky barrier diodes has been published (11). It gave maximum sensitivities (at 4·89 GHz and a modulation frequency of 30 kHz) of -135 dBm and -150 dBm (3 x 10^{-17} and 10^{-18} W) and dynamic ranges of 92 dB and 110 dB for point-contact and Schottky barrier diodes respectively. Super-heterodyne detection is employed in many microwave measuring instruments: maximum sensitivities of -145 dBm have been reported (12, 13) and signals below 10^{-19} W can be detected by one system using two synthesised sources locked to a common 10 MHz crystal oscillator (14). Sensitivities of these orders approach the limit imposed by noise and are obtained with bandwidths of 0·1 Hz or less.

4. THERMAL DETECTORS

4.1. General

It is evident from sections 2 and 3 above that any RF power sensor can be regarded as an incoherent radiation detector. Some are used for this purposes with reflectometers, i.e. instruments for measuring voltage reflection coefficient. In this section we discuss power sensors depending on thermal effects: not only have they been used as incoherent detectors, but one type has been investigated experimentally for use in homodyne detection.

4.2.   Thermoelectric detectors

Three forms of thermoelectric detector have been used for incoherent detection; from each a DC output voltage proportional to $|V_s|^2$ is obtained (where V_s is the RF signal voltage). The first utilises a resistive load dissipating RF power to heat one junction of a thermocouple, the cold junction being maintained in thermal contact with the waveguide containing the resistive load. This type has been used over waveguide bandwidths at frequencies up to 40 GHz (15).

The second form also uses thermocouples consisting of two thin-film 100 ohm metallic loads connected in parallel on a dielectric substrate disposed across a coaxial line. Each load consists of a bismuth-antimony thermocouple, arranged so that they are thermally in series. These detectors are available in 7 mm diameter coaxial line for use at frequencies up to 18 GHz (16).

A third type which has been reported (11) uses a planar ohmic contact with silicon. Power incident on the contact causes a high electric field strength within the bulk silicon, producing localised heating of its electrons so that the more highly energised cause conduction. There is apparently no practical upper frequency limit for such a device and results of square-law operation have been obtained over the range 1 to 100 μW. Unlike the other (directly heated thermocouple) detectors, this type is likely to be highly sensitive to ambient temperature.

Commercial coaxial line thermocouples are available for frequencies from 10 MHz to 18 GHz covering ranges of 0·1 nW to 10 μW and 1 μW to 100 mW; another covers 50 MHz to 26 GHz with a range of 1 μW to 100 mW. The lowest claimed noise (with bandwidth and temperature drift unspecified) is 15 pW. Less than 1% (0·04 dB) deviation from square-law response is claimed: this is consistent with a reported deviation of 0·03 dB in 20 dB experienced with one type of thin-film thermoelectric detector (17). This order of deviation holds over about a 40 dB range and the response time varies from 100 ms to 5 seconds, depending on manufacturer.

4.3.   Bolometric detectors

The word 'bolometer' is generally used to cover all forms of temperature-sensitive resistor employed for detecting radiation. Probably the most commonly used is a thermistor (typically a 400 μm diameter bead of semiconductor material), for its small size enables it to be mounted complete with RF matching elements and RF chokes for leads for its DC resistance to be measured. Coaxial line thermistor mounts are available for use over the 10 MHz to 18 GHz band and waveguide mounts for waveguide frequency bands up to 100 GHz. The maximum RF power input is a few milliwatts and the VSWR is generally not greater than 2. The change of VSWR with power level is imperceptible because the resistance of the thermistor is maintained constant by controlling DC bias

current supplied to it from a self-balancing bridge circuit
(18). If the resistance of each arm of a Wheatstone bridge
circuit is R and DC voltage is supplied to maintain balance,
then the RF power absorbed by a thermistor biased to resistance R is $k(V_1^2 - V_2^2)/4R$, where V_2 and V_1 are the DC bridge
voltages with and without RF supplied, respectively, and k
is a proportionality factor ranging from about $0 \cdot 8$ to $0 \cdot 9$
over the frequency band. Many thermistor mounts contain a
second thermistor to sense changes of ambient temperature.
The two thermistors are each connected in a separate self-
balancing Wheatstone bridge (19) and the RF power is derived
from the difference between the DC voltages across the bridge
circuits. One form of commercially available power meter
(8, 20) for use with these temperature-compensated thermistor
mounts provides a DC output proportional to power for use
with, for example, a chart recorder, as well as facilities
for measuring DC voltages derived from those supplied to the
bridges. The time constant of the recorder output is typically 35 ms and this type of power meter has been used to
demonstrate that a square-law response is obtained over at
least a 30 dB range (21).

4.4. Pyroelectric detectors

A pyroelectric material develops an emf when it is subject-
ed to to a change of temperature: this phenomenon has been
investigated experimentally for detection at millimetre wavelengths (22). Tapered resistive films, deposited on opposite faces of a 9 µm thick sheet of polyvinylidene fluoride
(PVDF), are positioned along the axis of a waveguide to form
a near-matched load (VSWR $<$ $1 \cdot 4$ over 95 to 105 GHz). The heat
generated in the resistors causes the temperature of the PVDF
sheet to change when the RF source is amplitude modulated by
a 30 Hz square wave, and this results in a reported NEP of
between 8 and 20 nW/Hz$^{\frac{1}{2}}$. No significant deviation from
square law response occurs over the power range 20 nW to
10 µW and, when it is used as a homodyne detector with 10 mW
reference power, square-law response from about $0 \cdot 1$ pW to
1 mW is reported (23).

5. CONCLUSION

In this chapter we have provided an introduction to some
of the detectors and detection methods in current use for
microwave measurements. We have not covered the application
of statistical and correlation techniques for recovering
wanted signals from noise. With the increasing speed and
decreasing cost of analogue-to-digital converters and of
microprocessors, this may well become an increasingly
significant feature of microwave measuring instruments.

REFERENCES

1. Cowley, A.M., and Sorensen, H.O., 1966, 'Quantitative
   comparison of solid-state microwave detectors', IEEE
   Trans, MTT-14, 588-602.

2. Sorger, G.U., and Weinschel, B.O., 1959, 'Comparison of deviations from square law for RF crystal diodes and barretters', IRE Trans, I-8, 103-111.

3. Fong-Tom, R.A., and Cronson, H.M., 1983, 'Diode detector characteristics for a 94 GHz six-port application', IEEE Trans, MTT-31, 158-164.

4. Somlo, P.I., and Hunter, J.D., 1982, 'A six-port reflectometer and its complete characterisation by convenient calibration procedures', IEEE Trans, MTT-30.

5. Cullen, A.L., and An, T.Y., 1982, 'Microwave characteristics of the Schottky-barrier diode power sensor', Proc IEE part H, 129, 191-198.

6. Oliver, B.M., and Cage, J.M., 'Electronic measurements and instrumentation', McGraw-Hill, 1971.

7. Woods, D., 1962, 'A coaxial millivoltmeter/milliwattmeter for frequencies up to 1 Gc/s', Proc IEE part B, 109, Suppl no 23, 750.

8. Oliver, B.M., 1967, 'Which RF voltmeter ?', Hewlett Packard Application Note no 60.

9. Tucker, D.G., 1947, 'Synchrodyne receiver design', Electronic Engineering, 19, 241.

10. Smith, R.A., 1951, 'The relative advantages of coherent and incoherent detectors: a study of their output noise spectra under various conditions', IEE Monograph no 6.

11. Jaggard, D.L., and King, R.J., 1973, 'Sensitivity and dynamic range considerations for homodyne detection systems', IEEE Trans, IM-22.

12. Weinert, F.K., 1980, 'An automatic precision IF-substitution vector ratiometer for the microwave frequency range', IEEE Trans, IM-29, 471-477.

13. Weinert, F.K., and Weinschel, B.O., 1981, 'A dual-channel automatic vector ratio meter', IEE Colloquium Digest 1981/49, 4/1-4/5.

14. Warner, F.L., Herman, P., and Cummings, P., 1983, 'Recent improvements to the UK national microwave attenuation standards', IEEE Trans, IM-32, 33-37.

15. Lemco, I, and Rogal, B., 1960, 'Resistive film milliwattmeters for the frequency bands 8·2-12·4 Gc/s, 12·4-18 Gc/s and 26·5-40 Gc/s', Proc IEE part B, 107, 427-430.

16. Luskow, A.A., 1971, 'This microwave power meter is no drifter', Marconi Instrumentation, 13.

17. Warner, F.L., 'Microwave attenuation measurement', Peter Peregrinus Ltd, 1977.

18. Larsen, N.T., 1976, 'A new self-balancing DC-substitution RF power meter', IEEE Trans, IM-25.

19. Aslan, E.E., 1969, 'Accuracy of a temperature compensated precision RF power bridge', IEEE Trans, IM-18, 232-236.

20. Anon, 1979, 'Operating and service manual for power meter 432A', Hewlett Packard.

21. Abbott, N.P., and Orford, G.R., 1981, 'Some recent measurements of linearity of thermistor power meters', IEE Colloq Digest 1981/49, 5/1-5/4.

22. Iwasaki, T., Inoue, T., and Nemoto, T., 1979, 'A matched load PVF_2 pyroelectric detector for millimetre waves', IEEE Trans, IM-28, 88-89.

23. Iwasaki, T., and Nemoto, T., 1980, 'Homodyne detection at 100 GHz with a pyroelectric detector', IEEE Trans, IM-29, 190-192.

ADDITIONAL REFERENCES

24. Peterson, E., and Hussey, L.W., 1939, 'Equivalent modulator circuits', Bell Syst Tech J, 18, 32-48.

25. Peterson, L.C., and Hussey, L.W., 1945, 'The performance and measurement of mixers in terms of linear network theory', Proc IRE, 33, 458-476.

26. Pound, R.V., 'Microwave mixers', (Rad Lab Series no 16), McGraw-Hill, 1945.

27. Torrey, H.C., and Whitmer, C.A., 'Crystal rectifiers', (Rad Lab Series no 15), McGraw-Hill, 1948.

28. Warner, F.L., 'Detection of millimetre and submillimetre waves' Chap 22 in 'Millimetre and submillimetre waves', (ed Benson, F.A.), Iliffe, 1969.

29. King, R.J., 'Microwave homodyne systems', Peter Peregrinus Ltd, 1978.

Chapter 7

Power measurement

L. C. Oldfield

1. INTRODUCTION

1.1 Power and the Transmission of Energy

Power is defined generally as the rate at which energy
is transmitted. The SI unit of power is the watt which is
the power when energy is transmitted at the rate of one joule
per second. At microwave frequencies energy is transmitted
as a result of the generation and collapse of electric and
magnetic fields, so the rate of energy transmission at a
given instant is dependent upon time and the microwave
frequency. Instantaneous power is not a quantity of general
interest, so the term 'power' in this paper will be defined
as the rate of energy transmission averaged over a number of
complete microwave periods. As a means for ensuring that
this energy transmission is directed towards the object of
interest (the load), use may be made of an antenna, or a
guiding structure in which multiple reflections of the energy
occur. Energy is absorbed at boundaries having imperfect
reflectivity, and reflected at boundaries having imperfect
absorptivity. The problem of measuring power is thus as much
concerned with the transmission medium and load as with the
output of the generator. The transmission of power over short
distances through hollow or coaxial waveguides can often be
considered free of dissipation, but they may contain discontinuities, the effect of which is to cause local imbalance
in the energy exchange between electric and magnetic fields,
resulting in reflection loss. In sect. (3) it is shown how
allowance can be made for the effect on power measurement of
reflection (mismatch) loss.

1.2 The Fundamental Role of Power

If the transmission guide is uniform in cross-section
and is non-dissipative, it can be shown that the relative
values of the fields are determined by the cross-sectional
dimensions of the guide, and the magnitude and direction of
power flow by two complex variables formulated according to
the nature of the application (1). For analysis of modes
and propagation in a particular form of guide they will be
the amplitude and phase of electric and magnetic fields (or
possibly voltage and current in the case of coaxial guide).

In measurement science, where the interaction between hardware components is of particular interest, it is convenient to work in terms of two waves b and a. Wave b propagates towards the plane of reference and wave a away from it. In contrast to the field and voltage-current representations, the magnitudes of the variables are invariant and phase is a linear function of distance along the guide. The magnitudes and relative phases of these variables are defined by the incident and reflected power P_i and P_r and the reflection coefficient of the load Γ according to

$$|b|^2 = P_i \quad , \quad |a|^2 = P_r \quad , \quad a/b = \Gamma \qquad (1)$$

Since P and Γ are amenable to measurement, whereas fields and voltages cannot be measured at any given plane without destroying the uniformity of the guide, power and reflection assume roles of fundamental importance in microwave measurements. We note that this is in marked contrast to dc and lf measurements, where on account of their ease of measurement, current and voltage assume this role.

1.3   Definitions of Power

1.3.1 Average power

Average power P_{av} is the rate of energy transfer averaged over many periods of the lowest modulation frequency involved. For a pulse modulated signal, power must be averaged over many repetitions of the pulse. Average power is measured with a power meter having a response time long in comparison with the modulation period.

1.3.2 Pulse power

For pulse power P_p the rate of energy transfer is averaged over the pulse width τ, where τ is the time between half power points. This definition is suitable for pulses where the shape departs only marginally from rectangular. Pulse power is usually computed from measurements of duty cycle and average power according to $P_p = P_{av}/duty$ cycle where duty cycle is τ x repetition frequency.

1.3.3 Peak envelope power

Peak envelope power P_e is the maximum power that could be recorded by a power meter having a response time much less than the modulation period and much greater than the microwave period. Peak envelope power is used when the pulse shape is sufficiently non rectangular for the pulse power definition to give significant error. The measurement of peak envelope power is discussed in sect. (5.5).

1.3.4 Available power

Available power P_A is the maximum power that can be extracted from a generator. It is the power delivered by

the generator to a load in the absence of any dissipation or
reflection between the two (sect. (3.1)).

2.   SURVEY OF POWER STANDARDS AND SENSORS

Devices for power measurement may be classified into
three broad categories (2):

a.   Fundamental or primary standards where power is derived
directly in terms of the base units of the SI system.

b.   Substitution standards where power at the microwave
frequency is compared with power at some other frequency
(usually dc).

c.   Secondary standards which can be used for power
measurement when they have been calibrated by fundamental or substitution standards.

Devices in these categories must have adequate sensitivity, linearity with power across the working range and
stability with time and environmental conditions.   In
addition, it is desirable that they should have a short
reading time, a large dynamic range, a rugged electrical and
mechanical structure and small physical size.

2.1 Fundamental Standards

This class of standards is of interest to standards
laboratories, where a link must be established between the
base units and the microwave watt.   A direct implementation
of the definition of power demands the measurement of
frequency and the energy stored in electric or magnetic
fields.   In the fundamental methods discussed here, energy
is measured in terms of the work done on charged particles
due to their motion through electromagnetic fields.   With
the possible exception of the 'adiabatic' calorimeter, we
note that thermal methods cannot be fundamental in the strict
sense, because they involve measurements of calorimetric
parameters where it must be assumed that the thermal
characteristics at microwave frequencies are the same as
those for dc.

2.1.1 Torque-vane and radiation pressure wattmeters (3,4,5)

The torque-vane wattmeter has been used in many forms.
A typical example is the device (3), where torque is produced
in the suspension of a conducting vane in which current is
induced by the electric field in a waveguide T-junction.   The
magnitude and direction of the torque required to maintain
the vane in its rest position is dependent upon the magnitude
and direction of energy transfer.   A general method for the
calibration of this type of instrument in terms of length and
force measurements has been given by Cullen (6); it does not
demand knowledge of the electric field in the junction.
As a high power feed-through instrument (it requires
tens of watts to achieve its lowest reported uncertainty (7)

of around ± 2%) it is useful, but otherwise its sensitivity to vibration, to standing waves in the guide and to convection currents make it unsuitable for general use.

2.1.2 Hall effect wattmeter (8,9)

When a slice of semiconductor is placed across a waveguide (Fig.1), the electric field E causes a current to flow.

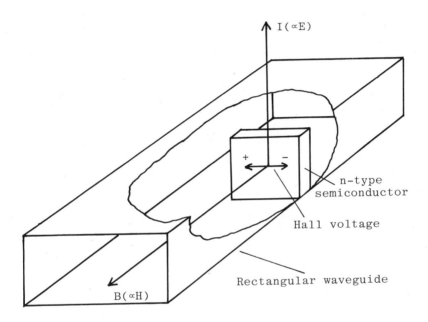

Fig.1 Hall effect in a waveguide

The current carriers are deflected by the magnetic field H in a direction mutually orthogonal to E and H. This produces the Hall voltage, the magnitude and polarity of which is determined by the magnitude and direction of net power transfer. Theoretical treatment of this device has not produced good agreement with practice, probably because the sensitivity of the effect is barely 5 μV.W^{-1}. Hall effect devices suffer from spurious voltages due to rectification in the semiconductor and also mismatch effects caused by the presence of the semiconductor in the guide. However, they are useful for high power feed-through applications where the load VSWR is large.

2.1.3 Electron beam wattmeter (10,11)

This device, like the torque vane devices, depends on the interaction between moving charges and electric fields. Since the electron beam requires no conducting path, the microwave problems associated with moving vanes are eliminated.

In the most recent development of this method (12) (Fig.2), a low intensity (about 100 nA) electron beam is

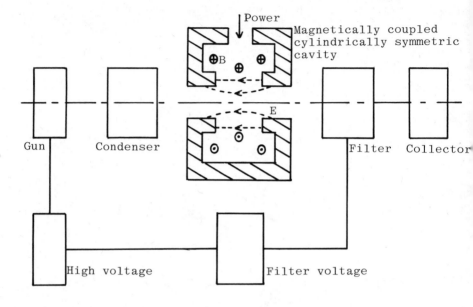

Fig.2   Electron beam power standard

produced by an electron gun and condenser system and is directed along the axis of cylindrical symmetry of a cavity excited at resonance by the power to be measured. The electric field across the cavity gap modulates the energy of electrons in accordance with the phase of their transit. The depth of this modulation is detected with an electrostatic filter in which the voltage is raised until electrons just fail to reach the collector. The voltage required for beam cut-off can be related to the rate of energy storage in the cavity.

Measurements are traceable to the standards of mass, length, time and the value of e/m_o (the ratio of electronic charge to rest mass). Although analysis and results from the device have indicated uncertainties of ± 0.5% in the 100 mW region when used at 9 GHz, it is on account of its narrowband limitation and cumbersome operation, likely to remain a standards laboratory equipment.

2.2   Substitution Standards

Since it is not possible in this lecture to give a comprehensive review of all the power standards in this category, the student is referred to the many excellent reviews that have been published (2,9,13-15).

2.2.1 Calorimeters

Calorimeters (Fig.3) depend on the conversion of input

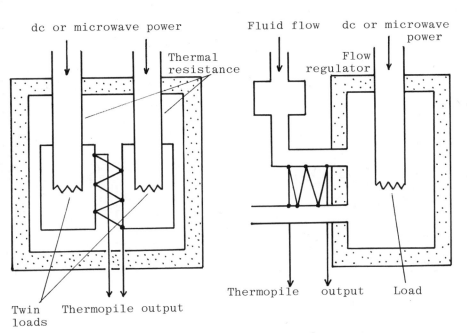

Fig.3   Twin load and fluid flow calorimeters

power to heat, i.e. they are terminating devices. They fall
into two broad categories: static calorimeters (16,17) and
fluid flow calorimeters (18). The simplest static device
consists of a single load isolated from the environment
(adiabatic calorimeter). Power is measured in terms of heat
capacity and rate of temperature rise. Although simple in
concept, this device is difficult to implement at power
levels below about 100 W. The twin load calorimeter does
not require the same environmental isolation because the
temperature rise is referred to a dummy load with similar
thermal properties. Power absorption results in an increase
of load temperature until a thermal balance is reached with
its surroundings. Power is measured in terms of the dc power
required to produce the same temperature rise. The technique
of substituting dc power for microwave power is useful
because it obviates the need for detailed knowledge of heat
losses and thermal capacities. For the most accurate work
the heat distributions in the dc and microwave elements must
be matched. Failure to meet this condition can result in
what is termed 'substitution error'.

Twin load calorimeters are used at the National Physical
Laboratory to establish the national standard of power in
14 mm coaxial waveguide for frequencies up to 8.2 GHz (16).
The uncertainty ascribed to 95% confidence is 0.2% at a level

of 10 mW.

The fluid flow calorimeter is intrinsically broadband
and is particularly useful at levels of several watts. If
the volumetric flow rate, specific heat, density and temperature rise are known, this apparatus may be calibrated
directly, but it is more often calibrated with dc power.
Flow calorimeters based on this principle are used at the
Electrical Quality Assurance Directorate to provide power
standards for the Ministry of Defence (18). The uncertainty
ascribed to the EQAD X-band flow calorimeter is ± 0.6% at a
level of 1 W.

2.2.2   Bolometers (barretter and thermistor)

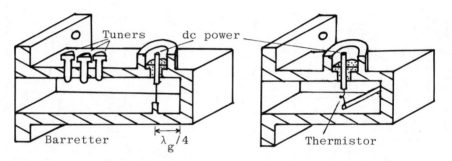

Fig.4   Waveguide mounted barretter and thermistor
        bolometers

A bolometer is an absorber (element) that has a large
temperature coefficient of resistance. This element may be
a thermistor, a fine wire (barretter) or a thin metallic
conducting film optimally placed within the waveguide with
respect to the peak microwave electric field (Fig.4a). The
sensitivity is typically 5-50 Ω.mW^{-1}.

The barretter technique is the most precise known for
measuring power at mW levels and is the one adopted by most
national standards laboratories for measuring power in hollow
waveguide. Barretters are operated by using dc power
substitution, so it is not necessary to know their temperature coefficient of resistance or to make first order
corrections for heat losses. The substitution consists of
adjusting by means of a self balancing bridge (Fig.5a) the
dc power that must be removed to balance the resistance
change caused by the incident microwave power. The dc power
removed is not a direct measure of the microwave power
because a proportion of this is absorbed by the element
mount. The performance of bolometric standards (element +
mount) of all types is defined by their reflection coefficient Γ and their effective efficiency η_e

$$\eta_e = \frac{\text{dc power substituted, } P_{dc}}{\text{total microwave power supplied at the input flange, } P_I}$$

(2)

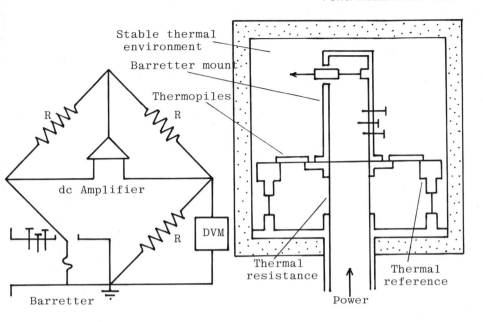

Fig.5   Operation and calibration of a barretter
        standard

where $P_I = P_{dc}$ + power dissipated in the mount structure,
P_d + difference between element dissipation at dc and microwave for the same element resistance, P_S + power leaked from
the mount at points between the input flange and the element,
P_L. The measurement of P_d, P_S and P_L is a standards laboratory task. For well designed standards P_L is negligible;
$P_d + P_S$ is derived from a technique that involves measurement
of the small temperature change that occurs at the reference
plane of a standard when it is operated with microwave power
rather than dc (Fig.5b). Microcalorimeters capable of
measuring temperature differences of 100 µ°C have been in
use at the National Bureau of Standards for some years (19,20).
These have been developed at the UK National microwave
standards laboratory (RSRE, Malvern) for calibration of tuned
barretter standards covering the frequency range 8.2-40 GHz
(21). Typical values of η_e for these standards is 90-99%
measured with an uncertainty of < ± 0.15%. Having been
calibrated, the barretter standard is used in conjunction
with a self balancing bridge and a precision voltmeter
(Fig.5a) according to

$$P = (V_0^2 - V_1^2)/(R.\eta_e) \qquad (3)$$

where R is the bridge resistance across which V_0 and V_1 are
measured for power off and on respectively.
     Thermistor bolometers (13) are arguably the most gener-
ally useful as power standards because they are electrically
and mechanically more rugged than barretters and are more

amenable to incorporation in a broad band mount structure
(Fig.4b). When used as broadband standards the correction
for mismatch can be significant, so the overall uncertainty
is greater than for the tuned barretter standard. The main
disadvantage of thermistors is their sensitivity to
environmental temperature, but this problem is greatly reduced
if a compensation thermistor is used, or if the results are
computed from thermistor voltage in response to alternate on-
off states of the measurement power. Thermistors have a much
longer time constant than barretters (approximately 0.1 s)
making them less suitable for measurement of peak envelope
power. Thermistor power sensors are being used at RSRE to
provide broad band national standards between 8.2 and 26.5
GHz and in the millimetre bands 50-75, 75-110 GHz.
     The thin film bolometer (9) consists of a temperature
sensitive resistive metal film deposited on dielectric
substrate. It has a shape and resistance determined by the
need to minimise the reflection coefficient of the element
and its mount structure. The thickness of the film is less
than a skin depth at the highest frequency to be used, and
the length is less than a wavelength. This ensures that the
current distribution is nearly independent of frequency, so
minimising any substitution errors due to different heating
effects for dc and microwave excitation. Thin film bolometers have been used experimentally at up to 100 mW as
terminating power sensors. Thin resistive films forming
lengths of waveguide wall (known as enthrakometers (22,9))
have been used for feedthrough power sensors. The attainable
stability and uniformity of thin resistive film deposition
does not yet appear to be sufficient for commercial production
of these devices.

2.2.3 Thermoelectric devices (9)

     A thermoelectric emf is generated when the junction of
two dissimilar metals is heated. There are two basic types
of thermoelectric device, distinguished by whether the power
is dissipated in the thermopile or in a load to which the
thermopile is attached (23). Directly heated sensors are
available for frequencies up to about 18 GHz, but the
indirect devices have been used beyond 40 GHz. Both types
are capable of being calibrated by substitution techniques
to uncertainties of ± 1%.
     A typical thermoelectric sensor (24) is constructed
using a thin film metallic load made up of bismuth and
antimony deposited on to a dielectric substrate in a way
that produces a number of junctions in series. The 'cold'
junctions are attached to the waveguide whereas the 'hot'
junctions are located in the air space within the waveguide.
     The dependence of thermopile voltage on incident power
departs significantly from linearity when the junction
temperature exceeds a certain limit, so although the thermal
properties of the sensors can be varied to produce devices
capable of covering a wide dynamic range (at least 10 nW to
3 W), the range of an individual sensor is only about two
decades. Typical response times vary between 0.1 and 5 s
according to design and operational range. Other disadvantages

of these sensors include their low sensitivity and the
difficulty of designing them to have a low reflection
coefficient over a broad band.   Their advantages include low
zero drift and sensitivity to environmental temperature, an
ability to withstand accidental overload and the simplicity
of the indicating equipment.

2.2.4 Superconducting quantum interference device (SQUID)
(25,26)
     The SQUID (Fig.6) is based on a loop of superconductor
split by a Josephson (weak contact) junction - that is a
junction that remains a normal conductor at superconducting
temperatures.   An arrangement that is often used is a spring
loaded point formed entirely from the superconductor.   The
weak contact is created by oxidisation of the point.

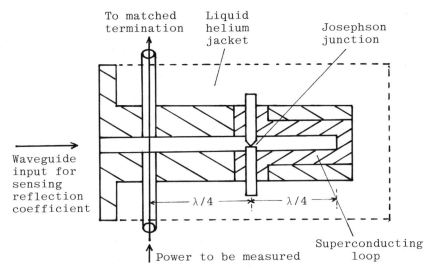

Fig.6   Microwave superconducting quantum
          interference device

     When a change of magnetic flux linkage through the loop
occurs the current induced is not a linear function of the
rate of change, but periodic with frequency 2 e/h where e is
the electronic charge and h is Planck's constant.   This is
an interference effect produced because the superconducting
electrons are not scattered and so can maintain coherence
between their wave functions.   Electrons can be considered
as quantised energy packets with wavelength λ = h/p where
p is the electron momentum generated by the changing flux
linkage.   As λ varies, the induced current passes through
maxima and minima according to the phase shift across the
junction resulting from the variations in path length around
the loop.
     In the device illustrated, the impedance at the wave-
guide port is periodic in response to the magnetic flux
linking the superconducting loop due to the microwave power

entering the coaxial port. By counting impedance minima, precise increments of power are measured, but due to the difficulty of measuring absolute flux linkage, a calibration with dc power is needed to enable the SQUID to be used as a power standard. This method is capable of great sensitivity (10^{-14} W) but is difficult to realise on account of the susceptibility of the junction to thermal and mechanical shock.

2.3   Secondary Standards

This section includes a selection of power sensors for which the substitution technique cannot conveniently be applied and the theoretical law is insufficiently well known for them to be used other than in a secondary role. They may be used as power standards when they have been calibrated against fundamental or substitution standards.

2.3.1   Crystal diode (9)

The current detected by a diode at low levels is approximately proportional to the square of the amplitude of the microwave electric field and thus to power. Diodes have a fast response time, a good dynamic range (μW to mW) and a high sensitivity (27) (5 mV.μW^{-1} into a high impedance load) but their stability with time and temperature is poor and the point contact types are susceptible to mechanical shock. It is difficult to produce a low reflection coefficient across a wide frequency range when the diode is used to terminate the guide, but when mounted in the wall of a waveguide to make a feedthrough power meter, the reflection coefficient can be very low.
Calibration procedures for diodes usually involve a means for improving the linearity of the indication. This may be achieved with appropriate circuitry, but the modern method in which the measured characteristic is function fitted on a computer, gives more flexibility, accuracy and working range.

2.3.2   Pyroelectric devices (28,29)

The pyroelectric effect occurs in materials, usually ferroelectric, which have a temperature-dependent electric polarisation. The voltage produced between two electrodes on a piece of pyroelectric material is proportional to the rate of change of temperature, so it is necessary to modulate the power to be measured. The power should be dissipated in the pyroelectric material because the thermal time constant of any attached absorber would seriously degrade the sensitivity. The problem of absorbing the power in a thin layer of pyroelectric material sets a lower limit to the frequency range that can be covered. Few pyroelectric detectors have been developed for frequencies below 300 GHz.

2.3.3   Golay cell (29)

This relies on pressure changes induced by heating in a

small cell of gas. The absorption of radiation is by a thin
metal film inside the cell, heat being transferred to the gas
by conduction. The gas cell is usually connected through a
small 'leak' to a larger reservoir of gas, so that the device
has a transient response like the pyroelectric device. The
Golay cell is most frequently used above 100 GHz.

3.   THEORY OF POWER MEASUREMENT AND COMPARISON OF SENSORS

This section covers the theory and techniques that must
be employed in direct power measurements or whenever one
power sensor is compared with another. Where it is necessary
to refer to a particular type of power sensor the thermistor
bolometer is cited on account of its widespread popularity.

3.1   Direct Power Measurement

The use of a terminating power meter to measure directly
the output of a microwave generator is the simplest form of
power measurement. The theory of this technique provides
the ground upon which more sophisticated use of power meters
is based.
Consider a generator connected to an absorbing load by
a section of lossless waveguide with standard cross-sectional
dimensions (Fig.7a). The reflection coefficients of the

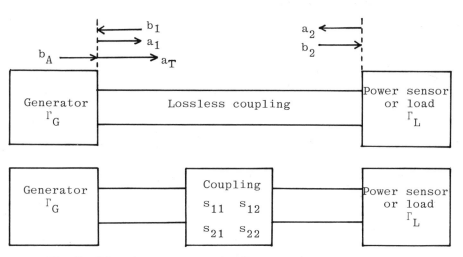

Fig.7   Direct measurement of generator power

generator and load, Γ_G and Γ_L, are defined with respect to
this standard waveguide. Let the AVAILABLE POWER (sect.
(1.3.4)) of the generator be P_A and the power emanating from
the output plane of the generator (TRUE POWER) be P_T (this
does not include power reflected from the load and re-
reflected at the generator). b_A and a_T are waves defined
such that $|b_A|^2 = P_A$ and $|a_T|^2 = P_T$ (sect. (1.2)), b_1, b_2
are waves incident upon generator and load respectively and
a_1, a_2 are the corresponding reflected waves. The wave

launched into the waveguide (and subsequently incident upon the load) is given by

$$b_2 = a_T + a_1 = a_T + b_1\Gamma_G$$

from which we obtain $P_T = |b_2 - b_1\Gamma_G|^2$. Since $b_1 = a_2 = b_2\Gamma_L$,

$$P_T = b_2^2 |1 - \Gamma_L\Gamma_G|^2 \qquad (4)$$

The power absorbed by the load is given by

$$P_L = |b_2|^2 - |a_2|^2 = |b_2|^2(1 - |\Gamma_L|^2) \qquad (5)$$

where $(1-|\Gamma_L|^2)$ is known as the MISMATCH LOSS FACTOR. From eqns. (4) and (5)

$$P_T = P_L \cdot |1-\Gamma_L\Gamma_G|^2 / (1-|\Gamma_L|^2) \qquad (6)$$

Generators are commonly specified in terms of their available power, because this is a quantity that is independent of the transmission system and load or power meter, and hence of multiple reflections that may exist.

By taking the difference between incident and reflected power at the generator output

$$P_T = P_A (1-|\Gamma_G|^2) \qquad (7)$$

hence the available power is given in terms of the load power by

$$P_A = P_L \frac{|1-\Gamma_L\Gamma_G|^2}{(1-|\Gamma_L|^2)(1-|\Gamma_G|^2)} \qquad (8)$$

From this expression we see that if Γ_L is the complex conjugate of Γ_G (equal magnitude but phase advanced by π), then $P_A = P_L$. Thus when the reflection coefficients of generator and load are conjugately matched, the power delivered to the load is the available power of the generator. Whilst this is an interesting result, it is of little use in practice because it is not possible, or even desirable, to have a lossless coupling between generator and load. Some isolation is necessary to prevent the operation of the generator from being affected by the characteristics of the load. This isolation is usually provided by a non reciprocal attenuator (isolator).

Let the properties of the real coupling be described in terms of scattering coefficients (Fig.7b) so that

$$b_1 = s_{11}(a_1 + a_T) + s_{12}a_2 \quad , \quad b_2 = s_{22}a_2 + s_{21}(a_1 + a_T)$$

$$a_1 = b_1\Gamma_G \qquad\qquad , \quad a_2 = b_2\Gamma_L$$

By elimination of a_1, a_2 and b_1 we find

$$b_2 = \frac{a_T s_{21}}{(1-s_{11}\Gamma_G)(1-s_{22}\Gamma_L) - s_{12}s_{21}\Gamma_G\Gamma_L}$$

Hence from eqns. (5) and (7)

$$P_A = \frac{P_L \left| (1-s_{11}\Gamma_G)(1-s_{22}\Gamma_L) - s_{12}s_{21}\Gamma_G\Gamma_L \right|^2}{|s_{21}|^2 (1-|\Gamma_G|^2)(1-|\Gamma_L|^2)} \qquad (9)$$

If an isolator with matched ports is used, s_{11}, s_{22} and s_{12} may be set to zero, so

$$P_A = P_L / [\, |s_{21}|^2 (1-|\Gamma_G|^2)(1-|\Gamma_L|^2)\,] \qquad (10)$$

We note that an attenuation measurement is required, but the phase relationship between Γ_G and Γ_L is now immaterial.

3.2 2-Port Comparison Measurements

If a load (or power sensor) and a power standard are connected alternately to the same generator, the relationship between the power absorbed by the load P_L and that absorbed by the power standard P_S is derived from eqn. (9)

$$\frac{P_L}{P_S} = \frac{1-|\Gamma_L|^2}{1-|\Gamma_S|^2} \cdot \left| \frac{(1-s_{11}\Gamma_G)(1-s_{22}\Gamma_S) - s_{12}s_{21}\Gamma_G\Gamma_S}{(1-s_{11}\Gamma_G)(1-s_{22}\Gamma_L) - s_{12}s_{21}\Gamma_G\Gamma_L} \right|^2 \qquad (11)$$

By using a well matched isolator between the generator and load, this expression reduces to

$$P_L/P_S = (1-|\Gamma_L|^2)/(1-|\Gamma_S|^2) \qquad (12)$$

3.2.1 Calibration factor of a power sensor

Let us suppose that we wish to calibrate a power sensor in terms of a power standard. If the standard has an effective efficiency η_S and its indication is I_S then $P_S = I_S/\eta_S$ and from eqn. (12) we obtain

$$\frac{I_u}{I_S} = \frac{\eta_u(1-|\Gamma_u|^2)}{\eta_S(1-|\Gamma_S|^2)} = \frac{K_u}{K_S} \qquad (13)$$

where the suffix u refers to the device to be calibrated, and K is the CALIBRATION FACTOR defined by the product of EFFECTIVE EFFICIENCY and MISMATCH LOSS FACTOR.

$$K = \eta(1-|\Gamma|^2) \qquad (14)$$

Given a matched generator and a standard for which K_S is known K_u for any unknown is simply the ratio of indications (the dc substitute power). A measurement of $|\Gamma|$ is necessary in order to find η.

3.3   3-Port Comparison Measurements

For the most accurate work, the 2-port comparator is limited by the need to measure the amplitudes and phases of several complex quantities (eqn. (11)) and in practice by the stability of the generator. The 3-port arrangement (Fig.8) alleviates both of these difficulties and also has the following important advantages:

a.   The sensitivity can be varied by choice of coupling factor (sects. (5.1,5.2)).

b.   The calibration is independent of the characteristics of the source.

c.   The equivalent generator reflection coefficient is usually small and can be tuned out if necessary.

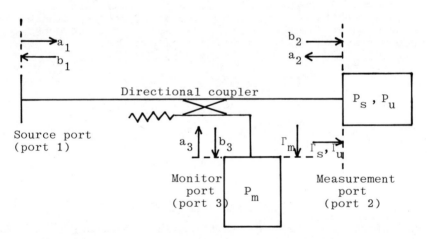

Fig.8   3-port power comparator

For the 3-port directional coupler the basic equations are

$$
\begin{pmatrix} b_1 \\ b_2 \\ b_3 \end{pmatrix} =
\begin{pmatrix} s_{11} & s_{12} & s_{13} \\ s_{21} & s_{22} & s_{23} \\ s_{31} & s_{32} & s_{33} \end{pmatrix} \cdot
\begin{pmatrix} a_1 \\ a_2 \\ a_3 \end{pmatrix} \qquad (15)
$$

If port 3 is terminated by a power sensor (the monitor) with reflection coefficient Γ_m, and port 2 with the standard, $a_3 = b_3 \Gamma_{m2}$ and $a_2 = b_2 \Gamma_{s2}$. By solving eqns. (15) for $P_s = |b_2|^2$ and $P_m = |b_3|^2$ we obtain

$$\frac{P_m}{P_s} = \left|\frac{s_{31}}{s_{21}}\right|^2 \left|\frac{1-\Gamma_s G_2}{1-\Gamma_m G_3}\right|^2 \frac{1-|\Gamma_m|^2}{1-|\Gamma_s|^2} \qquad (16)$$

where $\quad G_2 = s_{22} - \dfrac{s_{21}s_{32}}{s_{31}} \quad$ and $\quad G_3 = s_{33} - \dfrac{s_{31}s_{23}}{s_{21}}$

The three terms in this equation express respectively contributions from the 3-port, the interaction between the 3-port and the terminations, and the terminations.

If the monitor indication is I_{ms} when the standard is attached, and I_{mu} when the unknown is attached

$$\eta_u = \eta_s\left(\frac{I_{ms}}{I_{mu}} \cdot \frac{I_u}{I_s}\right)\left|\frac{1-\Gamma_u G_2}{1-\Gamma_s G_2}\right|^2 \frac{1-|\Gamma_s|^2}{1-|\Gamma_u|^2} \qquad (17)$$

Thus, the efficiency and reflection coefficient of the monitor are eliminated and the dependence on the 3-port is embodied in the single parameter G_2 which, though analogous to the generator reflection coefficient in eqn. (6), is not a reflection coefficient, hence the notation G rather than Γ. Accurate comparisons of power sensors cannot be carried out unless G_2 has been:

a. Made negligible by suitable design of the equipment.

b. Eliminated by using 180° phase standards.

c. Measured and allowed for.

Each of these options is viable and will be briefly discussed.

3.3.1 Making G_2 negligible

From the two terms defining G_2 (eqn. (16)) there arise two distinct requirements of the directional coupler for the elimination of G_2:

a. The output port must be matched ($s_{22} = 0$).

b. The directivity must be infinite (s_{31}/s_{32}).

Since no directional coupler is perfect, we must calculate the error in measuring η_u when it is assumed that G_2 is negligible. This is known as the MISMATCH ERROR. Assuming that the values of $|\Gamma_u|$ and $|\Gamma_s|$ are known, the error arises from ignorance of $|G_2|$ and of the phase relationship between G_2 and Γ_u, Γ_s (eqn. (17)).

Let $\eta_s \left(\dfrac{I_{ms} \cdot I_u}{I_{mu} \cdot I_s} \right) \dfrac{1 - |\Gamma_s|^2}{1 - |\Gamma_u|^2} = C_1$ (independent of G_2)

then $\qquad \eta_u = C_1 |(1 - \Gamma_u G_2)/(1 - \Gamma_s G_2)|^2$ (18)

The range of possible values of η_u is defined by the limits

$$\eta_u(\text{max}) = C_1 \left[\frac{1 + |\Gamma_u G_2|}{1 - |\Gamma_s G_2|} \right]^2 \quad , \quad \eta_u(\text{min}) = C_1 \left[\frac{1 - |\Gamma_u G_2|}{1 + |\Gamma_s G_2|} \right]^2$$

As $|\Gamma_s G_2|$ and $|\Gamma_u G_2|$ are both $\ll 1$, the rectangular % limits of η_u are

$$e\% \simeq \pm 50[(1 + |\Gamma_s G_2|)^4 - (1 - |\Gamma_u G_2|)^4]$$

Hence e% is given approximately by

$$e\% \simeq \pm 200 \ |G_2| \ (|\Gamma_s| + |\Gamma_u|) \tag{19}$$

Harris and Warner (30) indicate that the probability distribution governing a power measurement subject to mismatch error is U-shaped. Under these circumstances, the 95% confidence limits are given by

$$e'\% = 0.77e\% \tag{20}$$

As an example (31) we consider a typical high quality coupler for which $|G_2| = 0.016$, and a power standard where $|\Gamma_s| = 0.01$ (Table 1).

$|\Gamma_u|$	0.01	0.05	0.10	0.15	0.20	0.30
e'%	0.05	0.15	0.27	0.39	0.52	0.76

TABLE 1.   Mismatch error as a function of $|\Gamma_u|$

If a second high quality coupler and power sensor are added to make a back-to-back arrangement (32), approximate values of $|\Gamma|$ can be deduced from the forward and reverse power sensors. The error limits on the measurement of $|\Gamma|$ are found from the range of power ratios obtained when a short circuit connected at the output port is moved over $\lambda_g/2$. If eqn. (19) indicates that mismatch errors are unacceptably large when using the best couplers available, tuners may be used to improve the performance at discrete frequencies.

3.3.2 Eliminating the effect of G₂

We will show that given a precision waveguide spacer of length $\lambda_g/4$ (known as a 180° phase standard or quarter wave spacer) the mismatch error can be effectively eliminated (33, 34). The simplest implementation of this technique assumes that the $\lambda_g/4$ spacer is perfect, ie it is non dissipative and its insertion at the measurement port (Fig.8) does not affect $|\Gamma|$. The monitor is not disturbed during the measurements, so from eqn. (16) we define

$$C_2 = \left|\frac{s_{31}}{s_{21}}\right|^2 \frac{1-|\Gamma_m|^2}{|1-\Gamma_m G_3|^2}$$

giving

$$P_{ms}/P_s = C_2 |1-\Gamma_s G_2|^2/(1-|\Gamma_s|^2) \qquad (21)$$

If the $\lambda_g/4$ spacer is interposed between the coupler and the power standard, and the power ratio remeasured,

$$P'_{ms}/P'_s = C_2 |1+\Gamma_s G_2|^2/(1-|\Gamma_s|^2) \qquad (22)$$

hence $\qquad P_{ms}/P_s + P'_{ms}/P'_s = 2C_2(1+|\Gamma_s G_2|^2)/(1-|\Gamma_s|^2)$

For a standard where $|\Gamma_s| < 0.1$ and a moderate quality coupler where $|G_2| < 0.1$, $|\Gamma_s G_2|^2 < 0.0001$ and so may be neglected in comparison with unity. If the same assumption is made for the unknown power sensor

$$P_{mu}/P_u + P'_{mu}/P'_u = 2C_2/(1-|\Gamma_u|^2)$$

and the unknown efficiency is given by

$$\eta_u = \eta_s \left(\frac{1-|\Gamma_s|^2}{1-|\Gamma_u|^2}\right) \cdot \left(\frac{I_{ms}/I_s + I'_{ms}/I'_s}{I_{mu}/I_u + I'_{mu}/I'_u}\right) \qquad (23)$$

If copper waveguide is used, the losses in a $\lambda_g/4$ spacer are about 0.1%. However, since both standard and unknown are measured using the same spacer, the error in η_u is ≪ 0.1%. The effect of the extra flange joint when a $\lambda_g/4$ spacer is used can be eliminated by taking the first measurement with a spacer of length ℓ and the second with a spacer of length $\ell + \lambda_g/4$.

The $\lambda_g/4$ method is a useful one, but it has two significant limitations:

a. $|\Gamma|$ is not measured.

b. It can only be used at frequencies for which a $\lambda_g/4$

spacer is available.

The second limitation is removed, if in place of the $\lambda_g/4$ spacer, two other spacers are used where the lengths are each less than $\lambda_g/2$ at the highest frequency. By expressing eqn. (21) in the polar form

$$\frac{P_{ms}}{P_s} = \frac{C_2}{(1-|\Gamma_s|^2)} \cdot [1+R^2-2R\cos(\theta+\alpha)] \qquad (24)$$

where $R = |\Gamma_s G_2|$, $\theta = \angle(\Gamma_s G_2)$ and α is the electrical length of the spacer used, simultaneous equations describing measurements at three planes are solved (35) and $\Gamma_s G_2$ is derived, so removing the need to assume $|\Gamma_s G_2|^2 \ll 1$. If Γ_s is known, G_2 can be evaluated, as also can Γ_u by repeating the process with the unknown attached.

3.3.3   Measurement of G_2

The methods described here emerge as the natural development of those discussed in sect. (3.3.2). We accept that directional couplers are not perfect, so we seek to characterise them (measure G_2) so that the mismatch error can be avoided by mathematical correction. The computation involved is not trivial, but the benefits of the technique are considerable, in that accurate measurements may be made across a broad band without the need for mechanical adjustment.

For the cost of an extra power sensor and increased mathematical complexity, the spacer technique may be developed into the spacer reflectometer (36) (Fig.9a). This is not suitable for applications where rapid results are required, because it is necessary to measure the unknown by using at least three spacers in turn at the measurement port. However, it has the advantage that once $|\Gamma|$ has been measured, G_2 can be made small* by placing in front of the reverse power meter a well matched attenuator and then proceeding as described in sect. (3.3.2) using two spacers.

The inconvenience of the spacer reflectometer can be removed by adding a second directional coupler (Fig.9b) which allows the necessary phase shifts to be performed within the network. The multistate reflectometer (35,37) can be automated, and in the opinion of the author represents the current state of the art in power calibration. 6-port reflectometers **(as described in reference 38)** have been widely used for power calibration; they are in principle quicker in operation than the multistate, but are significantly more expensive to build and difficult to calibrate reliably over

*The definition of G_2 (eqn.(16)) applies to a 3-port, but since G_2 is the effective generator reflection coefficient at the measurement port, this parameter may be defined for any reflectometer. The accuracy of efficiency comparisons depends upon the absolute uncertainty in G_2, a quantity that is roughly proportional to G_2.

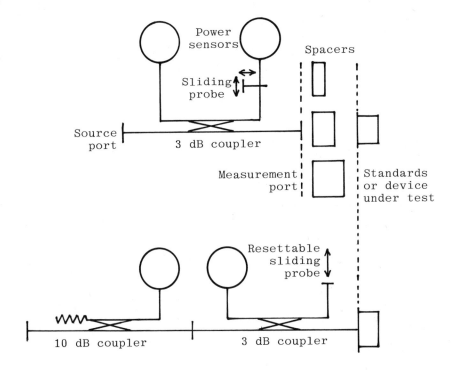

Fig.9(a) The spacer reflectometer and
     (b) The multistate reflectometer

a broad band (38).

Mainly on account of their versatility, speed and ability to measure amplitude and phase of reflection coefficient, computer corrected methods are preferred over mechanically corrected ones for power measurements.

4.   UNCERTAINTIES IN POWER MEASUREMENT

Uncertainty is defined as the 'assigned allowance for error'. Some of the common sources of uncertainty will be discussed, but no attempt will be made to deduce overall uncertainty because this will depend upon the method of power measurement being used. The technique for combining individual uncertainties is discussed in the lecture 'Uncertainty and Confidence'.

4.1 Measurement Uncertainty in Generator and Sensor Reflection Coefficients

In the past this has been the major source of uncertainty in power measurements, but with the advent of improved methods for measuring reflection coefficient, there is little excuse for incurring large mismatch errors. A power measurement is corrected for mismatch according to $|1-\Gamma G_2|^2/(1-|\Gamma|^2)$ (eqn.(21)), from which the RESIDUAL MISMATCH UNCERTAINTY can

be evaluated by differentiating with respect to Γ and G_2.
The resulting expression is a complicated one, so we consider
a particular example in which $|\Gamma| = |G_2|$, $\angle(\Gamma G_2) = \pi$ giving
a mismatch correction of approximately $(1+|\Gamma|^2)^3$. This when
differentiated with respect to $|\Gamma|$ gives the percentage
residual mismatch uncertainty e% in terms of the uncertainty
$\sigma_{|\Gamma|}$ in $|\Gamma|$.

$$e\% = 600\ \sigma_{|\Gamma|} \cdot \frac{|\Gamma|}{1+|\Gamma|^2} \tag{25}$$

$\sigma_{|\Gamma|}$ \ $|\Gamma|$	.05	.10	.20
.002	.06	.12	.23
.005	.15	.30	.58

TABLE 2.   Residual mismatch uncertainty as a function
of $\sigma_{|\Gamma|}$ and $|\Gamma|$

This table shows that e% rises with $|\Gamma|$ and $|G_2|$, so these
parameters should be kept small for the most accurate power
measurements.

4.2   Effective Efficiency of the Power Standard

This will be obtained from the manufacturer or a
standards laboratory. For X and J band calibrations at the
primary standard laboratory, the uncertainty will be $\simeq \pm 0.2\%$
to 95% confidence, at the master laboratory level it will be
$\simeq \pm 0.6\%$ and at the manufacturer's test facility may be
$\pm 2\%$.

4.3   Power Monitor Linearity

Where directional couplers with side-arm power monitors
are used for feedthrough power measurements or for comparing
power sensors, it should be ascertained that the monitor is
sufficiently linear over the required range. Recent work
with thermistor sensors (39) has shown linearity of 0.05%
over a 20 dB range.

4.4   Connector or Flange Repeatability

This is a major source of error, particularly for
coaxial systems. Many attempts have been made to assign
typical figures for the repeatability of different types of
coaxial connectors, but recent work shows that this can only
be done within very broad limits. Repeatability is a function
of connector tolerance, connector design, orientation of the
mating, nut torque, number of previous matings and any
mechanical strain caused by the connection. Moreover, since
connectors are easily damaged when mated to a damaged or

out-of-tolerance connector, a repeatability figure measured
for one pair may be very different when they are re-mated
following 'affairs' with other connectors. The uncertainty
to be quoted for connector repeatability when calibrating
power sensors should be that with respect to a high precision
standard connector, and should be assessed by measuring the
device several times using the connector in different orien-
tations. If this procedure is not followed, then worst case
uncertainties must be ascribed, and these can be sufficiently
large to swamp all other sources of uncertainty.

Waveguide flange repeatability is dependent upon the
flatness and surface finish of the flanges (40) so assessments
must be based upon the device being calibrated. Shims are
useful for producing a good quality connection between poor
quality flanges, but if they are used in a calibration, it
should be so stated because the difference between using and
not using shims can be as much as 0.2% for flanges in poor
condition.

4.5   Instrumentation

This includes all effects preventing the exact measurement of substituted dc power. If we confine the discussion
to closed loop systems, ie self balancing bridges where
reference oscillators are not required, the following effects
must be considered.

4.5.1 Drift or long term stability

This is usually determined by the thermal environmental
around the power meter bridge circuit and the thermal link
between laboratory and sensor. The remedy is to allow
sufficient warm-up of the power meter and to provide lagging
for the sensor.

4.5.2 Noise or short term stability

This is dependent upon the bridge circuits and the
voltmeter used for measuring the substituted power. Except
when measuring at very low power levels, this contributes
negligibly to the overall uncertainty.

4.5.3 Circulating earth currents

These can have a serious effect on power measurements,
particularly when more than one sensor is used on a single
piece of equipment. The remedy is to provide dc blocks
between all sensors.

4.5.4 Resistance and voltage uncertainty

For bolometric measurements, the bridge resistance must
be known (eqn. (3)). This is normally specified within close
limits by the manufacturer, but if different connecting
cables are used, it may be necessary to correct this value.
A good quality DVM with resistance range will be traceable to
National Standards, so uncertainties of this cause will

normally be negligible.

5.   MISCELLANEOUS TECHNIQUES

5.1 Very High Power

A directional coupler with appropriate coupling factor
may be used (sect. (3.3.1)). This technique is also used for
calibrating sensors with very low sensitivity. An alternative
technique uses a calibrated attenuator, but this is not
recommended because at high power the attenuator is liable
to be affected by its own dissipation.

5.2 Very Low Power

A directional coupler should be used (sect. (3.3.1)),
but the monitor is placed on the main arm and the side arm
provides the measurement port.

5.3 Calibrations Involving Waveguide Transitions

Methods for transferring the calibration from a power
sensor having a rectangular waveguide flange to one with a
coaxial connector, are given in (41,42). In the simpler
method (41), measurements are made using the transition at a
waveguide port and at a coaxial port. The unknown efficiency
is derived from simple ratios without regard to reflection
coefficients. The limits of uncertainty of a calibration by
this method are given as 1.0 to 1.4 per cent.
For more exacting work, three transitions are required
(43), but no assumptions about their properties are made.
The efficiency of each of the transitions is derived from an
experiment involving one of the others placed back-to-back.
The uncertainty is given as less than 0.1%. It should be
noted that these methods are also useful for calibrations
involving transitions between different sizes of hollow
waveguide.

5.4 Measurements in Multimode Waveguide (44)

The significant feature of the work cited is the use of
an oversize section of waveguide, where the propagation
characteristics approach those of free space and the impedance
and propagation constant are almost independent of mode. With
essentially plane wave propagation, power is proportional to
the average $|E|^2$ sensed by a number of probes arranged around
the periphery of the guide in one cross-sectional plane.

5.5 Peak Envelope Power (45)

This is defined in sect. (1.3.3). The majority of
methods for measuring P_e rely on comparing the maximum amplitude of the envelope with the amplitude of a reference, the
comparison being made on an oscilloscope supplied by a diode
peak voltage detector (Fig.10). The detector and reference
are either calibrated in terms of an average power meter using

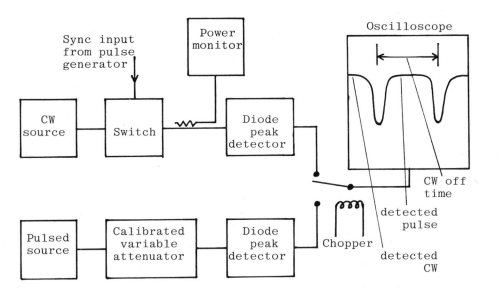

Fig.10   A 'notch' wattmeter

CW power, or this CW power is interrupted at the pulse
frequency to permit continuous comparison on the oscilloscope
display with the pulse amplitude (notch wattmeter (46)).
    Commercial versions of both techniques are available to
cover mW-kW levels. The range of uncertainties claimed
depend upon the pulse duration and are typically in the
range ± 5 to 15 per cent.

6.   CONCLUSIONS

    It has not been possible within the space and time
allowed to present a treatise on power measurement in
general. Having been written from the National standards
laboratory, it is inevitable that the lecture is slanted
more towards standards than general measurement procedures,
but given a grounding in the techniques used for the most
precise measurements, and an outline and source references
covering other techniques, the student will be better able
to assess the accuracy and suitability of the large number
of power sensors and power measuring systems that are
available.

REFERENCES

1.   Engen, G.F., 'Advances in microwave measurement science',
     Proc. IEEE, 66, 374-383, April 1978.

2.   Steele, J. McA., Ditchfield, C.R. and Bailey, A.E.,
     'Electrical standards of measurement Part 2: rf and
     microwave standards', Proc. IEE, 122, 1037-1053, Oct 1975.

3. Barlow, H.E.M., 'A microwave electrostatic wattmeter', Proc. IEE, 110, 85-87, January 1963.

4. Cullen, A.L., Rogal, B. and Okamura, S., 'A wide-band double-vane torque-operated wattmeter for 3 cm microwaves IRE Trans. MTT-6, 133-136, April 1958.

5. Barlow, H.E.M. and Cross, P.H., 'Radiation pressure instrument for absolute measurements of power at 35 GHz', Proc. IEE, 117, 853-855, April 1970.

6. Cullen, A.L., 'A general method for the absolute measurement of microwave power', Proc. IEE, 99, Pt. IV, 112-120, April 1952.

7. Wizner, W., 'A new method for absolute measurement of microwave power', PhD thesis, Univ. of London, 1968.

8. Barlow, H.E.M. and Katoaka, S., 'The Hall effect and its application to power measurement at 10 GHz', Proc. IEE, 105, Pt B, 53-60, January 1958.

9. Carlin, H.J., 'Measurement of power', Handbook of microwave measurements, 1, Chapter 3, 135-226, Eds. Sucher, M. and Fox, J., Wiley, New York, 1963.

10. Thomas, H.A., 'Microwave power measurements employing electron beam techniques', Proc. IRE, 45, 205-211, February 1957.

11. Oldfield, L.C. and Ide, J.P., 'An electron beam method for absolute measurement of microwave power', IEE Colloquium Digest, 2, 'The measurement of power at higher microwave frequencies', January 1979.

12. Oldfield, L.C. and Ide, J.P., 'An electron beam technique for realisation of the microwave watt', RSRE Report 82016, Ministry of Defence, RSRE, Malvern, Worcestershire (Report not yet completed).

13. Rumfelt, A.Y. and Elwell, L.B., 'Radio frequency power measurements', Proc. IEEE, 55, 837-850, June 1967.

14. Hewlett Packard Application Note 64-1, 'Fundamentals of rf and microwave power measurements', August 1967.

15. Lane, J.A., 'Microwave power measurements', Peregrinus, London, 1972.

16. Fantom, A.E., 'Improved coaxial calorimetric rf power meter for use as a primary standard', Proc. IEE, 126, 849-854, September 1979.

17. Vowinkel, B. and Röser, H.P., 'Precision measurement of power at millimetre- and sub-millimetre wavelengths using a waveguide calorimeter', Int. J. IR and mm waves, 3, 471-487, July 1982.

18. Abbott, N.P., Reeves, C.J. and Orford, G.R., 'A new waveguide flow calorimeter for levels 1-20 W', IEEE Trans. IM-23, 414-420, December 1974.

19. MacPherson, A.C. and Kerns, D.M., 'A microwave microcalorimeter', Rev. Sci. Instrum., 26, 27-33, January 1955.

20. Engen, G.F., 'A refined X-band microwave microcalorimeter' J. Research NBS, 63C, 77-82, July-September 1959.

21. Skilton, P.J., 'Developments in United Kingdom waveguide power standards', RSRE Report 80006, Ministry of Defence, RSRE, Malvern, Worcestershire, April 1980.

22. Collard, J., 'The enthrakometer, an instrument for the measurement of power in rectangular waveguides', J. IEE, 93, Pt IIIA, 1399-1402, No 9, 1946.

23. Lemco, I. and Rogal, B., 'Resistive film milliwattmeters for the frequency bands 8.2-12.4 Gc/s, 12.4-18 Gc/s and 26.5-40 Gc/s' Proc. IEE, 107B, 427-430, September 1960.

24. Luskow, A.A., 'This microwave power meter is no drifter', Marconi Instrumentation, 13, August 1971.

25. Sullivan, D.B., Adair, R.T. and Frederick, N.V., 'RF instrumentation based on superconducting quantum interference', Proc. IEEE, 66, 454-463, April 1978.

26. Gallop, J.C., 'The impact of superconducting devices on precision metrology and fundamental constants', Metrologia, 18, 67-92, June 1982.

27. Pratt, R.E., 'Very low level microwave power measurements' Hewlett Packard Journal, 27, 8-10, October 1975.

28. Blaney, T.G., 'Radiation detection at submillimetre wavelengths', J. Phys. E, 11, 856-881, September 1978.

29. Hadni, A., 'Pyroelectricity and pyroelectric detectors', Infrared and millimetre waves, 3, Chapter 3, 111-180, Ed. Button, K.J., Academic, New York, 1980.

30. Harris, I.A. and Warner, F.L., 'Re-examination of mismatch uncertainty when measuring microwave power and attenuation', IEE Proc. 128, Pt H, February 1981.

31. Ide, J.P., 'A broadband waveguide transfer standard for the dissemination of UK national microwave power standards', RSRE Memo 3390, Ministry of Defence, RSRE, Malvern, Worcestershire, January 1982.

32. Abbott, N.P. and Orford, G.R., 'A software programmable system for top echelon calibration of power meter thermistor mounts', EQAD Report SC102/79, Ministry of Defence, EQAD, Bromley, Kent, January 1979.

33.   Little, W.E. and Ellerbruch, D.A., 'Precise reflection
      coefficient measurements with an untuned reflectometer',
      J. Research NBS, 70C, 165-168, July-September 1966.

34.   Harris, I.A⦿, 'An accurate method for the comparison of
      two rf power meters which have similar or differing power
      ranges' Progress in radio science 1966-1969, 2, 249-254,
      URSI, 1971.

35.   Ide, J.P. and Oldfield, L.C., 'The theory and practice of
      a multistate reflectometer', RSRE Memo 3824, Ministry of
      Defence, RSRE, Malvern, Worcestershire, May 1985.

36.   Oldfield, L.C. and Ide, J.P., 'Measurement of complex
      reflection coefficients in W-band using a 4-port reflec-
      tometer and precision waveguide spacers', IEE Colloquium
      Digest, 53, 'Advances in s-parameter measurement at
      micro-wavelengths', May 1983.

37.   Oldfield, L.C., Ide, J.P. and Griffin, E.J., 'A multi-
      state reflectometer', IEEE Trans. IM-34, June 1985.

38.   Hill, L.D. and Griffin, E.J., 'An automatic stepped
      frequency 6-port reflectometer for WG22', Proc. IEE,
      Pt H, 77-81, April 1985.

39.   Abbott, N.P. and Orford, G.R., 'Some recent measurements
      of linearity of thermistor power meters', IEE Colloquium
      Digest, 1981/49.

40.   Skilton, P.J., 'A technique for determination of loss,
      reflection and repeatability of waveguide flanged
      couplings', IEEE Trans. IM-23, 390-394, December 1974.

41.   Engen, G.F., 'Coaxial power meter calibration using a
      waveguide standard', J. Research NBS, 70C, 127-138,
      April-June 1966.

42.   Skilton, P.J. and Fantom, A.E., 'A comparison of the
      United Kingdom national standards of microwave power in
      waveguide and coaxial lines', IEEE Trans. IM-27, 297-298,
      September 1978.

43.   Skilton, P.J., 'A technique for measuring the efficiency
      of waveguide-to-coaxial-line adaptors', IEEE Trans. IM-27,
      231-234, September 1978.

44.   Levinson, D.S. and Sleven, R.L., 'Power measurements in
      multimode waveguide', Microwave J, 59-64, October 1967.

45.   Hudson, P.A., 'Measurement of rf peak pulse power',
      Proc. IEEE, 55, 851-855, June 1967.

46.   Hudson, P.A., 'Implementation of the notch technique as
      an rf peak pulse power standard', NBS Tech Note 682, US
      Department of Commerce, NBS, Boulder, Colorado, July 1976.

Chapter 8

Attenuation measurement

F. L. Warner

1. INTRODUCTION

A loss of only 0.1 dB in an uncooled waveguide between an aerial and a receiver raises the system noise temperature by 7 K. An attenuation of 2 dB between a transmitter and its aerial causes 37% of the power to be wasted. A neglected loss of 0.01 dB in the output section of a liquid nitrogen noise standard causes a 1% error in the noise temperature. These three examples show that it is frequently necessary to carry out precise attenuation measurements in microwave work. Ten of the established methods for measuring microwave attenuation are described in section 5 of these notes. To provide essential stepping stones to that part of the paper, section 2 gives the basic definitions and equations related to attenuation and the next two sections discuss DC, AF, IF and RF attenuation standards which are used in some of the microwave attenuation measuring systems. Attenuation transfer standards, international attenuation intercomparisons and applications are briefly mentioned in sections 6, 7 and 8. Six-port, swept frequency and network analyser methods of measuring attenuation are described in later lecture notes.

2. BASIC DEFINITIONS AND EQUATIONS RELATED TO ATTENUATION

2.1 Insertion Loss and Attenuation

When a generator with a reflection coefficient, Γ_G, is connected directly to a load with a reflection coefficient, Γ_L, let the power dissipated in the load be denoted by P_1. Suppose now that a 2-port network is connected between the generator and the load, as shown in Fig. 1, and let this reduce the power dissipated in the load to P_2.

SIGNAL FLOW GRAPH FOR THE CONFIGURATION SHOWN IN FIG. I.

TWO-PORT NETWORK BETWEEN GENERATOR AND LOAD

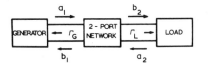
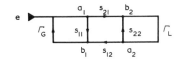

Fig. 1   Fig. 2

Then, by definition, the insertion loss of this 2-port network is given in decibels by:-

$$L_I = 10 \, \log_{10} \frac{P_1}{P_2} \qquad \qquad \dots (1)$$

Attenuation is defined as the insertion loss then Γ_G and Γ_L as well as upon the characteristics of the 2-port network; whereas attenuation is a property only of the 2-port network.

Using scattering coefficients s_{11}, s_{12}, s_{21} and s_{22} to define the 2-port network, the complex wave amplitudes shown in Fig. 1 are related as follows:-

$$b_1 = s_{11}a_1 + s_{12}a_2 \qquad \qquad \dots (2)$$

$$b_2 = s_{21}a_1 + s_{22}a_2 \qquad \qquad \dots (3)$$

The signal flow graph [1] for the configuration given in Fig. 1 is shown in Fig. 2.

Using the non-touching loop rule [1], we immediately get:-

$$\frac{b_2}{e} = \frac{s_{21}}{1 - (\Gamma_G s_{11} + s_{22}\Gamma_L + \Gamma_G s_{21}\Gamma_L s_{12}) + \Gamma_G s_{11}s_{22}\Gamma_L} \qquad \dots (4)$$

Denoting the assumed real characteristic impedance of the transmission system by Z_o, the power incident upon the load is $|b_2|^2/Z_o$ and the power reflected from the load is $|a_2|^2/Z_o$. Thus, the power dissipated in the load, P_2, is as follows:-

$$P_2 = \frac{|b_2|^2}{Z_o} - \frac{|a_2|^2}{Z_o} = \frac{|b_2|^2}{Z_o} (1 - |\Gamma_L|^2) \qquad \qquad \dots (5)$$

From (4) and (5) we get:-

$$P_2 = \frac{|e|^2 |s_{21}|^2 (1 - |\Gamma_L|^2)}{Z_o \left| 1 - (\Gamma_G s_{11} + s_{22}\Gamma_L + \Gamma_G s_{21}\Gamma_L s_{12}) + \Gamma_G s_{11}s_{22}\Gamma_L \right|^2} \cdot \qquad \dots (6)$$

The power, P_1, dissipated in the load when the generator is connected directly to it can be found immediately from equation (6) by letting $s_{11} = s_{22} = 0$ and $s_{21} = s_{12} = 1$.

Thus

$$P_1 = \frac{|e|^2 (1 - |\Gamma_L|^2)}{Z_o |1 - \Gamma_G \Gamma_L|^2} \qquad \qquad \dots (7)$$

Inserting (6) and (7) into (1), we get:-

$$L_I = 10 \, \log_{10} \frac{\left| 1 - (\Gamma_G s_{11} + s_{22}\Gamma_L + \Gamma_G s_{21}\Gamma_L s_{12}) + \Gamma_G s_{11}s_{22}\Gamma_L \right|^2}{|s_{21}|^2 |1 - \Gamma_G \Gamma_L|^2} \qquad \dots (8)$$

An expression for the attenuation, α, of the 2-port can now be found straight away from equation (8) by setting both Γ_G and Γ_L equal to zero.

Hence
$$\alpha = 10 \log_{10} \frac{1}{|s_{21}|^2} \qquad \ldots (9)$$

Attenuation is sometimes separated into two components, one associated with reflection and the other with dissipation.

Let P_I = the power incident upon the 2-port;
$\quad P_R$ = the power reflected by the 2-port back into the matched generator;
and P_L = the power dissipated in the matched load.

Then, the reflective component of the attenuation is given by:

$$\alpha_r = 10 \log_{10} \frac{P_I}{P_I - P_R} \qquad \ldots (10)$$

and the dissipative component of the attenuation is given by

$$\alpha_d = 10 \log_{10} \frac{P_I - P_R}{P_L} \qquad \ldots (11)$$

In this case, where both the generator and load are matched:

$$\frac{P_R}{P_I} = |s_{11}|^2 \text{ and } \frac{P_L}{P_I} = |s_{21}|^2 \qquad \begin{array}{l} \ldots (12), \\ (13) \end{array}$$

Therefore:-

$$\alpha_r = 10 \log_{10} \frac{1}{1 - |s_{11}|^2} \text{ and } \alpha_d = 10 \log_{10} \frac{1 - |s_{11}|^2}{|s_{21}|^2} \qquad \begin{array}{l} \ldots (14), \\ (15) \end{array}$$

Clearly
$$\alpha = \alpha_r + \alpha_d \qquad \ldots (16)$$

2.2 Mismatch Error If an attenuation measurement is made without the source and load being perfectly matched there will be an error in the result. This error, which is brought about by measuring insertion loss rather than attenuation, is known as the mismatch error[2] and it is given in decibels by:-

$$M = L_I - \alpha \qquad \ldots (17)$$

$$= 10 \log_{10} \frac{|1 - (\Gamma_G s_{11} + s_{22}\Gamma_L + \Gamma_G s_{21}\Gamma_L s_{12}) + \Gamma_G s_{11} s_{22}\Gamma_L|^2}{|1 - \Gamma_G \Gamma_L|^2} \qquad \ldots (18)$$

All of the variables in this last equation are complex quantities and all phase relationships are possible; so the mismatch uncertainty can lie anywhere between the limits given by the following equation:-

$$M_{limit} = 20 \log_{10} \frac{1 \pm |\Gamma_G s_{11}| \pm |s_{22}\Gamma_L| \pm |\Gamma_G s_{21}\Gamma_L s_{12}| \pm |\Gamma_G s_{11}s_{22}\Gamma_L|}{1 \mp |\Gamma_G \Gamma_L|} \qquad \ldots(19)$$

Two recent papers[138, 139] give expressions for mismatch uncertainty when all phase angles except those of the source and load are known. Fig. 3 shows a family of mismatch uncertainty curves calculated from equation (19). If the unknown is symmetrical and has a VSWR of 1.05, the source and load VSWRs must be less than 1.005 to keep the mismatch uncertainty below 0.001 dB when no phase angles are known.

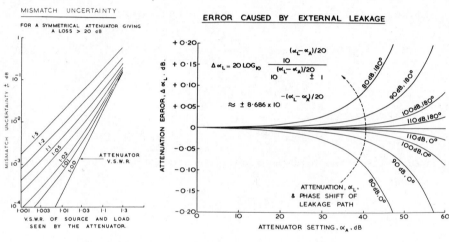

Fig. 3   Fig. 4

2.3 Error due to Leakage   Another source of error in any attenuation measurement is leakage. Let α_A denote the attenuation is dB through the attenuator under test and let α_L denote the attenuation in dB through a leakage path shunting this attenuator. Then it is easily shown that the error in the attenuation measurement due to leakage lies between the limits given by the equation in Fig. 4. A general family of leakage error curves is also shown in this figure. To keep the leakage error below 0.001 dB, α_L must be 80 db greater than α_A.

3. DC, AF AND IF VARIABLE ATTENUATION STANDARDS

The three widely used devices that fall into this category are: the Kelvin Varley voltage divider, the inductive voltage divider and the IF piston attenuator.

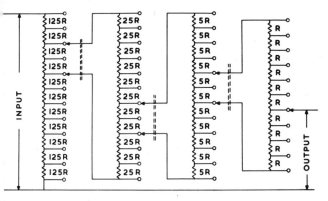

A 4 DECADE KELVIN-VARLEY VOLTAGE DIVIDER.

Fig. 5

3.1 Kelvin Varley Voltage Dividers
The Kelvin Varley voltage divider was first described in 1866[3]. Fig. 5 shows a 4 decade divider of this type[4-7], which enables the ratio of the output voltage to the input voltage to be varied from 0 to 1 in steps of 1 part in 10^4. Eleven equal resistors are used in each decade except the last which contains ten.

The switches always span two resistors and the values of the resistors in each succeeding decade become smaller by a factor of five. A brief study of Fig. 5 will show that this arrangement gives exact decimal switching, when it is unloaded.

Although decimal switching is normally used, the Kelvin Varley principle can be readily applied to a switching system with any numerical base, N. We then require N resistors in the last stage, (N+1) resistors in each of the other stages and the values of the resistors in each succeeding decade must be smaller by a factor N/2. Thus, in a binary Kelvin Varley voltage divider, N = 2 and all resistors have the same value[8].

When a Kelvin Varley divider has an infinite load, its input resistance, R_{in}, stays constant as the switch settings are varied. For the divider shown in Fig. 5, R_{in} = 1250 R. However, the output resistance does vary with the switch settings and it has a maximum value slightly greater than $R_{in}/4$ when the source resistance is zero. The frequency response of a Kelvin Varley divider is similar to that of a simple RC low pass filter. When good constructional techniques are employed, the 3 dB bandwidth is in the region of 100 kHz when R_{in} = 100 kΩ and 10 MHz when R_{in} = 1 kΩ. Kelvin Varley dividers with up to 7 decades are now commercially available.

In a Kelvin Varley divider, errors are caused by departures from the nominal resistance values and by resistance in the switch contacts and connecting leads. When all possible precautions are taken and ultra-stable resistors of the highest obtainable precision are used, the error in the output voltage at any setting can be made less than 10^{-6} x the input voltage.

3.2 Inductive Voltage Dividers   The inductive voltage divider (IVD), sometimes called a ratio transformer, forms an exceptionally accurate variable attenuation standard at audio frequencies[9-20]. Fig. 6 shows the circuit diagram of a 7 decade IVD. It consists of 7 very accurately tapped auto-transformers which are connected together with high class switches in such a way that the ratio of the output voltage to the input voltage can be varied from 0 to 1 in steps of 1 part in 10^7.

SIMPLIFIED CIRCUIT DIAGRAM* OF A 7 DECADE RATIO TRANSFORMER

$$\left(\text{SET TO GIVE A VOLTAGE RATIO OF } 0\cdot3162277 = 10 \text{ dB} \right)$$

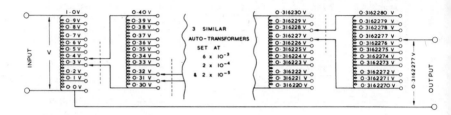

Fig. 6

Precision IVD's can be designed to operate at any frequency between about 10 Hz and 100 kHz and greatest accuracy is obtained at a frequency of about 1 kHz.

Many different ways of making the auto-transformers have been investigated. Good results can be achieved by taking 10 exactly equal lengths of insulated copper wire from the same reel and twisting them together in either a uniform or random manner. The resulting 'rope' is arranged to form a single layer around a toroidal core that is wound like a clock spring with very thin insulated Supermalloy tape. Nine appropriate pairs of wires from the 10 strand cable are then joined together so that 10 series-aiding coils are obtained. Supermalloy [21] (79% Ni, 15% Fe, other elements) is used because it has a very high permeability (> 100,000) and an extremely low hysteresis loss (< 1 $J/m^3/Hz$ at 0.5 tesla).

In an IVD, errors are caused by: inequalities in the series resistances and leakage inductances, inhomogeneities in the magnetic cores, distributed admittances between the windings, loading by the later decades and impedances in the connecting leads and switch contacts. With great care, all of these errors can be made extremely small.

The error, ε, in an IVD is defined as follows:-

$$\varepsilon = \frac{V_{out} - DV_{in}}{V_{in}} \qquad \qquad \cdots (20)$$

where V_{in} is the input voltage, D is the indicated ratio and V_{out} is the output voltage.

An 8 decade IVD is commercially available, for which $\varepsilon = \pm 4 \times 10^{-8}$. Several firms manufacture 7 decade IVD's and most of these have an ε value lying in the range $\pm 10^{-6}$ to $\pm 10^{-7}$.

At 1 kHz, a well designed IVD has an input impedance, z_{in}, well above 100 kΩ and an output impedance, z_{out}, of a few ohms.

When an IVD is set to give an attenuation of α dB, it is easily shown from equation (20) that the error $\Delta\alpha'$ in dB is given by:

$$\Delta\alpha' = -8.686 \ \varepsilon \ \text{antilog}_{10} \ (\alpha/20) \qquad \ldots(21)$$

Two IVD's can be used in cascade. The loading of the first one by the second then causes an additional error. The overall error is minimum when they are both set to the same ratio. Under these conditions, the error $\Delta\alpha''$ for a total attenuation of α is given by:

$$\Delta\alpha'' = -17.372 \ \varepsilon \ \text{antilog}_{10} \ (\alpha/40) + 20 \ \log_{10} \left| \frac{z_{in} + z_{out}}{z_{in}} \right| \qquad \ldots(22)$$

Fig. 7 shows how $\Delta\alpha'$ and $\Delta\alpha''$ vary with α under worst case conditions. A typical value of 0.0001 dB was taken for the last term in equation (22). For high values of attenuation it is clearly very advantageous to use two IVD's in cascade. When $\varepsilon = \pm 10^{-7}$, it can be seen that the cascade arrangement gives an error of less than 0.001 dB at 100 dB.

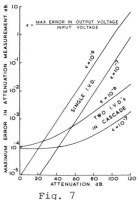

ERROR DUE TO INDUCTIVE VOLTAGE DIVIDER IMPERFECTIONS.

Fig. 7

3.3 Piston Attenuators   The piston attenuator was first described by Harnett and Case[22] in 1935. During the last 30 years, it has been widely used in microwave attenuator calibrators based on the IF substitution principle, which will be described in section 5. The essential parts of a piston attenuator are shown in Fig. 8. The attenuation is varied by altering the separation between the two coils. The circular tube acts as a waveguide beyond cut-off, and the launching system is designed so that only one mode is excited in it. The voltage induced in the pick-up coil falls off exponentially as the separation between the two coils is increased.

With perfectly conducting walls, it is easily shown from elementary waveguide theory that an increase in the coil separation from z_1 to z_2 produces an attenuation change in dB given by

$$\alpha_p = 8.686 \times 2\pi(z_2 - z_1) \left\{ \left(\frac{S_{nm}}{2\pi r} \right)^2 - \frac{1}{\lambda^2} \right\}^{\frac{1}{2}} \qquad \ldots(23)$$

where r is the radius of the cylinder, λ is the free space wavelength of the applied signal and S_{nm} is a constant which depends upon the mode that is excited (see Table 1).

When high accuracy is required, the skin depth must be taken into account and equation (23) must then be replaced by a much more complicated equation which has been derived by Brown[23].

Most of the piston attenuators used in IF substitution systems operate at either 30 MHz or 60 MHz in the H_{11} mode and have an internal diameter lying somewhere in the range 20 to 50 mm. In most practical cases,

$$\frac{1}{\lambda^2} \;\; << \;\; \left(\frac{S_{nm}}{2\pi r}\right)^2$$

so, for any given mode, the attenuation change produced by a displacement equal to the radius is a constant (see Table 1).

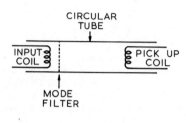

PISTON ATTENUATOR

CIRCULAR TUBE

Mode	Value of S_{nm}	α_p for displacement equal to radius dB
H_{11}	1.841	15.99
E_{01}	2.405	20.89
H_{21}	3.054	26.53
E_{11} & H_{01}	3.832	33.28

Fig. 8   Table 1

Differentiating equation (23), we get

$$\frac{\Delta\alpha_p}{\alpha_p} \approx - \frac{\Delta r}{r} \qquad\qquad \dots (24)$$

Thus the tolerance on the internal radius must be ±1 part in 10^4 to keep the attenuation error down to ±0.001 dB per 10 dB. To avoid erratic behaviour, the plunger which carries the movable coil must be insulated from the cylinder wall[24]. The mode filter can take the form of either a row of metal strips or a disc of very high dielectric constant material[24]. A well designed metal strip filter will attenuate the E_{01} mode by about 80 dB and the H_{11} mode by less than 0.5 dB. If the attenuation is reduced below about 30 dB, loading effects occur and there is a departure from the relationship given by equation (23).

Commercially available piston attenuators have an accuracy of about 0.01 dB per 10 dB. Extremely accurate piston attenuators have been developed by Yell[25, 140-142] Bayer[143, 144] and the Weinschel Engineering Company[145]. In each of these instruments, a laser interferometer is employed to measure the piston displacements.

4.   RF VARIABLE ATTENUATION STANDARDS

Many different types of coaxial and waveguide variable attenuators have been described in the literature. However only two of them - the waveguide rotary vane attenuator and the microwave piston attenuator - are well suited for use as microwave reference standards.

4.1 Rotary Vane Attenuators   The rotary vane attenuator was invented simultaneously by E A N Whitehead of Elliott Brothers (London) Ltd and A E Bowen of the Bell Telephone Laboratories; and the first description of it was given by Southworth[26]. The essential parts of a rotary attenuator are shown in Fig. 9. The two end vanes are fixed in a

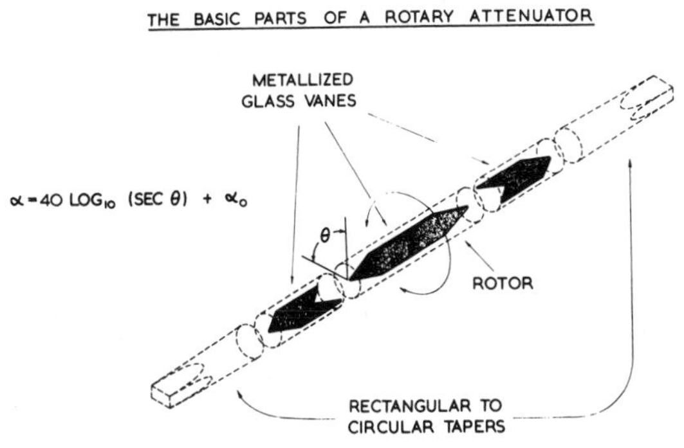

THE BASIC PARTS OF A ROTARY ATTENUATOR

METALLIZED
GLASS VANES

$\alpha = 40 \, LOG_{10} \, (SEC \, \theta) + \alpha_0$

ROTOR

RECTANGULAR TO
CIRCULAR TAPERS

$\frac{\alpha - \alpha_0}{dB}$	θ DEGREES
0	0.000
3	32.712
10	55.782
20	71.565
30	79.757
40	84.261
50	86.776
60	88.188

Fig. 9

direction perpendicular to the incident electric vector and the attenuation is varied by rotating the central section. For an ideal attenuator of this type the attenuation, α, in dB is given by the equation in Fig. 9, where θ is the angle of the central vane relative to the end vanes and α_0 is the insertion loss of the instrument when $\theta = 0$.

The rotary attenuator has many advantages over the other types of variable waveguide attenuators which are still in use; the attenuation it provides is almost independent of frequency, there is scarcely any change in phase through it as the attenuation is varied, its temperature coefficient is almost zero and it has a low VSWR under all conditions. Its biggest disadvantage is that the attenuation through it varies very rapidly with θ at high values of attenuation (see the table in Fig. 9).

The principal sources of error in a rotary attenuator are: end vane misalignment, insufficient attenuation in the central vane, internal reflections, an incorrect setting of the rotor angle and imperfections in the gear wheels that drive the rotor. Equations and curves for four of these sources of error are given in Figs. 10-13. In these equations, $\alpha_A = \alpha - \alpha_0$ and Γ denotes the reflection coefficient at each discontinuity (27).

Detailed analyses of the rotary vane attenuator have been given by Hand[27], Mariner[28], James[29], Larson[30-32, 146], Otoshi & Stelzried[33], Warner[147] and Guldbrandsen[148, 149].

Extremely accurate rotary attenuators with optical angular measuring systems have been described by Little et al[34] and Warner et al[35] The attenuator described in this last reference has an 11" diameter optical grating attached to its rotor and the angle of the rotor is displayed to a resolution of 0.001° on a row of numerical indicator tubes. According to measurements made at 10 GHz against the British national standard, this rotary attenuator follows the 40 $\log_{10}$ (sec θ) law to within $\pm$ 0.0015 dB up to 16 dB. The mismatch error is large enough to account for the very small discrepancies that were obtained. Several automated rotary attenuators have been made at RSRE (150-152). A precision rotary attenuator is an excellent device for producing an attenuation change of 0.001 dB to check the short term jitter and long term drift of a microwave attenuation measuring equipment.

Rotary attenuators are available from many commercial firms and a typical accuracy specification on a mass-produced one is:- "±2% of reading in dB or 0.1 dB, whichever is greater".

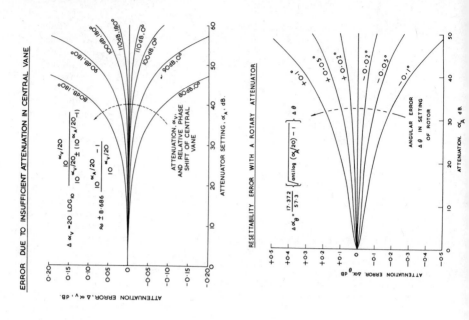

ERROR DUE TO INSUFFICIENT ATTENUATION IN CENTRAL VANE

$$\Delta \alpha_V = 20 \, LOG_{10} \frac{10^{\alpha_V/20}}{10^{\alpha_V/20} \pm (10^{\alpha_A/20}-1)}$$

$$\approx \pm 8.686 \frac{10^{\alpha_A/20}-1}{10^{\alpha_V/20}}$$

ATTENUATION, α_V, AND RELATIVE PHASE SHIFT OF CENTRAL VANE

ATTENUATOR SETTING, α_A, dB.

ATTENUATION ERROR, $\Delta \alpha_V$, dB.

RESETTABILITY ERROR WITH A ROTARY ATTENUATOR

$$\Delta \alpha_\theta = \frac{17.372}{57.3} \left\{ antilog \left(\alpha_A/20 \right) - 1 \right\} \Delta \theta$$

ANGULAR ERROR $\Delta \theta$ IN SETTING OF ROTOR

ATTENUATION, α_A, dB.

ATTENUATION ERROR $\Delta \alpha$, dB.

END VANE MISALIGNMENT ERROR.

$$\Delta \alpha_\epsilon = 20 \, LOG_{10} \frac{\cos^2 \theta \, \cos^2 \epsilon}{\cos (\theta - \epsilon) \, \cos (\theta + \epsilon)}$$

$$\approx 8.686 \left(\frac{\epsilon}{57.3} \right)^2 \left\{ 10^{(\alpha_A/20)} - 1 \right\}$$

LAST VANE
CENTRAL VANE
FIRST VANE

ATTENUATION, α_A, dB.

ATTENUATION ERROR, $\Delta \alpha_\epsilon$, dB.

ERROR DUE TO INTERNAL REFLECTIONS

$$\Delta A_R = 20 \, LOG_{10} \frac{1}{1 \pm 2 \Gamma^2 \, SIN^2 \theta} \approx \pm 17.372 \, \Gamma^2 \left\{ 1 - 10^{-A_A/20} \right\}$$

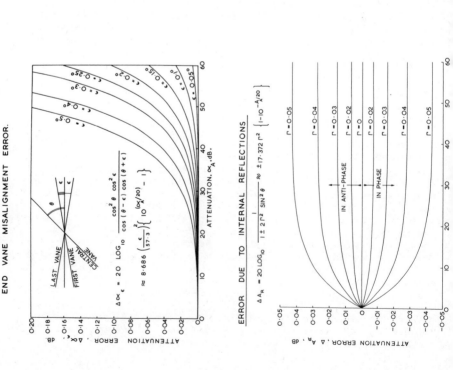

IN ANTI-PHASE

IN PHASE

ATTENUATION ERROR, ΔA_R, dB

4.2 Microwave Piston Attenuators   Most of the information given in
section 3.3 about IF piston attenuators also applies to microwave piston
attenuators. However, at microwave frequencies, the last term in
equation (23) cannot be neglected; so the attenuation rate varies with
frequency. Furthermore, multi-turn input and pick-up coils cannot be
used. A microwave piston attenuator has an enormous dynamic range and a
linear decibel scale; but its construction places extreme demands on
mechanical accuracy and it has a high residual attenuation, unless a
departure from linearity is accepted at the low end of the attenuation
scale. Several microwave piston attenuators have been described in the
literature[36-42]. A successful early one, that was used in the TRE S-
band Signal Generator Type 47, is shown in Fig. 14(a). The straight
filament in the evacuated glass tube launches the evanescent wave and it
is also used as a power indicator. The coupling loop in the klystron is
rotated until it just glows. The H_{11} mode is used. The sub-miniature
70 Ω resistor acts a pick-up loop and also enables the attenuated signal
to emerge from a "matched source". The inner tube that contains the
70 Ω resistor is moved back and forth by a rack and pinion and is
connected to the output socket with a flexible cable.

When great accuracy is needed or the frequency is very high, flexible
cables cannot be used. Ditchfield[42, 153] and Taylor[154] have eliminated them by using a sliding source attached to a sliding tube in a
precision 9 GHz H_{11} mode piston attenuator, which has been developed at
RSRE for use as a primary attenuation standard (see Fig. 14b). The

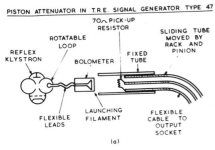

central circular section acts as a
waveguide beyond cut-off. The
launching and output sections take
the form of circular waveguides
filled with Stycast Hi-K. These
propagate the H_{11} mode but are beyond cut-off for the E_{01} mode. The
cut-off wavelengths for these two
modes are given by

$$(\lambda_c)_{H_{11}} = 1.706 \ d \ \sqrt{\varepsilon} \qquad \ldots (25)$$

$$(\lambda_c)_{E_{01}} = 1.306 \ d \ \sqrt{\varepsilon} \qquad \ldots (26)$$

where d and ε denote respectively the
diameter and dielectric constant of
the Stycast Hi-K rods. After choosing the central diameter to give the
required attenuation rate, ε was
chosen so that the free space wavelength of the input signal lies
between these two cut-off wavelengths.
Thus there is no need for special
mode filters. This piston attenuator
is mounted on a massive granite slab,
it is evacuated and the displacements of the input rod are measured with
a laser interferometer. The attenuation rate is 2.5 dB/mm and the
estimated accuracy is ±0.02 dB in 150 dB. The bore has been used as a
36 GHz cavity to determine the electrical diameter (from resonance

Fig. 14

measurements) and the skin depth (from Q measurements).

Dielectric rods have been used in a similar manner by other workers. A 24 GHz piston attenuator containing polystyrene rods has been described by Griesheimer[37], polystyrene rods were also used by Gordon Smith[38] in a 6 mm piston attenuator and Munier[41] employed an ebonite rod in a 3.3 GHz piston attenuator.

5.   METHODS USED TO MEASURE ATTENUATION

Many different methods are used to measure microwave attenuation and they can be grouped as follows [43-45, 155, 156]

(a)   power ratio methods;
(b)   comparison with an accurate microwave attenuator (RF substitution);
(c)   comparison with an intermediate frequency attenuator (IF substitution);
(d)   comparison with an audio frequency attenuator (AF substitution);
(e)   methods not requiring an attenuation standard;
(f)   methods based upon reflection coefficient measurements;
(g)   the shuttle-pulse method;
(h)   methods based upon Q measurements;
(i)   methods based upon the Josephson effect;
(j)   miscellaneous methods.

5.1 Power Ratio Methods   Fig. 15 shows a very simple system that can be used for measuring attenuation by the power ratio method. It employs a solid state microwave source and a thin-film thermo-electric (TFT) power meter[46]. Since a microwave amplitude stabilizer is not used, the solid state source should be operated from a highly stabilized power unit and its bias lead should be surrounded by lossy material to eliminate microwave leakage. Good frequency stability (a few parts in 10^6 per day) can be achieved by using an impatt oscillator which is locked to a high Q invar cavity[47].

SIMPLE SYSTEM FOR ATTENUATION MEASUREMENT BY THE POWER RATIO METHOD

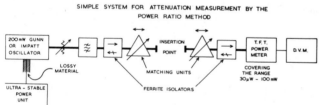

Fig 15

.Thin film thermo-electric power meters detect signals lying anywhere in the microwave spectrum; so the microwave band-pass filter is needed to eliminate harmonics present in the output from the microwave source. The matching units are adjusted until an almost perfect match (VSWR < 1.005) is seen looking in either direction from the insertion point. The isolators are used so that the matching conditions are not affected if any changes are made elsewhere. At RSRE a high resolution swept-frequency reflectometer[48] is normally used to carry out these matching operations. If a TFT power head covering the range 30 µW to 100 mW is used, maximum dynamic range can be obtained by employing a microwave source with an output of about 200 mW, since the loss through the filter, the matching units and the two isolators will be in the region of 3 dB.

Let the digital voltmeter reading be V_1 when the flanges at the insertion point are bolted directly together and V_2 when the "unknown" is inserted. Then, the attenuation through the "unknown" is given by:-

$$\alpha = 10 \log_{10} \frac{V_1}{V_2} \qquad \qquad \ldots (27)$$

With this system, only relative powers are needed so the efficiency of the TFT head need not be known. The lead between the TFT head and the TFT amplifier is flexible and can be moved about without detriment to the accuracy.

Careful measurements made on a simple system of this type at RSRE gave results which were in error by only 0.002 dB at 3 dB and 0.028 dB at 20 dB. These errors arise because the output EMF of a TFT head is not exactly proportional to the microwave input power.

The points mentioned in this section about precise matching, harmonic filtering, leakage suppression etc apply to all methods of attenuation measurement. A novel matching technique has been described recently by Hollway and Somlo[49]. A length of precision waveguide is introduced at the insertion point and an attenuator-reflector combination is moved back and forth inside it by a nylon thread. The matching unit adjacent to the reflector is adjusted until the transmitted power stays constant. The attenuator and reflector are then interchanged and the other matching unit is adjusted in the same way. A panoramic receiver containing a swept YIG filter is a very good instrument for detecting microwave harmonics. Leakage from an attenuation measuring system can be traced rapidly with a small horn and a superheterodyne receiver tuned to the source frequency.

With the system shown in Fig. 15, an increase, ΔP, in the output power, P, of the microwave oscillator between the two parts of a complete attenuation measurement will result in an attenuation error given by:

$$\Delta\alpha = -4.343 \ \Delta P/P \quad \text{decibels.} \qquad \ldots (28)$$

Thus, an increase of only 0.1% in the output power will give an error of -0.0043 dB.

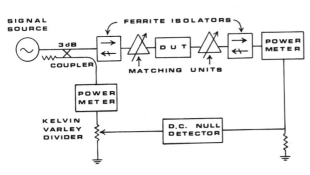

Fig. 16

To overcome this disadvantage, Stelzried, Reid and Petty[50,51] devised the dual channel equipment shown in Fig. 16. Two power meters are employed and a Kelvin Varley voltage divider is used to obtain a null both before and after inserting the "unknown". With this equipment short term jitter is about 0.0004 dB peak to peak and

and the long term drift is typically 0.0015 dB per hour.

Engen and Beatty[52] have described a power ratio method of measuring attenuation which gives uncertainties rising from 0.0001 dB to 0.06 dB over the range 0.01 dB to 50 dB. This remarkable performance is obtained by stabilizing the output from the microwave oscillator with a servo system containing a bolometer in an ultra-stable water bath, a DC amplifier and a ferrite variable attenuator. The power that is transmitted through the "unknown" is measured on a self balancing bolometer bridge which also has its vital components immersed in the ultra-stable water bath. The long term stability of this equipment is 0.0002 dB over a 3 hour period.

An equipment very similar to that described by Engen and Beatty has been set up at the Physikalisch-Technische Bundesanstalt in West Germany. A very detailed analysis of the errors in it has been given by Bayer[53]. He concludes that its systematic error in dB is given by:-

$$\Delta\alpha \approx 10\text{-}4 \ (1 + \alpha_u) \qquad\qquad \ldots(29)$$

where α_u is the attenuation in dB through the device under test.

An automated coaxial dual-channel power-ratio attenuator calibrator has been developed at RSRE by Hepplewhite. This contains a 10 W source and two self balancing power meters that employ Wollaston wire bolometers. The voltages across the two bridges (with the RF on and off) and the frequency are recorded automatically on punched paper tape and the results are worked out on a central computer. By taking a large number of readings rapidly in this way, drift errors are eliminated without resort to a water bath. The fixed coaxial components are rigidly clamped to a steel table and the ones that have to be moved about are attached to pneumatic supports[54]. A power ratio attenuator calibrator controlled by a desktop computer was described in 1978[157].

5.2 Comparison with an Accurate Microwave Attenuator   For this method of measurement, the standard is usually either a precision rotary vane attenuator or a very accurate microwave piston attenuator which is operated at the same frequency as the device under test. Several different configurations can be used. The simplest is a series connection of the source, the unknown, the standard attenuator, a receiver and a level indicator (with matched isolators on both sides of each attenuator). As the attenuation in the unknown is increased, the attenuation through the standard is reduced so that the receiver output stays constant. To achieve high accuracy with this arrangement, it is essential to have both an ultra-stable source and an ultra-stable receiver.

This disadvantage can be overcome by placing the unknown and the standard attenuator in parallel with each other and using a null detector. Magic tees or 3 dB couplers can be employed to split and recombine the microwave power. To achieve a balance in a system of this type, it is necessary to have a constant loss phase shifter in one of the channels. Unfortunately, it is very difficult to obtain such a device. If a superheterodyne receiver is used with either of these arrangements and two buffered precision rotary attenuators are used in tandem as the reference standard, a dynamic range in excess of 100 dB can be achieved. Furthermore, the mixer diode law is unimportant.

DUAL CHANNEL SERIES R.F SUBSTITUTION SYSTEM

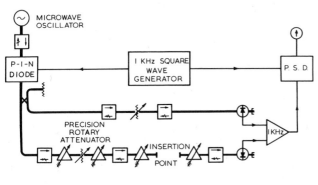

Fig. 17

Fig. 17 shows a dual channel series RF substitution system which was described recently by Larson and Campbell[55,56]. It overcomes the disadvantages of both the series and parallel arrangements described above; but the two microwave diodes in it are used as detectors, not mixers, so the dynamic range is reduced considerably.

The precision rotary attenuator which is used as the reference standard in this equipment is accurate to within ±0.002 dB up to 20 dB[34]. The drift of this equipment is typically 0.0002 dB over a period of 12 minutes.

5.3 Comparison with an Intermediate Frequency Attenuator This method has been widely used during the last 30 years. It gives good accuracy and a very large dynamic range (≈100 dB). It is particularly useful for calibrating attenuators in signal generators.

A series IF substitution configuration can be used[24,57], but the more popular arrangement is the parallel IF substitution system[58-67], of which a typical version is shown in Fig. 18. The 1 kHz square wave generator switches the microwave and 30 MHz oscillators on and off

PARALLEL I.F SUBSTITUTION SYSTEM

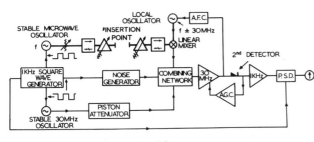

Fig. 18

alternately. The chopped microwave signal that emerges from the insertion point has its frequency changed to 30 MHz in a linear mixer. With the two flanges at the insertion point bolted directly together, the piston attenuator is adjusted until it delivers a chopped 30 MHz signal equal in amplitude to that from

the linear mixer. Under these circumstances, there is no 1 kHz component in the output from the 2nd detector and hence no output from the phase sensitive detector. The unknown is now inserted and the piston attenuator is adjusted until a fresh null is obtained. The piston attenuator used in an equipment of this type is normally calibrated directly in decibels; so the attenuation through the unknown can be obtained by simply subtracting the two piston attenuator readings. The automatic frequency control system is used to keep the mixer difference frequency almost equal to the frequency of the signal from the 30 MHz oscillator. The 30 MHz signal emerging from the linear mixer is more noisy than the signal emerging from the piston attenuator. To prevent this affecting the balance, some extra noise is fed into the combining network during the 500 μS periods when the 30 MHz oscillator is switched on. To obtain linear operation from the mixer, it is essential to keep the signal that enters it from the insertion point at least 30 dB below the local oscillator signal that is fed into it. In an equipment of this type, the most important source of error often arises in the displacement measuring system associated with the piston attenuator; eg if the diameter is chosen to give an attenuation rate of 40 dB per inch, an error of only 0.000025" in measuring the coil displacement will give an attenuation error of 0.001 dB.

Hollway and Somlo[68] have recently added several refinements to their parallel IF substitution system. These include: a switched reference path to reduce drift errors, a flexible waveguide in the measuring path to accommodate attenuators of different lengths, a steady flow of dry air through the system to remove moisture and automatic control of the IF oscillator power output for fine-balancing of the two channels.

An all-solid-state parallel IF substitution system was described in 1974 by Yell[69]. The Gunn oscillators in it have an amplitude stability of better than 0.002% per minute, the switching transients are gated out and the resolution is in the region of 0.0002 dB.

A fully automatic parallel I.F. substitution system, covering the range 1-18 GHz, was described by F K Weinert and B O Weinschel[158, 159] in 1976. It has a digital readout with a resolution of 0.001 dB and a single-step dynamic range of 110 dB. This equipment does not contain a piston attenuator. Its place has been taken by a box of extremely low temperature coefficient π type attenuators which are switched in and out automatically by very repeatable mercury-wetted reed relays. These π type attenuators are not standards in their own right. Each one is carefully calibrated against a very precise piston attenuator that is fitted with a laser interferometer. The errors are stored in a programmable read only memory that is incorporated in the receiver. These errors are then automatically subtracted in real time from the readout. The test and local oscillator signals are provided by totally independent, multi-band, all-solid-state, programmable, frequency stabilized sources(160). A barretter is used in the signal levelling loop to minimise harmonic errors. The I.F. amplifier is centred on 30 MHz. The signal and 30 MHz sources are switched on and off alternately by 20 kHz square waves. The switched attenuators cover the range 0-140 dB in 1 dB steps and the three least significant digits for the readout are obtained from a linear detector and an analogue to digital converter. With automatic balancing, the repeatability is better than ±0.01 dB at 60 dB. The absolute accuracy of the reference standard is ±(0.003 dB/10 dB + 0.005 dB).

The system described in the previous paragraph did not go into commercial production due to problems associated with pulling of the microwave signal source by the 20 kHz modulator. To overcome these problems, the Weinschel Engineering Company changed over to a CW method employing a vector cancellation principle and, in 1980, a 0.01-18 GHz dual channel IF vector substitution ratio meter, called the VM-4A, emerged[161-166].

This is the most sophisticated attenuator calibrator that is commercially available at the present time; so it will be described in some detail.   It has a resolution of O.OO1 dB and can perform measurements down to -145 dBm completely automatically.   A simplified block diagram of it is shown in Fig. 19.

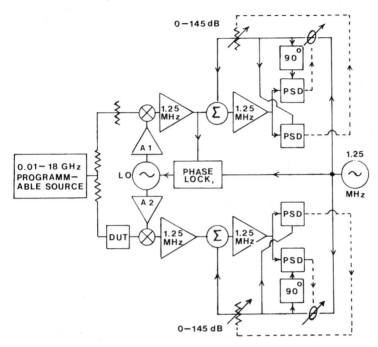

Fig. 19   Simplified block diagram of the Weinschel VM-4A

The programmable signal source delivers 20 mW throughout the band 0.01 to 18 GHz and the local oscillator is phase locked 1.25 MHz away from the signal frequency.   The device under test (DUT) is inserted in the lower channel.   The upper channel provides a strong signal for phase locking the local oscillator and it also provides immunity from source amplitude variations (since attenuation values are derived from date yielded by both channels).   RF leakage via the common local oscillator is eliminated by isolation amplifiers Al and A2.   Two pairs of mixers are used.   One set covers the range O.01-2.2 GHz and the other set operates over the band 1.9 to 18 GHz.   Two YIG tuned local oscillators are employed covering the ranges O.5-1.1 GHz and 1.9-4.6 GHz respectively.   Local oscillator frequencies below 500 MHz are generated by digital frequency divider circuits.   Local oscillator frequencies in the range 1-2.2 GHz are produced by frequency doublers.   Second harmonic mixing is used from 4-9.2 GHz and fourth harmonic mixing is used from 8-18 GHz, with an 8 dB increase in conversion loss.

The 1.25 MHz IF preamplifiers have low noise FET front ends and they are made exceptionally linear with negative feedback.   In each channel, the 1.25 MHz signal emerging from the IF preamplifier is nulled with a signal derived from the internal 1.25 MHz oscillator.   This latter signal has

its phase and amplitude controlled by two servo loops. One of them
twists the substituted signal until it is 180° out of phase with the IF
vector from the preamplifier. The second servo adjusts the amplitude of
the substituted signal until the cancellation is almost perfect. The
controlled attenuators in the amplitude servos are similar to those
described by Weinert and Weinschel in 1976[158, 159]. The servo loop
bandwidths are varied automatically from 500 Hz down to 0.2 Hz as the
error voltages fall from high to very low values. Adaptive digital
averaging reduces noise errors and provides short measurement times.

The measurement results are shown in dB and degrees on a 40 character
gas discharge display on the front panel. Prompting messages to the
operator also appear on this display. Two internal microprocessors
control all of the circuits in the VM-4A, process and display the data,
and perform self test functions. The exceptionally large dynamic range
of the VM-4A is due to the use of coherent detection[212] instead of
envelope detection after the IF amplification.

When the VM-4A is used with an external computer, mismatch errors can be
eliminated by using a self calibration procedure (see chapter 9   on
Microwave Network Analysers for this Vacation School). The accuracy
of the VM-4A is discussed further in Section 9.

A series IF substitution system developed by the Micro-Tel Corporation
was described in 1978[167] and a more advanced version of it (designated
1295-1-2) is now available[168]. This latest model covers the frequency
range 0.01-40 GHz, using frequency doubling and tripling techniques[169]
above 18 GHz. It has a resolution of 0.001 dB up to 70 dB and above
this value, the resolution is 0.01 dB. It contains a 30 MHz resistive
reference attenuator, which rises to 110 dB in 10 dB steps. Each step
can be independently adjusted from the front panel to coincide precisely
with a customer-supplied external 30 MHz precision attenuator. An IF
bandwidth of 15, 100 or 500 kHz can be selected and a bolometer is used
to detect the IF signal. This system can be operated either manually or
under computer control and its specified accuracy, excluding mismatch
uncertainty, is as follows:--

Attenuation step, dB	Accuracy dB
10	±0.04
30	±0.1
50	±0.2
80	±0.5
100	±0.8

A parallel 50 kHz substitution system was developed at RSRE in 1981[170-172].
Its block diagram is shown in Fig 20.

The signals from the two synthesized sources can be varied in 1 kHz
steps over the range 0.05-18 GHz. These two sources are always operated
with a 50 kHz separation and, since they are both locked to the same
10 MHz crystal controlled oscillator, an extremely stable 50 kHz intermediate frequency is always produced.

The attentuation reference standard in this system is a high class 7

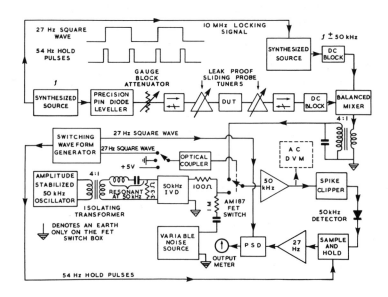

Fig 20   RSRE 0.05-18 GHz parallel 50 kHz substitution attenuator
calibrator

decade 50 kHz inductive voltage divider (IVD) that was designed and
made by T A Deacon of the National Physical Laboratory.

An AM187 break-before-make FET switch, driven by a 27 Hz square wave, is
used to compare the output signal from the mixer with that from the IVD.
At each setting of the device under test (DUT), the IVD is adjusted until
the output from the phase sensitive detector is zero and the changes in
attenuation through the DUT are then readily deduced from the IVD
settings.

To achieve great rigidity, many of the microwave components in this
system are clamped magnetically to an 8' x 3' cast iron table weighing
more than a ton.

The 2 DC blocks, the 50 kHz isolating transformer and the optical
coupler proved to be essential to eliminate circulating earth currents.
Noise balancing is achieved by superimposing some extra noise on the
IVD output signal. Switching spikes are eliminated by a sample and hold
circuit after the 50 kHz detector. To obtain a check on switching
errors, provision is also made for manual switching of the AM187 and,
for this mode of operation, an AC digital voltmeter (shown dotted) is
used as a level indicator.

This equipment gives national standards laboratory accuracy (see section
9), its dynamic range is about 100 dB and its drift rate is usally less
than 0.002 dB per hour. An order of magnitude reduction in the drift
rate (to < 0.0002 dB per hour) has recently been achieved by changing
over to the drift compensated series 50 kHz substitution system shown
in simplified form in Fig. 21. The drift is reduced by switching back

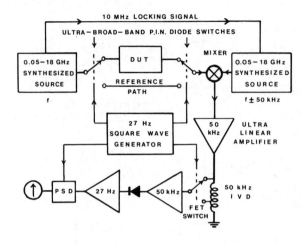

and forth at 27 Hz between the DUT and a stable microwave reference path, using two ultra-broadband PIN diode switches. The 50 kHz IVD is switched in and out of circuit at 27 Hz in antiphase with the DUT and adjusted until a null is obtained at the output.

Fig 21   RSRE drift-compensated series 50 kHz substitution attenuator calibrator

5.4   Comparison with an Audio Frequency Attenuator   The simplest AF substitution system is shown in Fig. 22. The microwave oscillator is square wave modulated at 100 Hz and the bolometer is used as a square law detector. Its output is passed through a precision audio attenuator and a low noise amplifier and it is then rectified and displayed on a meter. With the two flanges

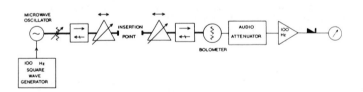

Fig. 22

at the insertion point bolted directly together, the audio attenuator is adjusted until a convenient reading is obtained on the output meter. The "unknown" is then inserted and the audio attenuator is readjusted until the same output as before is obtained. Due to the square law nature of the detector, the microwave attenuation in decibels through the "unknown" is equal to half the increment on the audio attenuator in decibels. A system of this type has been described by Korewick[70, 71] and a dual channel version of it, which is unaffected by amplitude variations in the microwave source, has been described by Sucher and Fox[72] and also by Pakay[73]. Dual channel equipments of this type are commercially available and a typical one has a dynamic range of 25 dB and an accuracy of ±0.005 dB over the range 0-1 dB. The accuracy falls off to ±0.1 dB over the range 20-25 dB. In systems of this type, the microwave power fed into the bolometer must be kept at least 20 dB below the bias power to achieve the required square law characteristic [74].

Another AF substitution system is shown in Fig. 23. This system has been developed to a very high degree by Clark [75] at NRC, Ottawa. The output from the synthesizer is split and fed into harmonic generators and single-sideband modulators which produce two microwave signals differing in frequency by 10 kHz. One of these signals is amplitude stabilized, passed through the unknown and then fed into a linear mixer. The other one is used as a local oscillator signal. Thus, a 10 kHz intermediate frequency is produced. At each setting of the unknown, the IVD is adjusted until the rectified output is equal to the voltage of the mercury battery, as indicated by a null reading on the recorder. The changes of attenuation in the unknown are then readily deduced from the settings of the IVD. With this equipment, the short term jitter is about 0.0003 dB peak to peak and the estimated systematic error varies from 0.0005 dB at 3 dB to 0.01 dB at 80 dB.

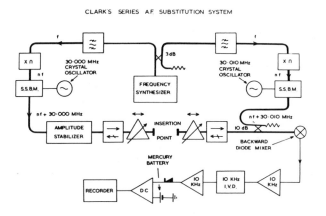

CLARK'S SERIES A.F. SUBSTITUTION SYSTEM

Fig. 23

A third system that contains an audio frequency reference standard was devised by Schafer and Bowman [76] in 1962. Their arrangement is called a modulated sub-carrier system and it has been analysed further by Little [77], Oseiko et al [78] and Kalinski [79]. Four modulated sub-carrier systems, in waveguide sizes 11A, 16, 18 and 22 were designed in 1968 at RRE for use as the British national standards. A block diagram of the X band one is shown in Fig. 24. To obtain the largest possible dynamic range without causing attenuator damage, a 1 W double-cavity klystron is used as the source and its output is split into two channels. The signal passing through the sub-carrier channel is amplitude modulated sinusoidally at 1 kHz with a DC coupled PIN diode envelope feedback system which stabilizes both the mean output power and the modulation depth. The amplitude modulated signal is passed through a level-setting attenuator and then through the attenuator under test which is sandwiched between carefully matched isolators. After this, it is divided between two linear homodyne balanced mixers. The unmodulated signal in the carrier channel is fed via a rotary phase shifter and PIN diode leveller into a magic tee where it is similarly split into two parts, one of which is fed directly into the upper balanced mixer while the other is fed via a 90° phase shifter into the lower balanced mixer. When there is a phase difference, θ, between the unmodulated wave and the carrier component of the amplitude modulated wave (at the upper balanced mixer) it is easily shown that the amplitudes of the 1 kHz signals emerging from the upper and lower balanced mixers are

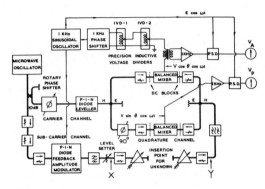

Fig. 24

proportional to cos θ and sin θ respectively. Immediately before each attenuation measurement, cos θ is maximised by adjusting the rotary phase shifter until sin θ and hence V_p are zero. This arrangement makes the adjustment of θ simpler and quicker because cos θ in the vicinity of its maximum value. At each setting of the attenuator under test, the second IVD is varied until its output is equal to the output from the upper balanced mixer; this condition being indicated by a zero value for V_A. The changes in attenuation are then readily deduced from the settings of IVD-2. The measuring equipment is nulled initially by adjusting IVD-1.

The WG16 system at RSRE has a short term jitter of about 0.0003 dB peak to peak and its zero drifts by about 0.001 dB in half an hour. Its accuracy can be judged from the results in Table 4 (see section 7).

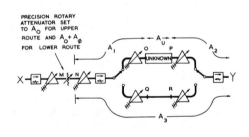

Fig. 25

To measure fixed devices with a loss exceeding 60 dB, the components between X and Y in Fig. 24 are replaced by the configuration shown in Fig. 25[80] and the attenuation in the level setter is reduced by $A_\phi \approx A_u/2$.

The six tuners are adjusted so that very good matches are seen looking to the left from M, O and Q and looking to the right from N, P and R with the switches in the appropriate positions. Switching over from the upper route to the lower route gives an attenuation change in dB of :-

$$A_{ml} = (A_o + A_1 + A_u + A_2) - (A_o + A_\phi + A_3)   \quad ... (30)$$

which is measured in the usual way. The unknown is then replaced by a piece of waveguide of the same length with an attenuation A_w. With the precision rotary attenuator now left fixed at $A_o + A_\phi$, switching over as before gives an attenuation change:-

$$A_{m2} = (A_1 + A_w + A_2) - A_3 \qquad \ldots (31)$$

Thus,

$$A_u = A_\phi + A_w + A_{m1} - A_{m2} \qquad \ldots (32)$$

A_ϕ is then measured in the usual way and $\bar{A}_w$ is determined by the technique described in section 5.6; so A_u can be found.

Further work on AF substitution techniques has been reported recently by Warner et al[150-152], Kawakami[173-174], Kalinski[175, 176] and Pakay and Torok[177].

5.5   Methods Not Requiring an Attenuation Standard   Several different ways of measuring microwave attenuation without an attenuation standard have been described in the literature[81-84]. Because of length restrictions, only the method due to Laverick[82] will be discussed here. Fig. 26 shows the equipment needed for this method. The measurement procedure is as follows:-

(1)   With SW1 closed, SW2 open and the unknown set at maximum attenuation, A_1 and θ_1 are adjusted until the output is zero;

(2)   With SW1 open and SW2 closed, A_2 and θ_2 are adjusted until the output is again zero;

(3)   With both switches closed, the unknown and θ_1 are adjusted until the output is yet again zero.

LAVERICK'S SELF-CALIBRATING ATTENUATION MEASUREMENT SYSTEM

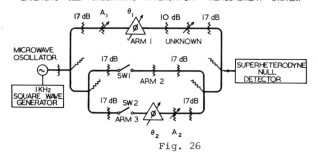

Fig. 26

Steps 2 and 3 are then repeated over and over again. The voltage ratios corresponding to successive movements of the unknown attenuator are 2, 3/2, 4/3, 5/4 etc. Table 2 summarises the complete process. This method requires a large number of waveguide

Number of operation	Output voltage from arm 1 before adjustment of unknown	Output voltage from arm 2 with SW1 closed	Output voltage from arm 3 after step 2	Combined output voltage from arms 2 and 3 during step 3	Output voltage from arm 1 after adjustment of unknown	Attenuation change in unknown dB.
1	+e	-e	-e	-2e	+2e	6.021
2	+2e	-e	-2e	-3e	+3e	3.522
3	+3e	-e	-3e	-4e	+4e	2.499
4	+4e	-e	-4e	-5e	+5e	1.938
5	+5e	-e	-5e	-6e	+6e	1.584
6	+6e	-e	-6e	-7e	+7e	1.339

TABLE 2

components and it suffers from several sources of error, which have been analysed by Laverick[82]. However, with care, it will give an accuracy of about ±0.02 dB up to 20 dB.

5.6 Methods Based Upon Reflection Coefficient Measurements   Several papers[85-90] have described methods of deducing attenuation from reflection coefficient measurements. The simplest method is to place the unknown between a standing-wave detector and a sliding short circuit, with the input end of the attenuator facing the short circuit. The magnitudes and phase angles of the reflection coefficients are measured for several different positions of the short-circuit and the results are plotted on polar graph paper. The points lie on a circle whose centre is usually displaced from the origin and, using the bilinear transformation, it can be shown that the radius of this circle is equal to

$$|s_{21}|^2 / (1 - |s_{11}|^2).$$

Thus, after finding this radius, the dissipative component of the attenuation can be found immediately from equation (15). A small correction may be needed to allow for the attenuation in the standing wave detector and the sliding short-circuit. To find the reflective component of the attenuation, the unknown is reversed, a perfectly matched load is connected to its output end, $|s_{11}|$ is measured and then equation (14) is used.

When measuring low values of attenuation by this technique, the Roberts-Von Hippel method of measuring large standing wave ratios[91] has proved to be very useful and a method due to Owens[92] is even better. Instead of using a standing wave detector, a tuned reflectometer can be employed to measure the reflection coefficients. At RSRE, this is done by replacing the components between X and Y in Fig. 24 with the components shown in Fig. 27[80]. Under normal measurement conditions, the amplitude modulated signal follows the route XABCDCBEFY and therefore passes through the unknown twice. The waveguide switch SW1 is a most important item. By switching to the fixed short circuit, the zero of the modulated subcarrier system can be reset

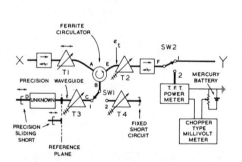

Fig. 27

rapidly at any time without unbolting or reconnecting any waveguide flanges. Measurements are made with the sliding short at positions $\lambda g/4$ apart, both with and without the unknown in position. The TFT power meter is used when the reflectometer is tuned and it is also used during one stage of the measurement procedure[80]. If great care is taken, low values of attenuation can be measured in this way to an accuracy of about ±0.001 dB. A combined reflectometer and power ratio technique for low loss measurements has been described by Skilton[178, 179].

5.7 The Shuttle Pulse Method   This method of measuring attenuation
was proposed initially by Ring[93] and the first description of it in
the published literature was given by King and Mandeville[94] in 1961.
Since then, it has been widely used by scientists working on millimetre
wave low-loss waveguide communication systems[95-102, 180]. A block
diagram of a typical shuttle pulse system is shown in Fig. 28. For most
of the time, the backward
wave oscillator (BWO) has
constant voltages applied
to its electrodes and it
provides a local oscillator
signal at a constant
frequency f_0. However,
flat-topped pulses with a
duration somewhere in the
range 10 nS to 1 μS and a
p.r.f. in the region of 10
kHz are applied to the
helix of this BWO and the
amplitude of these pulses
is chosen so that the
frequency is shifted from
f_0 to f_1 while they are
present. The ferrite
circulator directs the
output from the BWO,
through a rectangular to
circular transformer, to
the length of waveguide

SHUTTLE PULSE METHOD OF MEASURING
ATTENUATION

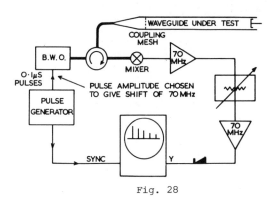

Fig. 28

under test, which is provided with a coupling mesh at the input end and
a short circuit at the far end. The coupling mesh is designed to transmit about 1% of the incident power. The pulses at frequency f_1 bounce
back and forth many times between the coupling mesh and the short
circuit. After each round trip, a fraction of this power passes back
through the coupling mesh and enters the mixer. Here, it is mixed with
the strong signal that is reflected by the front face of the coupling
mesh. The two way delay through the waveguide under test is arranged to
be longer than the pulse duration; so this strong signal is at a frequency f_0 when the train of pulses at frequency f_1 arrive. Mixer output
pulses at an intermediate frequency $f_1 - f_0$ are therefore produced and
these are amplified, detected and displayed on an oscilloscope.

Let α_1 denote the attenuation constant in dB per unit length of the waveguide under test, let L be the distance between the coupling mesh and
the short circuit and let A_m and A_s denote the losses in dB that occur
at each reflection from the mesh and short circuit respectively. Then
the total loss in dB that occurs in each successive two-way trip through
the test section is given by:-

$$A_t = 2\alpha_1 L + A_m + A_s . \qquad \qquad \dots (33)$$

The value of A_t can be determined from any adjacent pair of pulses on
the oscilloscope. Additional measurements are needed to find the values
of L, A_m and A_s; and then α_1 can be obtained from equation (33). The
shuttle pulse method gives an accuracy in the region of ±0.05 dB.

5.8 Determination of Attenuation from Q Measurements   Very accurate
measurements of the attenuation per unit length, α_1, in a uniform piece
of waveguide can be obtained by using the arrangement shown in Fig. 29.

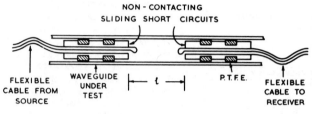

Q METHOD OF MEASURING ATTENUATION

NON - CONTACTING
SLIDING SHORT CIRCUITS

FLEXIBLE   WAVEGUIDE ℓ P.T.F.E.
CABLE FROM   UNDER
SOURCE   TEST

Two non-contacting
sliding short circuits
are placed inside the
waveguide so that a
cavity is formed. The
flexible coaxial cables
and adjustable coupling
loops are used to
couple power into and
out of the cavity. The
distance, ℓ, between
the two short circuits
is adjusted to several
different values that
give resonance and at
each of these settings,
the Q of the cavity is

FLEXIBLE
CABLE TO
RECEIVER

Fig. 29

measured.   (Sucher and Fox[103] describe many different techniques for
measuring Q and give 41 references to this subject.)

With slight rearrangement of the cavity equations given by Barlow and
Cullen[104], it is easily shown that the Q is given by:

$$Q = \frac{2\pi\ell\lambda_g}{\lambda^2(2 - |\Gamma_1| - |\Gamma_2|+2\alpha_1\ell)} \qquad \ldots (34)$$

where λ_g is the guide wavelength, λ is the free-space wavelength and Γ_1
and Γ_2 are the reflection coefficients looking into the left hand and
right hand short circuits. The values of Γ_1 and Γ_2 depend upon the
effectiveness of the choke systems, the conductivity of the end walls,
the frequency and the tightness of the coupling to the source and
receiver. Rearranging equation (34),

$$\frac{1}{Q} = \frac{\lambda^2}{2\pi\lambda_g} (2 - |\Gamma_1| - |\Gamma_2|) \cdot \frac{1}{\ell} + \frac{\lambda^2\alpha_1}{\pi\lambda_g} \qquad \ldots (35)$$

Thus by plotting $\frac{1}{Q}$ against $\frac{1}{\ell}$ for several successive resonance points,
a straight line is obtained whose intercept on the 1/Q axis is $\lambda^2\alpha_1/\pi\lambda_g$.
Thus, after finding λ and λ_g, α_1 can be found. By adopting this
procedure it is not necessary to measure Γ_1 and Γ_2.

Several workers[105-114] have determined attenuation values for waveguides from Q measurements and Skilton[115] has determined the losses in
various types of waveguide connectors by making Q measurements on
cavities containing up to 11 flanged couplings. Losses below 0.0005 dB
can be measured using this technique.

5.9 Method Based Upon the Josephson Effect   Certain metals such as
indium, lead, niobium and tin make a transition into a zero resistance
state when cooled below a few kelvin. They are then called super-

conductors. In 1962, Josephson[116] predicted that if two super-
conductors are separated by a very thin insulating layer and are joined
together elsewhere, a direct current would flow around the circuit
without any voltage being applied. He also said that a bias voltage,
V_B, applied across the barrier would generate an alternating current
with a frequency proportional to V_B (483.6 MHz per μV). Josephson's
predictions were soon proved experimentally. The Josephson effects have
now found many applications[117]. Josephson junctions are now employed
in ultra sensitive magnetometers and galvanometers, primary voltage
standards, voltage tunable oscillators, infra-red detectors, microwave
power and attenuation measuring equipments, etc.

A superconducting loop containing a Josephson junction is called a SQUID
(superconducting quantum interference device). The current that flows
round such a device depends upon the magnetic flux ϕ linking the loop
and is given by[181]

$$I_j = I_c \sin(2\pi\phi/\phi_o) \qquad \ldots (36)$$

I_c depends upon the nature of the junction and

$$\phi_o = h/2e = 2.0678538 \times 10^{-15} \text{ weber} \qquad \ldots (37)$$

where h is Planck's constant and e is the charge on an electron.

Let ϕ_x denote the component of ϕ which is due to an external source.
Then

$$\phi = \phi_x + LI_j \qquad \ldots (38)$$

where L is inductance of the loop. From (36) and (38), we get:-

$$I_j = I_c \sin\{2\pi(\phi_x + LI_j)/\phi_o\} \qquad \ldots (39)$$

If we add any integral multiple of ϕ_o to ϕ_x, the value of I_j is unchanged. The EMF around the loop is $d\phi/dt$. Thus the SQUID behaves like
a non-linear impedance which is a periodic function of ϕ_x.

Fig 30 shows how Kamper et al[118-123] use a SQUID to measure attenuation at 30 MHz.

A weak 9 GHz signal is used to indicate the periodic impedance variations of the SQUID as the 30 MHz attenuator under test is changed. A
point-contact niobium Josephson junction is formed across a section of
very low impedance X band waveguide that is lined with Babbitt metal
and short-circuited $\lambda_g/4$ behind the junction. The SQUID thus formed is
inductively coupled to the inner conductor of a 50 Ω coaxial line. The
30 MHz output current from the attenuator under test, a 1 kHz sine wave
and a DC bias current are passed through this inner conductor. The 9
GHz power reflected by the SQUID can be represented by a Fourier series
and written as:-

$$P = P_o + \sum_n P_n \cos\left\{ \frac{2\pi n}{I_o}(I_s \cos 2\pi f_s t + I_m \sin 2\pi f_m t + I_b) \right\} \qquad \ldots (40)$$

JOSEPHSON   ATTENUATOR   CALIBRATOR

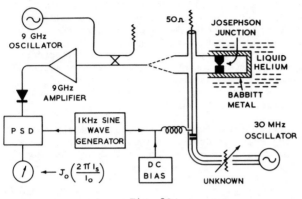

Fig. 30

where I_O is the current through the inner conductor that is needed to change the flux linking the SQUID by ϕ_O. Subscripts s, m and b refer respectively to the 30 MHz, 1 kHz and DC components of the current through this inner conductor.

This reflected power is passed through a 9 GHz tunnel-diode amplifier, rectified by a square-law detector and then fed through a phase sensitive detector. By averaging equation (40) over a complete RF cycle, then using Fourier analysis to find the amplitude of the 1 kHz component in the square law detector output and finally mixing this with the 1 kHz reference signal, the output from the phase sensitive detector is found to be:

$$V_{out} = K \sum_{n} P_n . J_o \left(\frac{2\pi n I_s}{I_o} \right) . J_1 \left(\frac{2\pi n I_m}{I_o} \right) \sin \left(\frac{2\pi n I_b}{I_o} \right) \quad \dots (41)$$

where K depends upon the gain of the receiving system and J_o and J_1 denote zero and first order Bessel functions. Even harmonic distortion (n = 2, 4, 6, etc) can be completely eliminated by adjusting the bias current until $I_b = 0.25\ I_o$ and this condition maximizes the desired output signal at n = 1. Third harmonic distortion can be eliminated by adjusting the amplitude of the 1 kHz signal until $I_m = 0.2033\ I_o$. This value makes the first order Bessel function zero. The fifth and higher order odd harmonics can be neglected. Therefore, when I_b and I_m have been correctly set, it follows that

$$V_{out} \propto J_o \left(\frac{2\pi I_s}{I_o} \right) \quad \dots (42)$$

Thus, it only requires a table of zero order Bessel function roots to determine the attenuation between any two nulls.

The roots of $J_o(x)$ occur at 2.40483, 5.52008, 8.65373 etc.

Let I_{s1}, I_{s2} and I_{s3} denote the first three values of the RF current that give a zero output. Then it follows that:

$$\frac{2\pi I_{s1}}{I_o} = 2.40483, \quad \frac{2\pi I_{s2}}{I_o} = 5.52008, \quad \frac{2\pi I_{s3}}{I_o} = 8.65373$$

Thus the change in attenuation needed to go from the first zero to the
second is

$$\Delta\alpha_{1-2} = 20 \ \log_{10} \ \frac{I_{s2}}{I_{s1}} = 20 \ \log_{10} \ \frac{5.52008}{2.40483} = 7.21722 \ dB$$

The change in attenuation needed to go from the second zero to the third
is

$$\Delta\alpha_{2-3} = 20 \ \log_{10} \ \frac{I_{s3}}{I_{s2}} = 20 \ \log_{10} \ \frac{8.65373}{5.52008} = 3.90516 \ dB$$

and so on. To reach 62 dB, one must cover 900 nulls. It is tedious to
count them by looking at the output meter; so electronic counters are
used to do this[123].

At NBS, agreement to within ±0.002 dB over a range of 62 dB was obtained
between a SQUID system and the normal calibration service[123]. A
similar agreement has been achieved at NPL[124]. Present SQUID systems
can be used up to about 1 GHz. With the readout signal at Q band, it may
be possible to carry out attenuation measurements up to X band.

Further work on the SQUID method of attenuation measurement has been
reported in eight recent papers[181-188].

5.10 Further Attenuation Measurement Techniques A novel time interval ratio technique for measuring attenuation has been developed at RSRE
by R H Johnson & P Cummings[189-191]. A fraction of the output from a
continuous wave source is fed straight into a thin-film thermo-electric
(TFT) sensor. The remainder of the signal is passed through the device
under test, then pulse modulated by a fast PIN diode and finally fed
into a second TFT sensor. At each setting of the device under test, the
PRF is varied (with constant pulse length) until the outputs from the
two TFT sensors are equal. The changes in attenuation are then deduced
from the PRFs which are provided by a precision time synthesizer
(HP 5359A). Heating and harmonic generation effects in the PIN diode
can cause small systematic errors in a system of this type; but
accuracies better than ±0.004 dB over a 23 dB range have been achieved.

In 1980, the Narda Microwave Corporation introduced a Microwave Multimeter[192-194]. Fig. 31 shows how the components in it are arranged
when attenuation measurements are required.

The microwave signal is provided by a varactor tuned solid state source.
The power that emerges from the device under test is detected by a zero-
bias Schottky barrier diode, which is followed by a chopper type DC
amplifier and a digital voltmeter. The instrument is controlled by an
internal microprocessor. The directional coupler shown in Fig. 31 plays
no role during attenuation measurements, but is used for reflection
coefficient measurements.

The diode detector is used with input signals of -50 dBm to +13 dBm and
its characteristic changes from square law to linear over this range. A
PROM (unique to each diode) is used to linearize the detector readings
to within +0.05 dB of the ideal square-law response. The variation of

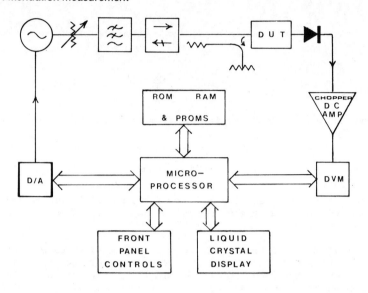

Fig. 31   Simplified block diagram of the Narda "Microwave Multimeter"

detector sensitivity with frequency is also stored in this PROM.   The
tuning of the microwave source is controlled by the microprocessor.   The
tuning curve is stored in a second PROM.

With appropriate units, this Microwave Multimeter can cover the frequency range 10 MHz to 18 GHz.   When used for attenuation measurements,
it has a dynamic range of 50 dB.   The instrumental uncertainty is
±0.15 dB and the mismatch uncertainty is, for example, ±0.25 dB when
measuring a 10 dB pad with input and output VSWRs of 1.2:1.

Three different voltage ratio techniques for measuring attenuation have
been devised recently at RSRE[170-172, 195].   These are shown in Figs
32A, 32B and 32C.

Each configuration makes use of the two microwave synthesized sources
and 50 kHz IF amplifier that have already appeared in Figs. 20 and 21.
The two sources are always operated with a 50 kHz separation and the
mixer is always operated in its linear region.   In the first and second
configurations, a precision AC digital voltmeter, with a frequency
response extending beyond 50 kHz,is used to measure the IF signal
voltages.   The arrangement shown in Fig. 32A has been widely used to
carry out automatic calibrations of programmable attenuators over a
20 dB range.   The addition of a GaAs FET power amplifier and a precision
gauge block attenuator results in the arrangement shown in Fig. 32B
which has a dynamic range of 90 dB.   When a device with high attenuation
is inserted for measurement, an appropriate amount of attenuation is
removed from the gauge block attenuator so that the weaker IF output
voltage stays within 20 dB of the datum level IF output voltage.   A
further increase in dynamic range to 150 dB has been achieved by replacing the DVM with a lock-in analyser (EG&G 5206) and inserting before it
a 50 kHz inductive voltage divider (IVD) as shown in Fig. 32C.   In this
case, when a very high attenuation DUT is inserted, appropriate amounts

of attenuation are removed from both the gauge block attenuator and the IVD to keep the two output voltages within 20 dB of each other. The simple equations needed to determine the attenuation, A_{dut}, in dB through the device under test are given in Fig. 32, where:-

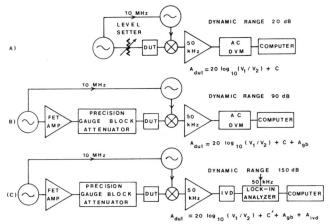

$$A_{dut} = 20 \log_{10}(V_1 / V_2) + C$$

$$A_{dut} = 20 \log_{10}(V_1 / V_2) + C + A_{gb}$$

$$A_{dut} = 20 \log_{10}(V_1 / V_2) + C' + A_{gb} + A_{ivd}$$

Fig. 32   Simplified block diagrams of 3 voltage ratio configurations used at RSRE

V_1 = IF output voltage at the datum setting;
V_2 = IF output voltage at the calibration setting;
C = correction factor (in dB) for non linearity in the 50 kHz amplifier and DVM;
C' = correction factor (in dB) for non linearity in the 50 kHz amplifier and lock-in analyser;
A_{gb} = attenuation (in dB) removed from the gauge block attenuator when the DUT is inserted;
A_{ivd}= attenuation (in dB) removed from the IVD when the DUT is inserted.

The correction factor (C or C') is determined about once a month over a 20 dB range, using a very stable 50 kHz oscillator followed by a 50 kHz IVD, and is then expressed as a function of V_2, (V_1 is always kept at 1 volt rms). The error due to imperfect correction for non-linearity is less than 0.0005 dB.

A computer program controls the DUT (if possible), takes readings from the DVM or lock-in analyzer, calculates C or C' and A_{dut} and then prints out the required results.

The 90 dB dynamic range configuration is shown in much more detail in Fig. 33.

The GaAs FET power amplifier (Avantek AWT-18057) gives an output power of 100 mW over the range 8 to 18 GHz.

The gauge block attenuator consists of three isolated switched couplers in cascade. The DUT is inserted between carefully matched isolators.

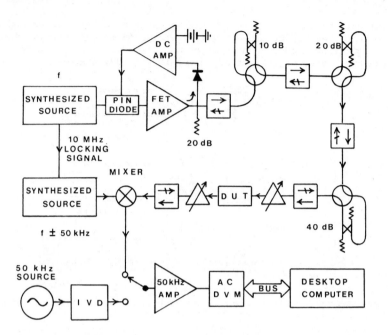

Fig. 33   RSRE voltage ratio system with 90 dB dynamic range

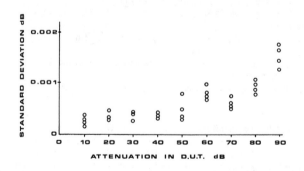

Fig. 34   Standard deviation versus attenuation over a 90 dB range using the configuration in Fig. 33.

The total systematic uncertainty of this system is estimated to be 0.001 dB per 10 dB up to 90 dB. A special DUT, consisting of a switched rotary vane attenuator in cascade with a switched coupler, was used to determine the repeatability of this system. The results obtained at 10 GHz are shown in Fig. 34. Each small circle denotes the standard deviation of 10 repeated measurements at the same attenuation value. All the standard deviations are seen to be below 0.0005 dB up to 40 dB, below 0.001 dB up to 70 dB and below 0.002 dB up to 90 dB.

This system is not plagued by switching spikes, modulator instabilities, harmonic errors, circulating earth currents, mixer-to-mixer leakage or AFC problems. The attenuation reference standard (50 kHz IVD) is only used for a short time about once a month so the wear and tear on it is very low. Furthermore, over a 90 dB dynamic range, noise balancing is not necessary[195].

A bridge method for the accurate measurement of waveguide wall loss was described recently by R V Gelsthorpe and R G Bennett [196-197]. Their bridge is shown in Fig. 35. A series of nulls is obtained by adjusting A2 and SC2. The upper short circuit SC1 is then moved by a quarter of a guide wavelength and a second series of nulls is obtained by again adjusting A2 and SC2. From the attenuator settings and sliding short settings it is now a straight forward matter [196] to calculate the attenuation in the waveguide under test. This measurement procedure eliminates the mismatch error, without a need for any precise matching.

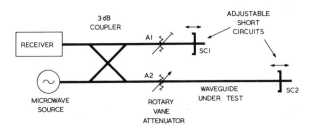

GELSTHORPE AND BENNETT'S BRIDGE METHOD

FOR THE ACCURATE MEASUREMENT OF WAVEGUIDE

WALL LOSS

FIG. 35

An RSRE implementation of this bridge is shown in Fig. 36.

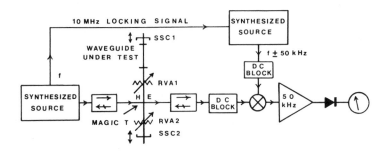

Fig. 36   RSRE rf substitution magic T bridge

The use of two synthesized sources makes this technique very attractive because the narrow band superheterodyne receiver stays perfectly in tune at the null positions, provides great sensitivity for accurate balancing and rejects harmonic signals. At the nulls, the AM noise from the signal source is cancelled out and the residual jitter at the output is less than 0.00001 dB. A recording of a 0.0001 dB change obtained at RSRE by moving precision rotary van attenuator, RVA1, from 0° to 0.194° is shown in Fig. 37.

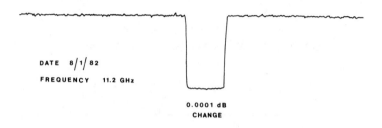

RECORDING OF A 0.0001 dB ATTENUATION CHANGE ON THE
R.S.R.E. RF SUBSTITUTION MAGIC-TEE WAVEGUIDE BRIDGE

DATE 8/1/82

FREQUENCY   11.2 GHz

0.0001 dB
CHANGE

Fig. 37

Six-port techniques have become very popular in recent years and dual six-port systems can be used to measure attenuation.

Further information on microwave attenuation measurement will be found in fourteen additional published papers[125-133, 198-203] and four recent books (155, 204-206).

Student's t-factors are needed whenever the uncertainty range associated with a set of attenuation measurements has to be determined for a specified confidence level. New expressions and some more precise values for Student's t-factor have been given in a recent paper by I A Harris and F L Warner[207].

6.   ATTENUATION TRANSFER STANDARDS

In order to find out how the results obtained with different attenuation measuring systems compare with each other, it is necessary to have a set of low temperature-coefficient attenuation transfer standards which are extremely stable, very rugged, readily portable and well matched. Furthermore, if variable attenuators are used as transfer standards, they must possess a very high degree of resettability. Most commercially available variable attenuators fail to meet this requirement; so fixed coaxial attenuators[134] or directional couplers in either coax or waveguide are often used as attenuation transfer standards. An inline directional coupler[135] is more satisfactory than a conventional one.

A very good attenuation transfer standard can be made by combining a directional coupler with a very repeatable waveguide switch as shown in Fig. 38. In one position of the switch the input port is connected straight through to the output port. In the other switch position, the input port is connected to the main arm of the directional coupler, which is terminated at the far end with a matched load. The power, which is coupled into the side arm of the directional coupler, is fed through the other arc of the switch to the output. A simple transfer standard of this type can, of course, only provide one value of attenuation; but many variations on this theme are possible, eg by combining 4 directional couplers with 4 waveguide switches, it is possible to make a transfer standard which covers the range 0 to 75 dB in 5 dB steps.

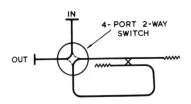

4- PORT 2-WAY SWITCH

DIRECTIONAL COUPLER WAVEGUIDE SWITCH
ATTENUATION TRANSFER STANDARD

Fig. 38

The repeatability of top grade waveguide switches is better than ±0.001 dB. The type of transfer standard shown in Fig. 38 is not affected by connector losses and it can be measured a large number of times very quickly without unbolting or reconnecting any waveguide flanges and without moving about any parts of the attenuation measuring system.

A rotary attenuator containing a very accurate optical angular measuring system[34, 35] would make an excellent transfer standard, but such a device is very expensive. At RSRE, a series of special rotary attenuators has been developed for use as transfer standards. These attenuators contain trains of very high class anti-backlash spur gears and they have an effective scale length of 180 metres.

The backlash in them is less than 0.001° and the resetting accuracy is better than 0.001°.

Foote and Hunter[136] have improved the resetting accuracy of a commercially available rotary attenuator by a factor of 20 by fitting it with a 180:1 precision Spiroid gear set.

A novel low value attenuation transfer standard has been described recently by Somlo and Mörgan[137]. The attenuation through the main arm of a directional coupler is changed by switching one of the side-arm terminations from a short circuit to a match. Another novel low value attenuation transfer standard has been described by Bayer and Stumpe[208].

7. INTERNATIONAL ATTENUATION INTERCOMPARISONS

Several international attenuation intercomparisons have been carried out and, to give some idea of the discrepancies that occur in such an exercise, Table 3 shows the mean values obtained in 1973 at RRE and the National Bureau of Standards at Boulder, Colorado, on 5 different attenuation transfer standards.

The results obtained in an international intercomparison of low values of attenuation ($\leqslant$ 1 dB) at 10 GHz have been reported by Bayer[209].

Transfer Standard	RRE value dB	NBS value dB
Italian 3 dB coupler	2.603	2.6026
UK 3 dB switched coupler	3.5879	3.5893
Italian 10 dB coupler	10.479	10.4773
UK 20 dB switched coupler	20.4878	20.4870
Canadian 20 dB coupler	20.572	20.573

TABLE 3

8.   APPLICATIONS OF MICROWAVE ATTENUATION MEASURING EQUIPMENTS IN INDUSTRY AND MEDICINE

The attenuation through wet insulating materials is very dependent upon the amount of water that is present; so microwave attenuation measuring equipments are now being widely used in industry for the measurement and control of moisture content. J Kalinski[175] has described some of the results that have been achieved with this technique in glass-works, coal mines, concrete precasting plants, copper ore processing plants, green fodder drying plants, etc. In sand, for example, moisture contents over the range 3 to 9% can be measured to an accuracy of 0.1%. The material under investigation is passed in a uniform manner between two horns that form part of the attenuation measuring equipment. M Furusawa et al[210] have described, briefly, an X band equipment for measuring the moisture content of tobacco shreds in a cigarette factory.

I Yamaura[211] at the Electrotechnical Laboratory in Tokyo has measured attenuation through the human torso over the frequency range 1.8-2.7 GHz.

Changes of attenuation caused by heartbeats and breathing were observed.

9.   CONCLUSIONS

IF substitution systems have been very widely used in the last 36 years for attenuation measurements up to 100 dB and a very popular configuration has been parallel IF substitution, with a piston attenuator as the reference standard, and envelope detection of the IF signal (see Fig. 18).

When a dynamic range well in excess of 100 dB is required, it is necessary to change over from envelope detection to coherent detection[212] and two systems that fall within this category are: the Weinschel VM-4A vector IF substitution system (see Fig. 19) and the RSRE voltage ratio system that employs both RF and IF gauge block attenuators (see Fig. 32C).

Equipments that have been used for extremely accurate high dynamic range attenuation measurements in national standards laboratories include:-

(a)   refined versions of the parallel IF substitution system shown in Fig. 18;
(b)   the parallel 50 kHz substitution system shown in Fig. 20;
(c)   the series AF substitution system with 10 kHz IF shown in Fig. 23;
(d)   the modulated sub-carrier system shown in Fig. 24;

(e)   the voltage ratio plus gauge block system shown in Fig. 33.

When every possible precaution is taken, accuracies approaching ±(0.001 dB per 10 dB + 0.001 dB) can be achieved up to 100 dB. At low values of attenuation, the main sources of error are: imperfect matching at the insertion point, mixer non-linearity, connector uncertainties, zero drift and short term instrumental jitter. At very high values of attenuation, the main sources of error are: noise, leakage, spurious signals, circulating earth currents and accumulated errors in the reference standard.

In a standards laboratory, both sides of the insertion point are normally very well matched (VSWR < 1.005) before an attenuation measurement is made and if the DUT has VSWRs less than 1.05 the mismatch uncertainty is about ±0.001 dB (see Fig. 3). Broad band matching is employed on both sides of the insertion point in some of the commercially available attenuator calibrators and the mismatch uncertainty can then exceed ±0.1 dB. A notable exception is the VM-4A (used with an external computer) where the mismatch uncertainty can be almost eliminated by a calibration and computer correction technique. Due to this technique, the accuracy of the VM-4A approaches that of its own reference standard which is specified as ±(0.003 dB per 10 dB + 0.005 dB). Thus, the accuracy of the VM-4A is only about 3 times poorer than the very best accuracy that can be achieved in a national standards laboratory.

Workers who possess only a small amount of equipment and need to make attenuation measurements over a 30 dB range should consider using either the single power meter technique shown in Fig. 15 or the AF substitution technique shown in Fig. 22.

For the measurement of low values of attenuation, good results can be obtained with a small amount of equipment by using either the reflection coefficient method (see Section 5.6) or the dual channel power ratio method (see Fig. 16). When very low values of attenuation need to be measured with great accuracy, either the Q measurement technique (see Section 5.8) or a bridge method (Fig. 35 or Fig. 36) is recommended.

If very low zero drift is needed over a long period of time, either the system shown in Fig. 21 or the power ratio system described by Engen and Beatty[52] should be used.

The shuttle pulse method is a very popular one for measuring the attenuation in the long lengths of helix waveguide used in millimetre wave communication systems.

The Josephson method suffers from the great disadvantage of needing liquid helium. However, it does not depend upon any precision machining and the only item needed to determine the results is a table of Bessel function roots; so it may eventually be adopted as a primary attenuation standard.

Finally, if moderate accuracy is acceptable and the required dynamic range is less than 50 dB, the versatile light weight Microwave Multimeter shown in Fig. 31 will give results very rapidly at many different frequencies.

10 .   REFERENCES

(1)   J.K. Hunton, IEEE Trans., MTT-8, 206-212, 1960
(2)   R.W. Beatty, J. Res. NBS, 52, 7-9, 1954.
(3)   C.F. Varley, Math. & Phys. Sect, Brit. Assoc. Adv. Sci., 14-15, 18
(4)   J.B. Kelley and H.H. Marold, Instruments & Control Systems, 33,
       626-628, 1960.
(5)   M.L. Morgan and J.C. Riley, IRE Trans., I-9, 237-243, 1960.
(6)   L.C. Fryer, AIEE Trans. (Communication & Electronics), 81, 128-
       135, 1962.
(7)   A.F. Dunn, IEEE Trans., IM-13, 129-139, 1964.
(8)   S.H. Neff, Amer. J. Phys., 38, 769-771, 1970.
(9)   A.D. Blumlein, British Patent No. 323037, 1928.
(10)   C.B. Pinckney, AIEE Trans., (Communication & Electronics), 78,
       182-185, 1959.
(11)   J.J. Hill and A.P. Miller, Proc. IEE, 109, 157-162, 1962.
(12)   T.L. Zapf, ISA Trans., 2, 195-201, 1963.
(13)   T.L. Zapf et al, IEEE Trans., IM-12, 80-85, 1963.
(14)   A.J. Binnie, J. Sci. Instrum., 41, 747-750, 1964.
(15)   W.C. Sze et al, IEEE Trans., IM-14, 124-131, 1965.
(16)   C.A. Hoer and W.L. Smith, J. Res. NBS, 71C, 101-109, 1967.
(17)   J.J. Hill and T.A. Deacon, Proc. IEE, 115, 727-735, 1968.
(18)   T.A. Deacon and J.J. Hill, Proc. IEE, 115, 888-892, 1968.
(19)   T.A. Deacon, Proc. IEE, 117, 634-640, 1970.
(20)   J.J. HIll, IEEE Trans., IM-21, 368-372, 1972.
(21)   O.L. Boothby and R.M. Bozorth, J. Appl. Phys., 18, 173-176, 1947.
(22)   D.E. Harnett and N.P. Case, Proc. IRE, 23, 578-593, 1935.
(23)   J. Brown, Proc. IEE, 96, pt. III, 491-495, 1949.
(24)   R.E. Grantham and J.J. Freeman, Trans. AIEE, 67, pt. I, 329-335,
       1948.
(25)   R.W. Yell, CPEM Digest, 108-110, 1972.
(26)   G.C. Southworth, "Principles & Applications of Waveguide Trans-
       mission", Van Nostrand, 374-376, 1950.
(27)   B.P. Hand, Electronics, 27, 184-185, 1954.
(28)   P.F. Mariner, Proc. IEE, 109B, 415-419, 1962
(29)   A.V. James, IRE Trans., I-11, 285-290, 1962.
(30)   W. Larson, IEEE Trans., IM-12, 50-55, 1963.
(31)   W. Larson, IEEE Trans., IM-14, 117-123, 1965.
(32)   W. Larson, IEEE Trans., IM-16, 225-231, 1967.
(33)   T.Y. Otoshi and C.T. Stelzried, IEEE Trans., MTT-19, 843-854, 1971
(34)   W.E. Little et al, J. Res. NBS, 75C, 1-5, 1971.
(35)   F.L. Warner et al, IEEE Trans., IM-21, 446-450, 1972.
(36)   L.G.H. Huxley,"Wave Guides", (Cambridge, 1947), Ch. 3.
(37)   C.G. Montgomery, "Technique of Microwave Measurements", (McGraw-
       Hill, 1947), Ch. 11.
(38)   A.C. Gordon-Smith, Wireless Eng., 26, 322-324, 1949.
(39)   A.B. Giordano, Proc. IRE, 38, 545-550, 1950.
(40)   H.M. Barlow and A.L. Cullen, "Microwave Measurements", (Constable,
       1950), Ch. 8.
(41)   J. Munier, J. Phys. Radium, 16, 429-430, 1955.
(42)   C.R. Ditchfield, CPEM Digest, 7-9, 1974.
(43)   R.W. Beatty, NBS Monograph No. 97, 1967.
(44)   D. Russell and W. Larson, Proc. IEEE, 55, 942-959, 1967.
(45)   C.S. Gaskell, Design Electronics, 6, 30-33, 1969.
(46)   A.A. Luskow, Marconi Instrumentation, 13, 30-34, 1971.
(47)   K. Wilson et al, Proc. European Microwave Conference A6/2:1-

A6/2:4, 1971.
(48) D.L. Hollway and P.I. Somlo, IEEE Trans., MTT-17, 185-188, 1969.
(49) D.L. Hollway and P.I. Somlo, IEEE Trans., MTT-22, 560-561, 1974.
(50) C.T. Stelzried and S.M. Petty, IEEE Trans., MTT-12, 475-477, 1964.
(51) C.T. Stelzried, M.S. Reid and S.M. Petty, IEEE Trans., IM-15, 98-104, 1966.
(52) G.F. Engen and R.W. Beatty, J. Res. NBS, 64C, 139-145, 1960.
(53) H. Bayer, Metrologia, 11, 43-51, 1975.
(54) K.L. Hepplewhite, Electronics Letters, 11, 575-577, 1975.
(55) W. Larson and E. Campbell, Digest of IEEE International Convention, 434-435, 1971.
(56) W. Larson and E. Campbell, NBS Tech. Note No. 647, 1974.
(57) H.L. Kaylie, IEEE Trans., IM-15, 325-332, 1966.
(58) G.F. Gainsborough,JIEE, 94, Pt. III, 203-210, 1947.
(59) A.L. Hedrich et al, IRE Trans., I-7, 275-279, 1958.
(60) B.O. Weinschel et al, IRE Trans., I-8, 22-31, 1959.
(61) B.O. Weinschel, Microwave J., 4, 77-83, 1961.
(62) R.J. Turner, Proc. IEE, Pt. B, Suppl. No. 23, 775-782, 1962.
(63) D.J. Walliker, Proc. IEE, Pt. B, Suppl. No. 23, 791-795, 1962.
(64) R.F. Clark and B.J. Dean, IRE Trans., I-11, 291-293, 1962.
(65) D.L. Hollway and F.P. Kelly, IEEE Trans., IM-13, 33-44, 1964.
(66) D.H. Russell, ISA Trans., 4, 162-169, 1965.
(67) R.W.A. Siddle and I.A. Harris, The Radio & Electronic Engineer, 35, 175-181, 1968.
(68) D.L. Hollway and P.I. Somlo, IEEE Trans., IM-22, 268-270, 1973.
(69) R.W. Yell, IEEE Trans., IM-23, 371-374, 1974.
(70) J. Korewick, IRE Trans., MTT-1, 14-21, 1953.
(71) J. Korewick, Electronics, 27, 175-177, 1954.
(72) M. Sucher and J. Fox, "Handbook of Microwave Measurements, Vol. 1", Polytechnic Press, 412-413, 1963.
(73) P. Pakay, Periodica Polytechnica, 18, 105-116, 1974.
(74) B.O. Weinschel, IRE Trans., I-4, 160-164, 1955.
(75) R.F. Clark, IEEE Trans., IM-18, 225-231, 1969.
(76) G.E. Schafer and R.R. Bowman, Proc. IEE, 109, Pt. B, Suppl. No. 23, 783-786, 1962.
(77) W.E. Little, IEEE Trans., IM-13, 71-76, 1964.
(78) A.A. Oseiko et al, Meas. Tech. (USA), No. 3, 394-396, 1970.
(79) J. Kalinski, IEEE Trans., IM-21, 291-293, 1972.
(80) F.L. Warner et al, IEEE Trans., IM-23, 381-386, 1974.
(81) C.M. Allred, NBS Tech. News Bull., 41, 132-133, 1957.
(82) E. Laverick, IRE Trans., MTT-5, 250-254, 1957.
(83) R.L. Peck, J. Res. NBS, 66C, 13-18, 1962.
(84) M.C. Davies, Proc. IEE,109, Pt. B, Suppl. No. 23, 796-800, 1962.
(85) W.T. Blackband and D.R. Brown, JIEE, 93, Pt. IIIA, 1383-1386, 1946.
(86) A.L. Cullen, Wireless Engineer, 26, 255-257, 1949.
(87) R.W. Beatty, Proc. IRE, 38, 895-897, 1950.
(88) G.A. Deschamps, J. Appl. Phys., 24, 1046-1050, 1953.
(89) J.H. Vogelman, Electronics, 26, 196-199, 1953.
(90) A.F. Pomeroy and E.M. Suarez, IRE Trans., MTT-4, 122-129, 1956.
(91) S. Roberts and A. von Hippel, J. Appl. Phys., 17, 610-616, 1946.
(92) R.P. Owens, Proc. IEE, 116, 933-940, 1969.
(93) D.H. Ring, unpublished paper.
(94) A.P. King and G.D. Mandeville, BSTJ, 40, 1323-1330, 1961
(95) J.A. Berry, Marconi Review, 28, 22-28, 1965.
(96) W.H. Steier, BSTJ, 44, 899-906, 1965.
(97) D.T. Young and W.D. Warters, BSTJ, 47, 933-955, 1968.

(98)   M. Shimba, Electronics & Communications in Japan, 52A, 53-58, 1969.
(99)   N. Lacey and I.S. Groves, IEE Conference Publication No. 71, 137-141, 1970.
(100)   M. Shimba et al, IEEE Trans., IM-21, 215-219, 1972
(101)   S.C. Moorthy, IEEE Trans., IM-22, 311-314, 1973.
(102)   R.P. Bomer, IEEE Trans., IM-23, 386-389, 1974.
(103)   M. Sucher and J. Fox, "Handbook of Microwave Measurements", (Polytechnic Press, 1963), Vol. 2, Ch. 8.
(104)   H.M. Barlow and A.L. Cullen, "Microwave Measurements", (Constable, 1950), Ch. 3.
(105)   J.A. Young, Proc. IEE, 106B, Suppl. No. 13, 62-65, 1959.
(106)   A.E. Karbowiak and R.F. Skedd, Proc. IEE, 106B, Suppl. No. 13, 66-70, 1959.
(107)   D.G. Keith-Walker, Proc. IEE, 106B, Suppl. No. 13, 71-75, 1959.
(108)   R. Hamer and R.J. Westcott, Proc. IEE, 109B, Suppl. No. 23, 814-819, 1962.
(109)   P.I. Somlo, Proc. IRE Australia, 23, 585, 1962.
(110)   J.K. Chamberlain, Electronic Engng., 38, 579-581, 1966.
(111)   G.H.L. Childs, IEE Conference Publication No. 71, 269-274, 1970.
(112)   W.J. Clapham, IEE Conference Publication No. 71, 303-306, 1970.
(113)   J. Uhlir, IEEE Trans., MTT-20, 38-41, 1972.
(114)   A.D. Olver et al, Electronics Letters, 9, 424-426, 1973.
(115)   P.J. Skilton, IEEE Trans., IM-23, 390-394, 1974.
(116)   B.D. Josephson, Physics Letters, 1, 251-253, 1962.
(117)   B.W. Petley, "An Introduction to the Josephson Effects", (Mills & Boon, 1971).
(118)   R.A. Kamper and M.B. Simmonds, Appl. Phys. Lett., 20, 270-272, 1972.
(119)   R.A. Kamper et al, Proc. Applied Superconductivity Conference, 696-700, 1972.
(120)   R.A. Kamper et al, Proc. IEEE, 61, 121-122, 1973.
(121)   R.A. Kamper et al, NBS Tech. Note 643, 1973.
(122)   R.A. Kamper et al, NBS TEch. Note 661, 1974.
(123)   R.T. Adair et al, IEEE Trans., IM-23, 375-381, 1974.
(124)   J. McA. Steele et al, Proc. IEE, 122, No. 10R, 1037-1053, 1975.
(125)   J.A. Fulford and J.H. Blackwell, Rev. Sci. Inst., 27, 956-958, 1956.
(126)   P.D. Lacy and K.E. Miller, Hewlett Packard J., 9, 1-6, 1957.
(127)   A.P. Hook, Brit. Comm. and Electronics, 7, 922-923, 1960.
(128)   T. Iwase, Rept. Electrotech. Lab. (Tokyo), 28, 13-16, 1964.
(129)   R.W. Beatty, Proc. IEEE, 53, 642-643, 1965.
(130)   P.I. Somlo, Proc. IREE Australia, 30, 13-15, 1969.
(131)   T. Nemoto et al, IEEE Trans., MTT-17, 396-397, 1969.
(132)   T. Nemoto, Electronics & Communications in Japan, 52B, 68-76, 1969.
(133)   W. Larson et al, IEEE Trans., MTT-18, 112-113, 1970.
(134)   A.A. Luskow, Marconi Instrumentation, 14, 81-86, 1974.
(135)   W. Larson, IEEE Trans., MTT-12, 367-368, 1964.
(136)   W.J. Foote and R.D. Hunter, Rev. Sci. Inst., 43, 1042-1043, 1972.
(137)   P.I. Somlo and I.G. Morgan, IEEE Trans., MTT-22, 830-835, 1974.
(138)   F.L. Warner, IEE Proc., 127, Pt. H, 66-69, 1980.
(139)   I.A. Harris and F.L. Warner, IEE Proc., 128, Pt. H, 35-41, 1981.
(140)   R.W. Yell, CPEM Digest, 54-56, 1978.
(141)   R.W. Yell, IEEE Trans., IM-27, 388-391, 1978.
(142)   R.W. Yell, IEE Colloquium Digest No. 1981/49, pp. 1/1-1/5, 1981.
(143)   H. Bayer, CPEM Digest, 453-456, 1980.

(144) H. Bayer, IEEE Trans., IM-29, 467-471, 1980.
(145) Microwaves, 17, 12, March, 1978.
(146) W. Larson, NBS Monograph 144, 1975.
(147) F.L. Warner, Ch.4 in Reference 155.
(148) T. Buldbrandsen, IEEE Trans., IM-28, 59-66, 1979.
(149) B. Guldbrandsen and T. Guldbrandsen, IEE Proc., 128, Pt. H, 46-52, 1981.
(150) F.L. Warner, D.O. Watton, P. Herman and P. Cummings, CPEM Digest, 162-164, 1976.
(151) F.L. Warner, D.O. Watton, P. Herman and P. Cummings, IEEE Trans., IM-25, 409-413, 1976.
(152) F.L. Warner, D.O. Watton, P. Herman and P. Cummings, IEE Colloquium Digest No.1977/15, pp. 2/1-2/4, 1977.
(153) C.R. Ditchfield, Ch. 4 in Reference 155.
(154) J.L. Taylor, CPEM Digest, 457-462, 1980.
(155) F.L. Warner, "Microwave Attenuation Measurement", (Peter Pereginus, 1977).
(156) F.L. Warner, Electronics & Power, 27, 371-374, 1981.
(157) HP Application Note, 64-2, 1978.
(158) F.K. Weinert and B.O. Weinschel, CPEM Digest, 94-96, 1976.
(159) F.K. Weinert and B.O. Weinschel, IEEE Trans., IM-25, 298-306, 1976.
(160) B.O. Weinschel, Microwave systems News, 1, 61-68, April, 1977.
(161) B.O. Weinschel and F.K. Weinert, Microwave Systems News, 10, pp. 94, 97, 101, 102, 104-106, May, 1980.
(162) F.K. Weinert, IEEE MT&T Symposium Digest, 442-444, 1980.
(163) F.K. Weinert, CPEM Digest, 463-467, 1980.
(164) F.K. Weinert, IEEE Trans., IM-29, 471-477, 1980.
(165) F.K. Weinert, Microwave Journal, 24, pp. 51-53, 56-57, 60, 85, April, 1981.
(166) F.K. Weinert and B.O. Weinschel, IEE Colloquium Digest No. 1981/49, pp. 4/1-4/6, 1981.
(167) Microwave Journal, 21, 40-41, May, 1978.
(168) Micro-Tel Leaflet, "1295 precision attenuation measurement receiver", March, 1982.
(169) Microwaves, 21, 99, August, 1982.
(170) F.L. Warner and P. Herman, IEE Colloquium Digest No. 1981/49, pp. 2/1-2/6, 1981.
(171) F.L. Warner and P. Herman, CPEM Digest, pp. C3-C5, 1982.
(172) F.L. Warner, P. Herman and P. Cummings, IEEE Trans, IM-32, 33-37, 1983.
(173) T. Kawakami, CPEM Digest, 21-23, 1978.
(174) T. Kawakami, IEEE Trans, IM-27, 33-38, 1978.
(175) J. Kalinski, J. Microwave Power, 13, 275-281, 1978.
(176) J. Kalinski, IEEE Trans, IECI-28, 201-209, 1981.
(177) P. Pakay and A. Torok, The Radio and Electronic Engineer, 47, 315-319, 1977.
(178) P.J. Skilton, CPEM Digest, 97-99, 1976.
(179) P.J. Skilton, IEEE Trans, IM-25, 307-311, 1976.
(180) R.P. Bomer, IEE Conf. Publ. No. 146, 113-116, 1976.
(181) I.A. Harris, Appendix 8 in Reference 155.
(182) R.A. Kamper, Microwave J., 19, 39-41 and 52, April 1976.
(183) B.W. Petley, K. Morris, R.W. Yell and R.N. Clarke, Electronics Letters, 12, 237-238, 1976.
(184) B.W. Petley, K. Morris, R.W. Yell and R.N. Clarke, NPL Report DES 32, 1976.
(185) N.V. Frederick, D.B. Sullivan and R.T. Adair, IEEE Trans, MAG-13,

361-364, 1977.
(186) Microwaves, 16, 12, June, 1977.
(187) D.B. Sullivan, R.T. Adair, and N.V. Frederick, Proc. IEEE, 66, 454-463, 1978.
(188) H. Seppa, CPEM Digest, pp. R3-R5, 1982.
(189) R.H. Johnson, IEE Colloquium Digest No. 1981/49, pp. 3/1-3/4, 1981.
(190) R.H. Johnson, UK Patent GB 2054172, 26 January, 1983.
(191) R.H. Johnson and P. Cummings, IEE Colloquium Digest No. 1983/53, pp. 16/1-16/9, 1983.
(192) Microwaves, 19, 86-87, March, 1980.
(193) R.P. Coe, Microwave Journal, 24, 73-74, 76, 79, April, 1981.
(194) Microwave Systems News, 11, 107, November 1981
(195) F.L. Warner and P. Herman, IEE Colloquium Digest No. 1983/53, pp. 17/1-17/7, 1983.
(196) R.V. Gelsthorpe and R.G. Bennett, Proc. IEE, 124, 589-592, 1977.
(197) R.V. Gelsthorpe and R.G. Bennett, IEE Conf. Publ. No. 152, 113-115, 1977.
(198) J. Kalinski, Proc. 5th European Microwave Conference, 223-227, 1975.
(199) J. McA. Steele, C.R. Ditchfield and A.E. Bailey, Proc. IEE, 122, No. 10R, 1037-1053, 1975.
(200) R.J. King and R. Knudsen, Electronics Letters, 12, 560-562, 1976.
(201) R.F. Clark, IEEE Trans., IM-25, 126-128, 1976.
(202) P.I. Somlo, IEEE Trans., IM-27, 76-79, 1978.
(203) G.F. Engen, Proc. IEEE, 66, 374-384, 1978.
(204) T.S. Laverghetta, "Microwave Measurements and Techniques", (Artech House, 1976).
(205) R.J. King, "Microwave Homodyne Systems", (Peter Peregrinus, 1978).
(206) T.S. Laverghetta, "Handbook of Microwave Testing", (Artech House, 1981).
(207) I.A. Harris and F.L. Warner, Proc. IEE, 125, 902-905, 1978.
(208) H. Bayer and D. Stumpe, Metrologia, 18, 187-192, 1982.
(209) H. Bayer, Metrologia, 16, 141-147, 1980.
(210) M. Furusawa et al, J. Microwave Power, 11, 196, 1976.
(211) I. Yamaura, IEEE Trans., MTT-25, 707-710, 1977.
(212) R.A. Smith, Proc. IEE, 98, Pt. 4, 43-54, 1951.

Chapter 9

Microwave network analysers

F. L. Warner

1   INTRODUCTION

A microwave network analyser is a swept or stepped frequency equipment
that measures both the real and imaginary parts of all four s parameters
of a 2-port device. Microwave network analysers have completely revo-
lutionised the microwave measurement field. The earliest successful
equipments of this type were developed by Elliott Brothers (London) Ltd
in the mid-nineteen fifties and these have been briefly described by
Kinnear[1]. In the early nineteen sixties, work on pulse compression
and phased array radars focused attention on the phase characteristics
of microwave components and several other firms started to develop
swept frequency equipments for measuring both phase and magnitude. In
1967, two very advanced systems of this type appeared on the commercial
market. Using appropriate combinations of swept sources and resolvers,
one of these systems was designed to cover the range 100 MHz to
12.4 GHz[2] and the other one was even more ambitious, being designed
to give coverage without any gaps all the way from 2 MHz to 40 GHz[3].
Both of these systems, and the earlier Elliott ones, provide a polar
display of the magnitude and phase angle of transmission or reflection
coefficients against frequency. With a Smith chart graticule over the
polar display, the normalized input impedance or the normalized input
admittance of a network at any frequency in the band can be immediately
seen. In the network analysers introduced in 1967, rectangular displays
can also be obtained showing the variation with frequency of transmission
gain or loss in dB, transmission phase, return loss in dB or reflection
phase.

In 1972, another very advanced network analyser became commercially
available and was described by Gorss[4]. This instrument covers the
range 400 kHz to 500 MHz, has two measurement channels, a characteristic
impedance of either 50 Ω or 75 Ω at the insertion points and a dynamic
range of greater than 115 dB. In addition to displaying all of the
quantities mentioned earlier two at a time, it will also give a direct
sweep display of group delay against frequency.

All of the swept frequency systems mentioned up to now suffer to some extent from mismatch, tracking and directivity errors, which limit the overall accuracy to about ±2%. To eliminate these errors and at the same time gain many other advantages, various firms and universities developed computer-controlled microwave network analysers (5-35, 52-71). In an equipment of this type, the frequency of the signal source is controlled by the computer and is stepped instead of swept across the required bands. At the beginning of each working day, various impedance standards are connected to the equipment and the systematic internal errors are measured at each frequency of interest and stored in the computer. Measurements are then made on the unknown devices under the guidance of a stored computer programme. At each frequency, the computer corrects the raw measurement data from the device under test, using the stored information about the internal systematic errors, and then prints out and displays the corrected results.

In section 2, the various microwave and circuit configurations which are used in network analysers are described; and section 3 gives the full theory of one technique that can be used to correct for internal systematic errors in a computer controlled network analyser. Additional sections deal with: other calibration procedures, errors arising in a microwave network analyser and the evaluation of uncertainties.

2   DESCRIPTIONS OF NETWORK ANALYSERS

2.1   General   In a microwave network analyser, the signal which is

transmitted through or reflected from the device under test (called the unknown) is compared in amplitude and phase with a reference signal using:-

(a)   a single sideband system;
(b)   a twin-channel superheterodyne system;
(c)   a modulated sub-carrier system;
(d)   a vector nulling system;
(e)   a homodyne system; or
(f)   a multi-port technique.

Mixers are employed in systems (a) to (e). In a multiport network analyser, both amplitude and phase information are deduced solely from power measurements.

In this section, many different network analyser configurations that employ mixers will be described.

2.2   Single sideband systems   A simplified block diagram of the
arrangement described by Kinnear[1] in 1958 is shown in Fig. 1. This is a single-sideband system. It was developed before backward wave oscillators became commercially available; so the microwave source in it is a motor-driven reflex klystron with automatic adjustment of the reflector voltage. The amplitude of the microwave signal is kept constant with an electronically controlled waveguide attenuator. Part of the output is tapped off by the first directional coupler, changed in frequency by 200 Hz in a motor driven rotary phase shifter and then

fed into the balanced mixer. The rest of the microwave signal is fed
through a rotary attenuator and the second directional coupler to the
"unknown". External waveguide links are connected in position so that
either a fraction of the reflected signal from the unknown, or the
signal that is transmitted through it, can enter the balanced mixer.
Both the amplitude and phase of the 200 Hz signal that emerges from the
mixer are precisely related to the amplitude and phase of the microwave
signal from the unknown. The required polar display is obtained by
following the 200 Hz amplifier with two phase sensitive detectors that
have 200 Hz sine and cosine reference waveforms applied to them. The
mathematical expressions included in Fig. 1 will help to explain the
operation of this system.

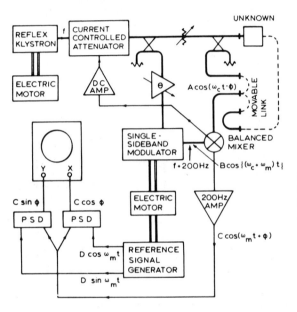

Fig. 1 Single sideband
network analyser de-
scribed by Kinnear in
1958

2.3   Twin channel superheterodyne systems   A simplified block
diagram of a non-computer-controlled twin-channel superheterodyne
system is shown in Fig. 2.

The microwave signal source is a backward wave oscillator which is
swept over the required range by the saw-tooth generator. Part of its
output is tapped off by the first coaxial coupler and fed into the
reference channel. A similar fraction of its output is tapped off by
the second coaxial coupler and applied to the device under test.
Depending upon the position of the reflection/transmission switch, the
power which is either reflected from or transmitted through the
unknown is fed into the test channel. The test and reference channels
contain identical double superheterodyne receivers, with 1st and 2nd
intermediate frequencies of 20 MHz and 200 kHz respectively. The first
local oscillator is phase locked 20 MHz away from the microwave signal
source. The phase lock loop has to be designed so that a phase locked
condition is always maintained while the BWO is swept over ranges up to

an octave wide. The signals entering the mixers from the test and
reference channels are kept about 30 dB below the signals from the
first local oscillator. Under these conditions, the two IF signals
have the same relative amplitudes and phases as the microwave reference
and test signals. The AGC system keeps the signal level constant at the
output of the reference channel and varies the gain of both 20 MHz IF
amplifiers by the same amount. This arrangement prevents the results
being affected by power variations in the microwave signal source. The
various displays mentioned in the introduction are obtained by using a
peak detector, a log converter[36], a phase detector, a 90° phase
shifter and two phase-sensitive detectors. The mathematical expressions
in Fig. 2 will clarify the operation of these circuits. The 200 kHz
switchable attenuator shown in Fig. 2 is used to avoid overloading on
strong signals. A network analyser of this type described by Anderson
and Dennison[2] gives an attenuation measurement accuracy of about
±0.1 dB per 10 dB and, in a 6 hour stability test, the total drift did
not exceed 0.05 dB and 0.2°.

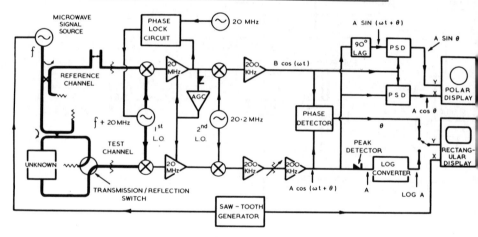

TWIN CHANNEL SUPERHETERODYNE NETWORK ANALYSER

Fig. 2

In a computer-controlled network analyser of this type[5, 6, 10] a
punched paper tape programme is fed into the computer and instructions
are then given to the operator by a teleprinter. The microwave
oscillator frequencies and the reflection/transmission switch are
controlled by the computer in accordance with information typed on a
keyboard by the operator. The 200 kHz attenuator is automatically
controlled by the computer so that the phase-sensitive detectors always
operate under favourable conditions. The output signals from these two
detectors are fed into the computer sequentially via an analogue-to-
digital converter.

When the primary microwave source for a network analyser is a backward
wave oscillator, the frequency accuracy is about ±0.25%. In the more

expensive computer-controlled network analysers, the frequency accuracy
is improved several orders of magnitude by deriving the microwave signals
from a computer-controlled frequency synthesizer which starts with a
quartz oscillator having a stability of a few parts in 10^9 per day. This
superior type of source enables precise measurements to be made on high-
Q cavities and gives a general improvement in repeatability.

By calibrating a network analyser with impedance standards and then
using the computer to eliminate the internal systematic errors, a large
improvement in the overall accuracy is obtained. To give an example,
two 60 dB couplers were measured as carefully as possible against the
British national standard and the mean attenuation values obtained were
58.865 dB and 59.564 dB. They were then measured on a computer
controlled network analyser at RSRE and the results obtained were
58.84 dB and 59.56 dB respectively.

In addition to controlling and correcting the measurements, the
computer can also be used to carry out mathematical operations on the
output data, typical examples being:-

(a)   conversion of s parameters to h, y or z parameters;
(b)   calculation of the gains obtainable from microwave transistors
       whose s parameters have been measured;
(c)   conversion of reflection coefficients to standing wave ratios;
(d)   calculation of group delay from adjacent points in the phase/
       frequency characteristic.

The system shown in Fig. 2 resembles the Hewlett Packard 8410 110 MHz
to 12.4 GHz non-computer-corrected network analyser that was introduced in 1967[2]. The first mini-computer corrected network analyser -
the HP 8540A - was sold in 1967. An improved version of it - the
HP 8542A - covering the range 110 MHz to 18 GHz sold well over the
period 1969 to 1973[10-13]. This was followed by the HP 8542B which
had better phase sensitive detectors, improved software and a graphics
display system. In 1977, a less expensive semi-automatic computer-
corrected network analyser - the HP 8409A - was introduced[68].
This model contains solid state swept sources, an HP 9825A desktop
controller and an IEEE-488 interface bus. Further upgrading led in
1980 to the HP 8409B which employs an improved error correction
technique[56] and an HP 9845 desktop controller. In 1981 the HP 8409C,
equipped with a digital main frame sweep oscillator, was placed on the
market. Lastly, in 1981, a low cost 0.5-18 GHz computer corrected
network analyser - the HP 8408A - appeared[69, 70]. This uses an
HP 85 controller and a new test set that is switched manually between
transmission and reflection.

Workers in other firms have reported many improvements to HP microwave
network analysers and five of them will now be briefly mentioned.
W. Kennan of Avantek has extended the upper frequency limit of his
HP 8409B from 18 GHz up to 40 GHz[71]. Fig. 3 shows the arrangement
that he uses to cover the range 26.5 to 40 GHz. RF signals from the
HP 8409B covering the range 4.5-18 GHz are up-converted to cover the
range 26.5-40 GHz by mixing them with a 44.5 GHz signal from a
temperature stabilized Gunn oscillator. After transmission through or
reflection from the device under test, the Q band signal is converted
back again to its original frequency range (4.5-18 GHz) using the same

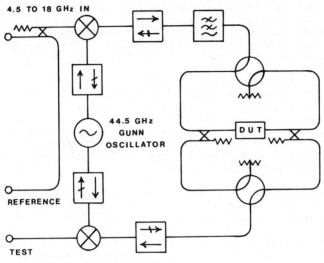

local oscillator
as before. The
reference channel
signal is fed
straight into the
network analyser.
The up-converter
and down-converter
used in this
system were
supplied by
Honeywell-
Spacekom.

In 1976, Yamaguchi
et al of TRW gave
a description of
a network analyser
that covers the
frequency range
55-65 GHz[72-74].
This is shown in
Fig. 4.

Fig. 3   Circuit used by Kennan with a HP 8409B
         to obtain coverage over the range
         26.5 - 40 GHz

A backward wave
oscillator sweeps
through the range

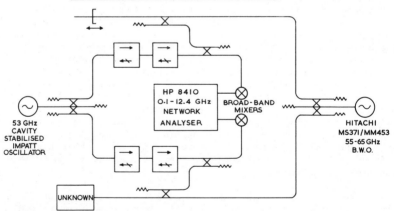

Fig. 4

55-65 GHz and a 53 GHz cavity stabilized IMPATT oscillator provides the
local oscillator signal. The IF outputs from the two broad band millimetre wave mixers thus sweep through the band 2-12 GHz. These IF signals
are fed straight into an HP 8410 network analyser, which phase locks
successfully throughout the entire 10 GHz sweep. The symmetrical wave-

guide configuration shown in Fig. 4 is only suitable for reflection coefficient measurements. The four ferrite isolators eliminate cross talk
between the test and reference channels and the sliding short is used to
equalize the lengths of the two paths.

B Perlman et al of RCA have made their HP 8409B more versatile and more
interactive by replacing the HP 9845 controller with an HP 1000F
minicomputer and adding to it a colour graphics display system[75].

The dynamic range of an HP 8409B at TRW has been raised from 60 dB to
80 dB by replacing the normal sources with low-phase-noise synthesized
sources and lengthening the averaging times[76].

D E Bradfield[77] has modified an HP 8410 so that transmission and reflection results can be observed simultaneously. This has been achieved
by using two PIN diode switches to connect the receiver to the reflected
signal on odd-numbered sweeps and to the transmitted signal on even-
numbered sweeps. The two PIN diodes are driven by the Q and $\bar{Q}$ outputs
from a JK flip-flop which is triggered by the blanking pulses from the
saw-tooth generator.

In a computer-
operated measurement
system developed at
the Bell Telephone
Laboratories by Davis
et al[19], circuit
drift errors are
eliminated by switch-
ing back and forth at
a 30 Hz rate between
the unknown and a
stable microwave
reference path. A
simplified block
diagram of this
system is shown in
Fig. 5.

COMPUTER OPERATED TWIN-CHANNEL SUPERHETERODYNE NETWORK ANALYSER

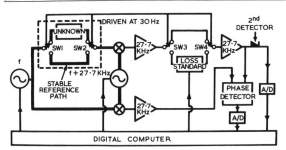

Fig. 5

MICROWAVE SWITCHING SYSTEM USED BY DAVIS,
HEMPSTEAD, LEED AND RAY.

Fig. 6

The loss standard
contains relay-
switched precision
attenuators which
cover the range 0 to
60 dB in 0.01 dB
steps. Switches SW3
and SW4 are driven in
synchronism with
switches SW1 and SW2.
The detected 27.7
kHz signal is fed,
via an analogue-to-
digital converter,
into the computer
which automatically
adjusts the loss
standard until the

30 Hz component in the 2nd detector output is almost zero. Thus, this arrangement is basically a driftless automatic IF substitution system. The microwave switching is performed with a 17½" diameter rotating vane. The part of the system which is enclosed by the dotted rectangle in Fig. 5 is shown in much greater detail in Fig. 6. The mechanical method of switching was adopted as it was found to be more stable and more reproducible than any electronic method of switching. This computer-controlled network analyser gives an amazingly good performance. Accuracies of transmission parameters for small loss are ±0.002 dB, ±0.01° and ±0.1 nS. This equipment is reported to be 10 times slower than the one described by Hackborn[5]. However, microwave measurements can be carried out on it 300 times faster than they could by using point-by-point techniques.

The twin-channel superheterodyne network analyser has enjoyed great success for 16 years but is now being challenged by the multi-port network analyser, which needs only one microwave source and offers higher accuracy with less frequent calibration. Nevertheless, the superheterodyne method still retains the following advantages:-

(a)   faster results;
(b)   greater dynamic range;
(c)   no harmonic errors;
(d)   provision of a fast-sweep quick-look capability;
(e)   no problems when measuring devices with gain.

At the present time, the total cost of the multi-port network analyser sold by Norsal is similar to that of the HP 8408A.

2.4   Modulated Sub-Carrier Systems   A simplified block diagram of a modulated sub-carrier network analyser[3] is shown in Fig. 7.

In this system, which stems from a paper by Cohn and Oltman[37], the microwave signal that is applied to the unknown is amplitude modulated at 200 kHz or thereabouts. The two signals which are applied to the left hand balanced mixer can be written as $E(1 + m \sin \omega_m t) \sin(\omega_c t + \theta)$ and $V \sin \omega_c t$. Combining these two signals vectorially, the amplitude of the resultant is found to be:-

$$V_R = \{V^2 + E^2(1 + m \sin \omega_m t)^2 + 2VE (1 + m \sin \omega_m t) \cos \theta\}^{\frac{1}{2}} \quad \ldots(1)$$

$$= V\{1 + r^2 + 2r \cos \theta\}^{\frac{1}{2}} \quad \ldots(2)$$

where

$$r = \frac{E}{V} (1 + m \sin \omega_m t) \quad \ldots(3)$$

In practice r is deliberately made much smaller than unity; so, neglecting r^2 in comparison with unity and then expanding by the binomial theorem and neglecting all terms except the first two, we get:-

$$V_R \approx V\{1 + r \cos \theta\} \approx V + E \cos \theta + Em \cos \theta \sin \omega_m t \quad \ldots(4)$$

Thus, after linear mixing with a conversion factor η and passage through a tuned amplifier with a gain G, the output of the upper channel is seen to be Em Gηcos θ sin $\omega_m t$ which can be written more simply as A cos θ sin $\omega_m t$.

Due to the presence of the quadrature splitter, the two signals which are applied to the right hand balanced mixer can be written as $E(1 + m \sin \omega_m t) \sin(\omega_c t + \theta)$ and V cos $\omega_c t$. Proceeding now in the same way as before, the output of the lower channel is found to be A sin θ sin $\omega_m t$. Thus, the two signals required for the polar display can be obtained very easily in this case by employing two phase sensitive detectors. The mathematical expressions in Fig. 4 will help to explain how log A and θ are obtained in this type of network analyser. A modulated sub-carrier network analyser has the great advantage of requiring only one microwave source and there is thus no need for the elaborate phase lock loop which is an essential part of a twin channel superheterodyne system. This advantage becomes even greater at millimetre wavelengths where electronically tunable sources are extremely expensive.

In 1967, the Wiltron Company[3] introduced a series of modulated sub-carrier network analysers with block diagrams similar to that shown in Fig. 7. Sufficient units were introduced to give coverage from 2 MHz to

MODULATED SUB-CARRIER NETWORK ANALYSER

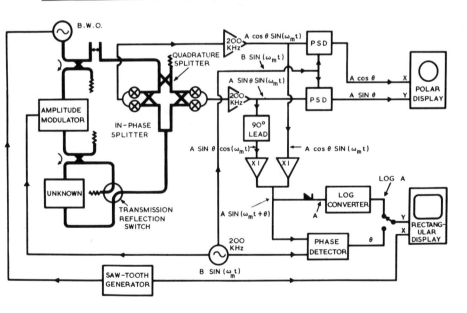

Fig. 7

40 GHz. In 1970, Wiltron brought out a computer programmed modulated sub-carrier network analyser for use with a time sharing computer terminal[14]. The Wiltron network analysers were not as successful commercially as those offered by HP and manufacture of them

was terminated in 1978. (Coverage of the range 0.1-18 GHz needed 8
Wiltron test sets but, with the HP systems, only 2 are required).

In the last few years, there has been a renewed interest in millimetre
waves and, with this part of the spectrum in mind, M A Wood of RSRE has
recently made a further detailed study of the modulated sub-carrier
network analyser[78-80]. He carried out his experimental work at 35 GHz,
often using the configuration shown in Fig. 8, which contains only one
balanced mixer. A 0° or 90° phase shift can be obtained in the carrier
channel by setting the upper waveguide switch to the appropriate
position. Several precise calibration procedures for the arrangement
shown in Fig. 8 have been developed by M A Wood.

M.A. WOOD'S VERSION OF THE MODULATED SUB-CARRIER NETWORK ANALYSER

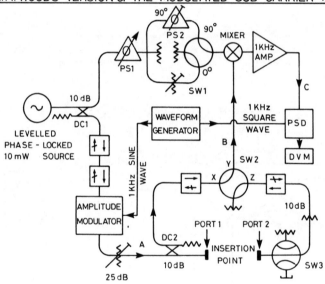

Fig. 8

Using one of these procedures, he calibrated a precision rotary vane
attenuator at 35 GHz over a 45 dB range and all of his results fell
wthin ±0.014 dB of those obtained on the configuration used for national
standards level measurements. Again operating at 35 GHz, measurements of
reflection coefficient modulus, $|\Gamma|$, were carried out at RSRE on three
1-port devices using both a calibrated homodyne network analyser and a
tuned reflectometer. The results obtained are given in Table 1.

Thus, M A Wood has demonstrated that a calibrated modulated-sub-carrier
network analyser is capable of giving very accurate results.

2.5 Vector nulling systems Over the period 1971 to 1976, a very
sophisticated 0.1 to 12.4 GHz microwave network analyser was developed
at NBS[81]. The aim was to make an equipment with an accuracy 10 times

Load number	|Γ| measured with homodyne system	|Γ| measured with tuned reflectometer
1	0.479 ± 0.002	0.481 ± 0.004
2	0.0941 ± 0.002	0.0937 ± 0.004
3	0.0107 ± 0.004	0.0108 ± 0.004

TABLE 1

better than that of commercially available automatic network analysers. Fig. 9 shows a grossly simplified block diagram of this NBS equipment. It contains two 0.1 to 12.4 GHz stepped microwave sources, which are phase locked 100 kHz apart. The output from one of these sources is fed to the unknown and either the signal that is reflected from it or the

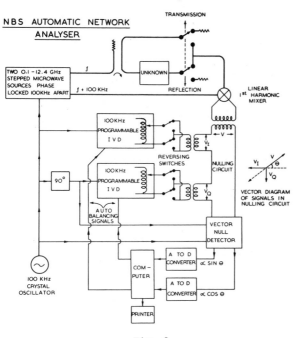

signal that is transmitted through it is mixed with the other microwave signal. The 100 kHz output from the mixer is automatically nulled by in-phase and quadrature 100 kHz signals from two computer-controlled inductive voltage dividers (see the vector diagram in Fig. 9). The reversing switches enable a null to be obtained when the mixer output lies in any quadrant. Careful intercomparisons at 3 GHz against the best attenuator calibrators at NBS have shown that this network

Fig. 9

analyser is accurate to within ±0.01 dB at 50 dB.

The same principle is used in the Weinschel VM-4A, which has already been described in the Attenuation Lecture Notes for this Vacation School.

2.6   Homodyne systems   In 1977, an easily implementable homodyne network analyser was described by Watanabe and Ashiki[82] of Shizuoka University in Japan. This is shown in Fig. 10. The output from the microwave source is split between two channels. The lower channel contains the unknown and a binary 90° phase shifter, that consists of a ferrite circulator, a fast P-I-N diode switch and a short circuit spaced

$\lambda_g/8$ behind the diode. The two signals are recombined in a balanced mixer. The real and imaginary parts of the unknown's transmission co-efficient, S_{21}, are measured sequentially and separated from each other by sample-and-hold circuits. The necessary switching circuits and their waveforms are shown in detail in Fig. 10. It is seen that the P-I-N diode switch is driven by a 200 kHz square wave. Since the microwave frequency may be swept, very rapid sampling is necessary and 200 kHz was chosen because it was close to the highest frequency at which all of the associated electronic circuits would perform satisfactorily. To avoid cross talk between the two output signals, it is seen from Fig. 10 that the samples are taken from the central parts of the non-phase-shifted and 90°-phase-shifted periods.

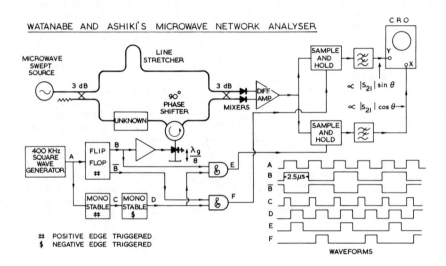

Fig. 10

When the unknown has a complex transmission coefficient denoted by $|S_{21}|e^{j\theta}$ and the line stretcher is correctly set, it is easily shown that the filtered outputs from the two sample and hold circuits are proportional to $|S_{21}|\cos\theta$ and $|S_{21}|\sin\theta$. Application of these two output signals to the X and Y plates of an oscilloscope immediately gives a polar display of the unknown's transmission coefficient.

Watanabe and Ashiki[82] tried out this very simple network analyser at X band and it performed as expected. A 4 GHz version of it, containing a broad band 90° phase shifter, was assembled and assessed at RSRE by M A Wood[78]. The results he obtained with it agreed very closely with the corresponding results yielded by an HP 8542A twin channel super-heterodyne network analyser. However, the DC component in the mixer output, drift in the DC amplifier and $1/f$ noise all proved to be trouble-some and, in later work, he preferred the modulated sub-carrier configuration shown in Fig. 8.

Further work on homodyne network analysers has been reported by Ostwald and Schiek[83-84].

SIGNAL FLOW GRAPHS FOR REFLECTION

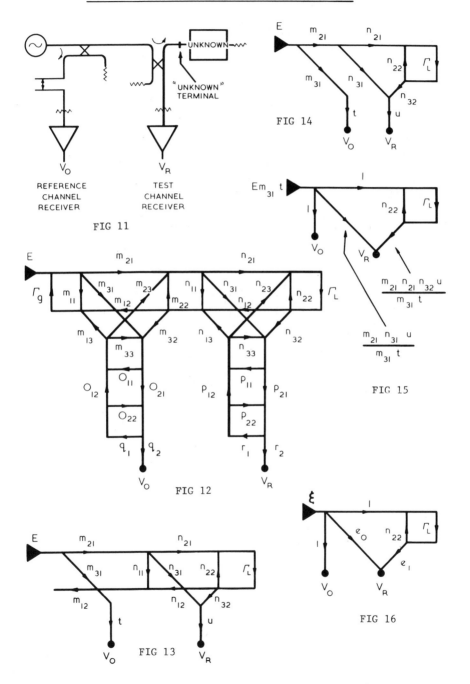

FIG 11

REFERENCE
CHANNEL
RECEIVER

TEST
CHANNEL
RECEIVER

FIG 14

FIG 15

FIG 12

FIG 13

FIG 16

SIGNAL FLOW GRAPHS FOR TRANSMISSION

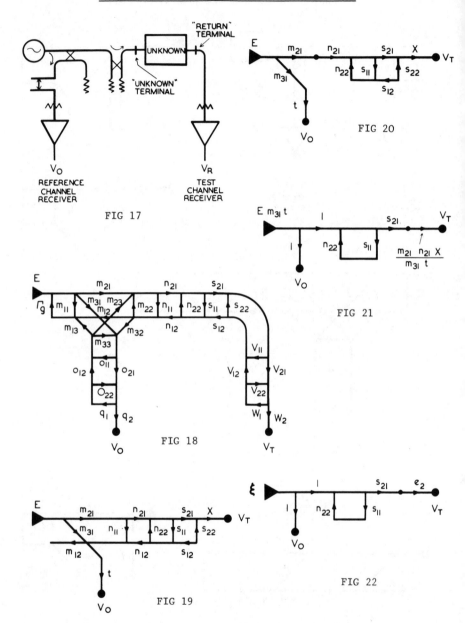

FIG 17

FIG 18

FIG 19

FIG 20

FIG 21

FIG 22

3. ELIMINATION OF INTERNAL SYSTEMATIC ERRORS

This section describes how the internal systematic errors are eliminated in a computer-controlled network analyser of the twin-channel super-heterodyne type. Firstly, let us consider the case when the reflection/transmission switch is set for reflection measurements. Then, for the purpose of this analysis, the block diagram given in Fig. 2 can be simplified to the configuration shown in Fig. 11.

Assuming that there are no leakage paths and also that the unused port of each directional coupler is perfectly matched, the signal flow graph[45-51] for Fig. 11 is shown in Fig. 12. This signal flow graph contains 24 unknowns and is difficult to analyse; so drastic simplifications will now be made. Let us assume that perfect matches are seen looking back into the source ($\Gamma_g = 0$) and downwards into O_{11} and p_{11}. Also, let us assume that the first coupler is perfectly matched at each port ($m_{11} = m_{22} = m_{33} = 0$) and has infinite directivity ($m_{32} = 0$). Then Fig. 12 simplifies to Fig. 13. From the rules for simplifying signal flow graphs[49, 51], it follows that:

$$t = \frac{O_{21}q_2}{1 - O_{22}q_1} \qquad \ldots (5)$$

$$u = \frac{p_{21}r_2}{1 - p_{22}r_1} \qquad \ldots (6)$$

Eliminating the paths which lead only to the matched source, Fig. 13 simplifies to Fig. 14. Using the non-touching loop rule[45, 47] the outputs from the reference and test channels are seen from Fig. 14 to be $Em_{31}t$ and

$$Em_{21} \left\{ n_{31}u + \frac{n_{21}\Gamma_L n_{32}u}{1 - n_{22}\Gamma_L} \right\} \text{ respectively.}$$

Fig. 14 still contains too many unknowns but it can be redrawn as shown in Fig. 15. Again using the non-touching loop rule, it is easily seen that Fig. 15 gives exactly the same output voltages as Fig. 14.

Let $Em_{31}t = \xi$, $\dfrac{m_{21}n_{31}u}{m_{31}t} = e_o$ and $\dfrac{m_{21}n_{21}n_{32}u}{m_{31}t} = e_1$ (7) (8) (9)

Then, we finish up with the signal flow graph shown in Fig. 16 which contains only 3 unknowns.

When the reflection/transmission switch is set for transmission measurements, we can use the simplified block diagram shown in Fig. 17. The signal flow graph for Fig. 17 is shown in Fig. 18, where s_{11}, s_{21}, s_{12} and s_{22} are the 4 scattering parameters describing the unknown. Proceeding now in a similar way as before, Fig. 18 can be progressively simplified to Fig. 22, where

$$x = \frac{v_{21}w_2}{1 - v_{22}w_1} \quad \text{and} \quad e_2 = \frac{m_{21}n_{21}x}{m_{31}t} \qquad (10), \ (11)$$

Thus, altogether, we finish up with 4 internal unknowns, e_o, e_1, e_2 and n_{22} whose values must be determined at each frequency of interest, before the s parameters of the unknown can be calculated.

The values of these 4 internal unknowns are found by carrying out a 4 stage calibration process.

From Fig. 16 we see that:-

$$\frac{V_R}{V_o} = e_o + \frac{\Gamma_L e_1}{1 - n_{22}\Gamma_L} \qquad \ldots (12)$$

and from Fig. 22, it follows that:-

$$\frac{V_T}{V_o} = \frac{s_{21}e_2}{1 - n_{22}s_{11}} \qquad \ldots (13)$$

The four stages of the calibration process are as follows:-

(a)   Reflection with matched termination   First of all, Γ_L is made equal to zero by placing a high class matched termination on the "unknown" terminal. Then, from equation (12), we get:-

$$\frac{V_R}{V_o} = M_1 = e_o \qquad \ldots (14)$$

(b)   Reflection with direct short   Secondly, Γ_L is made equal to -1 by placing a direct short circuit on the "unknown" terminal. From equation (12) we now get:-

$$\frac{V_R}{V_o} = M_2 = e_o - \frac{e_1}{1 + n_{22}} \qquad \ldots (15)$$

(c)   Reflection with offset short   The next operation involves placing an offset short of length, ℓ, on the "unknown" terminal. We then have

$$\Gamma_L = -e^{-j2\beta\ell} = \Gamma_s \qquad \ldots (16)$$

Substituting this value into equation (12), we get

$$\frac{V_R}{V_o} = M_3 = e_o + \frac{\Gamma_s e_1}{1 - n_{22}\Gamma_s} \qquad \ldots (17)$$

Knowing the values of ℓ and the phase constant β, the value of Γ_s is easily calculated by the computer at each frequency of interest.

(d)   Transmission with through connection   Finally, the "unknown" terminal is connected directly to the "return" terminal and a transmission measurement is made. In this case, $s_{11} = s_{22} = 0$ and $s_{12} = s_{21} = 1$; so, from equation (13) it follows that:-

$$\frac{V_T}{V_o} = M_4 = e_2 \qquad \qquad \dots (18)$$

From equations (14), (15), (17) and (18) we get

$$e_o = M_1 \qquad \qquad \dots (19)$$

$$e_1 = \frac{(M_1 - M_2)(M_3 - M_1)(1 + \Gamma_s)}{\Gamma_s(M_3 - M_2)} \qquad \qquad \dots (20)$$

$$e_2 = M_4 \qquad \qquad \dots (21)$$

$$n_{22} = \frac{(M_3 - M_1) + \Gamma_s(M_2 - M_1)}{\Gamma_s(M_3 - M_2)} \qquad \qquad \dots (22)$$

When an unknown 2 port is connected to the "unknown" terminal and a perfectly matched termination is placed behind it, $\Gamma_L = s_{11}$. Hence, from equation (12) we get:-

$$\frac{V_R}{V_o} = M_{rf} = e_o + \frac{s_{11}e_1}{1 - n_{22}s_{11}} \qquad \qquad \dots (23)$$

Rearranging, we get

$$s_{11} = \frac{M_{rf} - e_o}{e_1 + n_{22}(M_{rf} - e_o)} \qquad \qquad \dots (24)$$

When a transmission measurement is made with the unknown 2-port in position:

$$\frac{V_T}{V_o} = M_{tf} = \frac{s_{21}e_2}{1 - n_{22}s_{11}} \qquad \qquad \dots (25)$$

Thus $s_{21} = \dfrac{M_{tf}(1 - n_{22}s_{11})}{e_2} \qquad \qquad \dots (26)$

The other two s parameters, s_{22} and s_{12}, can now be found in a similar manner by reversing the unknown and repeating both the reflection and transmission measurements.

All of the parameters used in this section are complex quantities having both real and imaginary parts. This fact makes the computers task much more complicated than it may appear to be at first sight.

4. OTHER CALIBRATION PROCEDURES

The procedure described in section 3 is only one of many that can be used. In the procedure described by Hand[12], a leakage path is included and it is not assumed that there is a perfect match looking into the receiver. Thus, there are 6 unknowns in the final signal flow graphs and a 6 stage calibration procedure has to be carried out. Instead of using a fixed matched termination, a sliding one is employed and this is measured at 4 different positions. The computer then determines the centre of the circle that passes through the 4 measured values. This operation eliminates errors that would otherwise be caused by the small residual reflection of the sliding termination. After completing this operation, the other 5 stages of the calibration procedure are: reflection with a direct short, reflection with an offset short, reflection and transmission with a through connection and, finally, transmission without a through connection (both measurement ports terminated). This 6 stage calibration procedure results in quite complicated equations which were solved by iterative processes in the early computer-corrected microwave network analysers made by Hewlett Packard. However, Kruppa and Sodomsky[16], Davis, Doshi and Nagenthiram[27] and Rehnmark[28] have shown that explicit solutions can be found for these equations. By using explicit formulae instead of iterative methods, the computational efficiency can be improved.

Instead of performing the calibration with a matched termination, a direct short and an offset short, it can be carried out with:-

(a)   a direct short and two offset shorts;
(b)   a direct short, an offset short and an open circuit;
(c)   a matched termination, a direct short and a standard mismatch;
(d)   a direct short, an offset short and a sliding termination whose reflection coefficient need not be known;
(e)   a direct short and two sliding terminations - the first with a high reflection coefficient and the second with a low one, neither of which need be known.

Method (a) was proposed by Silva and McPhun[20] and it is particularly suitable for microstrip work and also for tasks where space restrictions rule out the use of a precision matched sliding termination. Care must be exercised when choosing the offset lengths as certain values can cause overloading in the computer.

Method (b) was devised by Shurmer[22] for carrying out measurements on microwave transistors mounted in a microstrip jig. The open circuit is produced by simply leaving the "transistor well" empty. A discontinuity capacitance is then present but this can be taken into account[38].

Gould and Rhodes[26] described method (c). Their standard mismatch is a commercially-available item with a V.S.W.R. of 1.5 ±5%, over the band 1 to 4 GHz. This calibration technique does not require precise measurements of the signal frequency as the three standards are frequency insensitive and they are all used at the same plane.

Methods (d) and (e) were devised by Kasa[31] and refined by Engen[34]. Further calibration procedures have been described recently by Franzen and Speciale[85, 86] and Silva and McPhun[87-89]. The latest HP network

analysers employ a 12 term universal error model described by
Fitzpatrick[56].

Woods[35] has pointed out that the calibration procedure described by
Hand[12] overlooks the small impedance changes that occur when the
reflection/transmission switches are changed over. To avoid this
shortcoming, he has suggested that these switches should be eliminated
by using 3 receiving channels. He has carried out a completely rigorous
analysis of this proposed new scheme. By adopting it the random errors
caused by non-repeatability of the switches would be eliminated and
two 10 dB pads would become unnecessary; so the dynamic range would be
improved. Three 1-port standards would be needed to carry out the
calibration.

5.   ERRORS ARISING IN A MICROWAVE NETWORK ANALYSER

Several sources of error have already been mentioned and a complete
list for a twin channel superheterodyne network analyser is given
below:

(a)   imperfect matching at the insertion point
(b)   non-infinite directivity in the couplers;
(c)   departures from perfect tracking in the two channels;
(d)   leakage;
(e)   noise;
(f)   imperfections in the calibration standards (these are caused by
       incorrect diameters for the inner and outer conductors, an error
       in the length of the offset short, contact resistances between the
       inner and outer conductors, eccentricity in the coaxial lines,
       line losses and discontinuity capacitances);
(g)   imperfections in the instrumentation (these include frequency
       errors, power variations in the microwave signals, non-repeat-
       ability in the microwave switches, non-linearity in the mixers,
       gain and phase drifts in the IF amplifiers, DC drift in the AGC
       loop, errors in the 200 kHz attenuator, quadrature errors in the
       phase sensitive detectors, errors in the analogue to digital
       converter, etc);
(h)   impedance changes that occur when the reflection/transmission
       switches are changed over;
(i)   connector uncertainties.

In a computer-corrected microwave network analyser, we have seen that
errors (a), (b) and (c) are eliminated. If leakage paths are included in
the signal flow graphs, error (d) is also eliminated. Error (h) was
discussed in Section 4 and error (i) can be minimised by using precision
coaxial connectors such as the GR 900 or APC-7. Drift errors can be
eliminated by using the technique described by Davis et al[19]. Thus,
if all possible steps are taken, we are still left with errors (e) and
(f) and various instrumental errors.

Adam[6] has given equations for the magnitude and phase uncertainties
that occur when making reflection measurements on an HP 8542A network
analyser with phase locked sources. These are as follows:-

$$\text{magnitude uncertainty} = \pm\ (0.0015 + 0.005|s_{11}| + 0.003|s_{11}|^2) \qquad \ldots (27)$$

angle uncertainty $= \pm \{0.25^{\circ} + \tan^{-1} \frac{0.0015}{|s_{11}|} + 4 \tan^{-1}(0.005|s_{11}|)\}$ ...(28)

Hand[12] has made a detailed study of the errors caused by (e), (f) and
(g) in the HP 8542A. He used a random number generator to assign random
phase angles between O and 2π and random magnitudes (between specification limits) to all of the variables that contribute to these errors.
He fed these quantities into a computer programme that yields a set of
calculated s-parameters which are compared with the input values to
determine the errors. This process is repeated 100 times, with new
random magnitudes and phase angles each time. The ninth worst error is
then printed out and this gives the measurement uncertainty with a 96%
confidence level. For the results thus obtained, Hand's paper[12]
should be consulted.

Ridella[23, 30] has derived equations for the calibration repeatability,
the measurement repeatability, the residual mismatch etc; and he has also
reported the results that were obtained when various devices were
measured on 4 different automatic network analysers.

6. EVALUATION OF UNCERTAINTIES

A computer-corrected microwave network analyser is an extremely complex
instrument and, occasionally, users wonder if a small hidden fault is
producing any errors in the thousands of results that it prints out each
hour. A way of restoring faith is to measure on it each week a precisely
calculable 2 port standard and see if there are any significant differences between the measured and theoretical values. Four different
devices that have been considered for this role are:

(a)   a half-round inductive obstacle in a short length of rectangular
      waveguide[39];
(b)   a capacitive rod in a short length of rectangular waveguide[40];
(c)   a section of coaxial line or waveguide, a quarter wavelength long
      at mid-band, with a characteristic impedance different from the
      normal value[41, 42, 43].
(d)   a short length of dielectric-filled normal-sized waveguide[90].

Item (c) is the one that has been most
widely used. Due to length restrictions,
only the reflection characteristics of
the waveguide version of item (c) will
be discussed here. Consider the
configuration shown in Fig. 23, where
h_n denotes the normal waveguide height
and h_r and ℓ denote the height and
length of the 2-port standard. The
discontinuity capacitances must be
taken into account and their susceptances are denoted by B. Then,
neglecting losses, the admittance
looking into the 2-port standard is
seen to be:-

REDUCED HEIGHT WAVEGUIDE
REFLECTION COEFFICIENT STANDARD

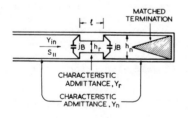

Fig. 23

$$Y_{in} = jB + Y_r \frac{(Y_n + jB) + jY_r \tan \theta}{Y_r + j(Y_n + jB) \tan \theta}$$ ...(29)

and

$$s_{11} = \frac{Y_n - Y_{in}}{Y_n + Y_{in}} = - \frac{1 - \alpha^2 - \alpha^2 b^2 + \dfrac{2\alpha b}{\tan \theta}}{1 + \alpha^2 - \alpha^2 b^2 + \dfrac{2\alpha b}{\tan \theta} - j\left(\dfrac{2\alpha}{\tan \theta} - 2\alpha^2 b\right)} \quad \dots(30)$$

where $\alpha = \dfrac{h_r}{h_n} = \dfrac{Y_n}{Y_r}$, $b = \dfrac{B}{Y_n}$ and $\theta = \dfrac{2\pi \ell}{\lambda_g}$ \hfill (31) (32) (33)

It is now a straight-forward matter to split equation (30) into real and imaginary parts or find expressions for $|s_{11}|$ and the reflection phase angle.

Marcuvitz[44] gives the following approximate expression for the normalized discontinuity susceptance:-

$$b \approx \frac{2h_n}{\lambda_g} \left\{ \frac{\alpha^2 + 1}{2\alpha} \log_e \frac{1 + \alpha}{1 - \alpha} + \log_e \frac{1 - \alpha^2}{4\alpha} + \frac{2}{A} \right\} \quad \dots(34)$$

where

$$A = \left(\frac{1 + \alpha}{1 - \alpha}\right)^{2\alpha} \cdot \frac{1 + \sqrt{1 - (h_n/\lambda_g)^2}}{1 - \sqrt{1 - (h_n/\lambda_g)^2}} - \frac{1 + 3\alpha^2}{1 - \alpha^2} \quad \dots(35)$$

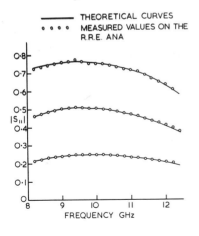

CALCULATED AND MEASURED VALUES
OF $|S_{11}|$ VERSUS FREQUENCY FOR
THREE 2-PORT TRANSFER STANDARDS

——— THEORETICAL CURVES
• • • • MEASURED VALUES ON THE
R.R.E. ANA

Fig. 24

Fig. 24 shows calculated and measured values of $|s_{11}|$ versus frequency for three 2-port transfer standards of this type. The theoretical curves were obtained from equations (30) to (35). The measured values were obtained on the HP 8542A automatic network analyser at RSRE.

7. CONCLUSIONS

The type of network analyser shown in Fig. 2 provides complete magnitude and phase information about one and two port networks. This information is provided almost instantaneously over an enormous frequency range, with an accuracy in the region of ±2%. Computer-controlled microwave network analysers are even more exciting instruments. By the use of very ingenious calibration and computer correction techniques, these automatic network analysers are able to give results which are almost as accurate as those obtained in the best microwave standards laboratories. The speed with which measurements can be made by a computer-controlled network analyser is almost unbelievable. Tasks which would take several days to perform using point-by-point techniques can be completed in a few minutes on them. The computer-controlled network analyser has revolutionized the microwave measurement field and is an excellent example of successful automation.

8. REFERENCES

(1)   J A C Kinnear, Brit. Commun. & Electronics, 5, 359-361, 1958.
(2)   R W Anderson and O T Dennison, HP Journal, 18, 2-10, 1967.
(3)   - Wiltron Company Catalogue, 2-5, 1968.
(4)   C G Gorss, IEEE Trans, IM-21, 538-543, 1972.
(5)   R A Hackborn, Microwave Journal, 11, 45-52, 1968.
(6)   S F Adam, IEEE Trans, IM-17, 308-313, 1968.
(7)   H V Shurmer, Electronics Letters, 5, 209, 1969.
(8)   W J Geldart et al, BSTJ, 48, 1339-1381, 1969.
(9)   S F Adam, "Microwave Theory and Applications", Prentice Hall, 351-457, 1969.
(10)   D K Rytting and S N Sanders, HP Journal, 21, 2-10, 1970.
(11)   W A Ray and W W Williams, HP Journal, 21, 11-15, 1970.
(12)   B P Hand, HP Journal, 21, 16-19, 1970.
(13)   B Humphries, HP Journal, 21, 20-24, 1970.
(14)   - Wiltron Company Catalogue,20-23, 1970.
(15)   H V Shurmer, Electronics Letters, 6, 733-734, 1970.
(16)   W Kruppa and K F Sodomsky, IEEE Trans, MTT-19, 122-123, 1971.
(17)   H V Shurmer, IEE Colloquium Digest 1971/13, pp. 4/1-4/6, 1971.
(18)   H V Shurmer, The Radio and Electronic Engineer, 41, 357-364, 1971.
(19)   J B Davis et al, IEEE Trans, IM-21, 24-37, 1972.
(20)   E F Da Silva and M K McPhun, Electronics Letters, 9, 126-128, 1973.
(21)   R T Davis, Microwaves, 12, 40-44 and 77, 1973.
(22)   H V Shurmer, Electronics Letters, 9, 323-324, 1973.
(23)   S Ridella, Proc European Microwave Conference, paper B.3.4, 1973.
(24)   J K Fitzpatrick, Microwave J, 16, 65-68, 1973.
(25)   S Arnold and I Whiteley, IEE Conf Publication No. 103, 125-130, 1973.
(26)   J W Gould and G M Rhodes, Electronics Letters, 9, 494-495, 1973.
(27)   O J Davies et al, Electronics Letters, 9, 543-544, 1973.
(28)   S Rehnmark, IEEE Trans, MTT-22, 457-458, 1974.
(29)   R W Beatty, Microwave J, 17, 45-49 and 63, 1974.
(30)   S Ridella, CPEM Digest, 51-53, 1974.
(31)   I Kasa, CPEM Digest, 90-92, 1974.
(32)   W E Little, CPEM Digest, 331-332, 1974.
(33)   I Kasa, IEEE Trans, IM-23, 399-402, 1974.
(34)   G F Engen, IEEE Trans, MTT-22, 1255-1260, 1974.
(35)   D Woods, Electronics Letters, 11, 403-404, 1975.
(36)   J F Gibbons and H S Horn, IEEE Trans, CT-11, 378-384, 1964.
(37)   S B Cohn and H G Oltman, IRE Int Conv Rec, 9, part 3, 147-150, 1961.
(38)   D S James and S H Tse, Electronics Letters, 8, 46-47, 1972.
(39)   D M Kerns, J Res NBS, 64B, 113-130, 1960.
(40)   L Lewin, J Res NBS, 72C, 197-201, 1968.
(41)   R W Beatty, Electronics Letters, 9, 24-26, 1973.
(42)   R W Beatty, CPEM Digest, 87-89, 1974.
(43)   R W Beatty, NBS Tech Note 657, 1974.
(44)   N Marcuvitz, "Waveguide Handbook", (McGraw-Hill, 1951), pp. 307-309.

(45)   S J Mason, Proc. IRE, 41, 1144-1156, 1953.
(46)   C S Lorens, MIT Quart Prog Rept, 97-102, 1956.
(47)   S J Mason, Proc IRE, 44, 920-926, 1956.
(48)   J K Hunton, IRE Trans, MTT-8, 206-212, 1960.
(49)   N Kuhn, Microwave Journal, 6, 59-66, 1963.
(50)   D M Kerns and R W Beatty, "Basic Theory of Waveguide Junctions
        and Introductory Microwave Network Analysis", Pergamon Press,
        118-127, 1967.
(51)   S F Adam, "Microwave Theory and Applications", Prentice Hall,
        86-106, 1969.
(52)   R W Beatty, NBS Monograph No. 151, 1976.
(53)   D Woods, Proc. IEE, 124, 205-211, 1977.
(54)   F L Warner, "Microwave Attenuation Measurement", (Peter Peregrinus
        1977) Ch. 11.
(55)   S F Adam, Proc. IEEE, 66, 384-391, 1978.
(56)   J Fitzpatrick, Microwave J., 21, 63-66, May, 1978.
(57)   E F da Silva and M K McPhun, Electronics Letters, 14, 832-834,
        1978.
(58)   M Hillbun, Microwaves, 19, 87-90, 92, January, 1980.
(59)   D Woods, IEE Proc., 127, Pt. H, 82-86, 1980.
(60)   J Fitzpatrick, Microwave Systems News, 10, 77, 78, 80, 82, 85-87,
        89, 93, May, 1980.
(61)   S V Bearse, Microwaves, 20, 15-18, 21, 22, January, 1981.
(62)   G R Cobb, Microwave J., 24, 63-68, April, 1981.
(63)   J Fitzpatrick and J Williams, Microwave Systems News, 11, 96, 97,
        99, 102, 104, June, 1981.
(64)   H V Shurmer, The Radio and Electronic Engineer, 51, 287-298, 1981.
(65)   G R Cobb, J Fitzpatrick and J Williams, Microwave Systems News,
        11, 27, 30, 33, 35, November, 1981.
(66)   M A Maury, Microwave J., 25, 18, 20, 22, 26, 28, April, 1982.
(67)   J Fitzpatrick, Microwave J., 25, 43, 44, 46, 48, 52, 54, 56,
        April, 1982.
(68)   HP Application Note 221, 1977.
(69)   Microwave Systems News, 11, 114, September 1981.
(70)   Microwaves, 20, 108, September 1981.
(71)   W Kennan, Microwaves, 20, 61, 63-68, September 1981.
(72)   G M Yamaguchi et al, CPEM Digest, 124-127, 1976.
(73)   G M Yamaguchi et al, IEEE Trans., IM-25, 424-431, 1976.
(74)   L T Yuan et al, IEEE Trans, MTT-24, 981-987, 1976.
(75)   B Perlman, D Rhodes and J Schepps, Microwave J., 25, 73-80, April,
        1982.
(76)   F G Mendoza et al, Microwave J., 25, 65, 66, 69, April, 1982.
(77)   D E Bradfield, Microwave J., 25, 154-155, September 1982.
(78)   M A Wood, IEE Proc., 128, Pt. H, 257-262, 1981.
(79)   M A Wood, Electronics Letters, 18, 1025-1026, 1982.
(80)   M A Wood, IEE Proc., 129, Pt. H, 363-366, 1982.
(81)   W E Little et al, CPEM Digest, 130-133, 1976.
(82)   K Watanabe and M Ashiki, IEEE Trans., IM-26, 309-312, 1977.
(83)   O Ostwald and B Schiek, IEEE Trans., IM-30, 152-154, 1981.
(84)   O Ostwald and B Schiek, Proc. 11th European Microwave Conference,
        617-621, 1981.
(85)   N R Franzen and R A Speciale, Proc. 5th European Microwave
        Conference, 69-73, 1975.
(86)   R A Speciale et al, Proc. 6th European Microwave Conference,
        210-214, 1976.
(87)   E F da Silva and M K McPhun, The Radio and Electronic Engineer,

48, 227-234, 1978.

(88) E F da Silva and M K McPhun, Microwave Journal, 21, 97-100, June, 1978.

(89) E F da Silva, University of Warwick Ph.D Thesis, 1978.

(90) A L Cullen and S K Judah, Microwaves, Optics and Acoustics, 1, 120-124, 1977.

(91) F L Warner, "Microwave Attenuation Measurement", (Peter Peregrinus 1977), pp. 276-277.

(92) M A Wood, IEE Colloquium Digest No. 1983/53, pp. 15/1 - 15/6, 1983.

Chapter 10

Noise measurements

M. W. Sinclair

1   INTRODUCTION

The fundamental limitation to the sensitivity in any receiver
system is the electrical noise present within the system. In RF and
microwave measurements the system noise will often limit the accuracy
with which any particular parameter may be measured. Electrical noise
arises from many sources which may be broadly classified into two
groups:

a.   Sources external to the measurement system

b.   Sources within the system itself

The former include familiar areas such as noise from electrical
motors, dirty switch contacts, ignition systems and electrical storms.
In general the effect of these sources decreases rapidly with frequency
and in the RF and Microwave region – the so-called 'microwave window' –
internally generated noise is predominant. The sources within the
system are the active and passive devices which go to make it up.
Electrical noise from these sources consists of continuous electromagnetic radiations obeying certain statistical laws. Broadly speaking
the signals have random amplitude (a Gaussian distribution function)
and wide frequency coverage. It is with these internal sources we are
concerned in this lecture.

2   TYPES OF NOISE

2.1   Thermal Noise

Thermal or Johnson (1) noise is present in all systems. It arises
from the random motion of electrons within an electrical conductor, this
random motion being a function of the physical temperature of the
conductor. Since any movement of electrons constitutes a current,
between any two points within the conductor a voltage will exist. This
voltage has a Gaussian amplitude distribution function and, in the RF
and Microwave region, a uniform spectral distribution. This latter
property gives rise to the often used expression 'white noise' when
describing the output of a thermal noise generator. Nyquist (2) derived
a formula for the output noise voltage from a thermal source

$$\overline{v_n^2} \;=\; 4kTBR \qquad\qquad (1)$$

where $\overline{V_n^2}$ = mean square open circuit noise voltage

   T = absolute temperature of the source

   R = resistance of the source

   B = bandwidth of the system

   k = Boltzmann Constant (1.38×10^{-23} joules/K)

Whilst the 'noise voltage' is a familiar parameter found in the specification of audio frequency systems, 'noise power' is more common and in fact more appropriate when considering RF and microwave systems. It is usual to consider the power available from the source

$$\text{Available power } P = \frac{\overline{V_n^2}}{4R} = kTB \text{ watts} \tag{2}$$

Thus we see that the available power is independent of the source resistance and directly proportional to physical temperature. Another parameter commonly referred to in the literature is the 'available power spectral density', where

$$\text{Available Power Spectral Density, } S = \frac{P}{B} = kT \text{ watts/unit band} \atop \text{bandwidth} \tag{3}$$

It is necessary to point out here that the expressions derived above are only approximate, and in fact the full expression for available power spectral density is (3)

$$P = k\phi$$

where

$$\phi = T \cdot p(f)$$

and

$$p(f) = H(e^H - 1)^{-1}$$

$$H = \frac{hf}{kT}$$

Here, h = Planck constant, f = frequency. The symbol ϕ has been referred to as the 'quantum noise temperature' (3). In general the factor $p(f)$ is close to unity except where T is very low and f very high, or in mathematical terms

$$p(f) \doteqdot 1 \text{ when } hf \ll kT$$

Thermal noise is present in both active and passive devices, and thus each component within any transmission system is a potential source of noise which will contribute to the overall system noise to a greater or lesser extent.

2.2   Shot Noise

Active devices will in general contribute some thermal noise to a system but do have other sources. By far the most important of these is shot noise (4). When a current flows in an active device the instantaneous value of current varies about the mean value because it emanates from a source where electrons are emitted in a random manner. Considering the case of a thermionic diode operating in the temperature limited region (ie all electrons emitted from the cathode reach the anode), the mean square value of the current is

$$\overline{i_N^2} = 2e\, I_D\, B \qquad (4)$$

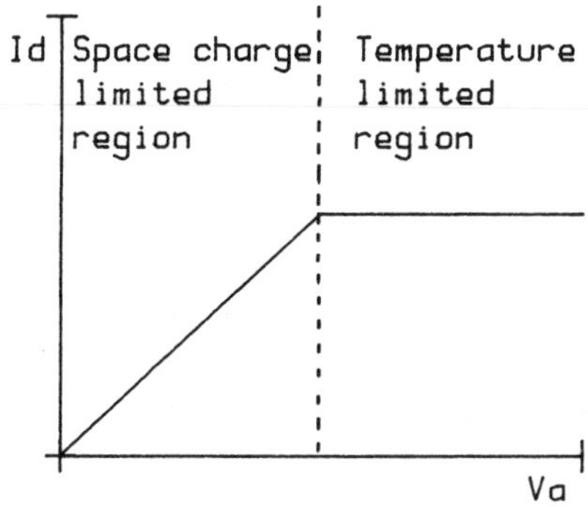

where I_D = mean current

e = electronic charge

B = bandwidth

If this current flows through an external resistance R, the power available at its terminals due to this current will be

$$P = \tfrac{1}{2}\, e\, I_D\, BR \qquad (5)$$

At frequencies above 300 MHz this value is modified due to the finite transit-time effects of the electrons (reduces P) and inter-electrode capacitance and series inductance (increase P). The overall effect is an increase in output at high frequency (4, 5). Equation (5) may then be written as

$$P = \tfrac{1}{2}\, e\, I_D\, BR\, C(f) \qquad (6)$$

where $C(f)$ is a frequency dependent factor which must be determined experimentally. Active devices may also generate other types of noise which will merely be named here as induced grid noise, partition noise and flicker noise (6).

Figure 1 (labels): Id; Space charge limited region; Temperature limited region; Va

3   DEFINITIONS

In many cases it is found convenient to express noise power in terms of a temperature. The basis of this is of course the thermal noise generator output power kTB. Any noise generator may be represented by an equivalent thermal noise generator which produces the same amount of available power as the source. If this power is $kT_e B$, then T_e is the equivalent noise temperature of the source. As will be seen later, it is also possible to define the noise generated in a receiver in terms of an equivalent noise temperature.

In many technical papers equalities of noise power are often reduced to equalities of noise temperature, where the common factors k and B have been dropped from the expressions.

A term frequently encountered in noise measurement is the excess noise ratio or ENR. This has been defined (3) in relation to a resistive termination (thermal noise source) at 290K (17°C), this sometimes being referred to as the Standard Noise Temperature, T_o

$$\text{ENR} = \log_{10} \frac{T_e - T_o}{T_o}\ \text{dB} \qquad (7)$$

where T_e is the equivalent noise temperature of the source.
There is some confusion amongst noise source manufacturers in the use of this definition. A common method of calibrating noise sources involves tuning the receiver input for a perfect match (ie Z_o of the transmission line). The noise powers delivered into this perfectly matched load are compared, and the unknown defined in terms of an "effective noise temperature", T_e'. The ENR is then calculated using this parameter.

Thus

$$ENR' = 10 \log_{10} \frac{T_e'-T_o}{T_o}$$

The effective noise temperature T_e' and the equivalent (available) noise temperature T_e are related by the expression

$$T_e' = T_e(1 - |\Gamma_s|^2)$$

where $|\Gamma_s|$ is the reflection coefficient modulus of the source. For low reflection sources the difference between the two ENR values is small. However, some solid state sources have a relatively high reflection coefficient, and differences between ENR and ENR' can be significant as shown in figure 2.

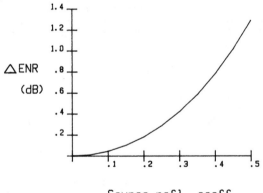

$\triangle$ ENR
(dB)

Source refl. coeff.

Figure 2

The noise performance of receivers may be defined in terms of an <u>equivalent (input) noise temperature</u> or perhaps more commonly <u>noise factor</u> or <u>noise figure</u>. The noise factor is defined(7) at a specified input frequency as the ratio of:

1. the total noise power per unit bandwidth at a corresponding output frequency available at the output port when the noise temperature of its input termination is standard (290K) at all frequencies to

2. that portion of 1. engendered at the input frequency by the input termination at the standard noise temperature 290K. Here 2. includes only that portion of noise from the input termination which appears at the output via the principal-frequency transformation of the system. This is an important factor when considering the noise factor of a crystal mixer where more than one input frequency can mix with the local oscillator signal to give the same output frequency. We shall consider

this point again later.

If we now refer to the system depicted in Figure 3, from our definition the noise factor is

$$F = \frac{N_o}{GkT_oB} \qquad (8)$$

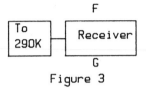

Figure 3

where F and G are the noise factor and gain of the receiver respectively and N_o is the total output power from the receiver.

$$N_o = GFkT_oB \qquad (9)$$

Now this comprises noise from the receiver and noise from the source termination. The latter is of course GkT_oB.

Hence the contribution from the receiver is

$$N_R = GFkT_oB - GkT_oB = (F-1)GkT_oB \qquad (10)$$

Note once again that the definition for noise factor requires that the source termination be at the standard temperature of 290K. This is inconvenient in the case of, for example, a satellite ground station where the input termination is the aerial pointing towards the sky. The source temperature can then be less than 10K. A more convenient measure of receiver noise for general use is the equivalent (input) noise temperature and is defined as (8):

The input termination noise temperature which, when the input termination is connected to a noise-free equivalent of the transducer (receiver), would result in the same output noise power as that of the actual transducer connected to a noise-free input termination.

In common with other similar definitions this is somewhat obscure in the reading. The concept is easier to understand if one first considers that all the output noise power contributed by the receiver itself is referred back to the input (ie divide by G). If we now imagine that all that noise power is being produced not within the receiver but by a resistive termination at some temperature T_e producing an equivalent amount of thermal noise power kT_eB, then T_e is the equivalent input noise temperature of the receiver. This approach leads naturally to a relationship between noise factor and equivalent noise temperature using equation (10).

From equation (10) the receiver contribution referred back to the input is

$$N_i = (F-1)kT_oB \qquad (11)$$

From our definition of equivalent noise temperature therefore

$$N_i = kT_eB \qquad (12)$$

and thus

$$T_e = (F-1)T_o \qquad (13)$$

Hence, knowing either of the receiver noise parameters we may calculate the other. Moreover, Figure 3 may be represented as shown in Figure 4. Using equations (9), (11) and (12) we have

$$N_o = G(kT_oB + kT_eB) \tag{14}$$

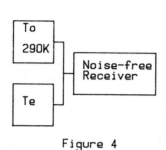

Figure 4

Thus the total input noise temperature of the system, sometimes referred to as the <u>operating noise temperature</u>, T_{op}, is given by

$$T_{op} = T_o + T_e \tag{15}$$

Note that from (14) the total power at the input port is the sum of the noise powers from the two individual sources. It is in fact true for the general case that when we have two or more completely independent (ie non-coherent) sources at any point in a system, then the total noise power at that point is the sum of the individual noise powers.

4   NOISE SOURCES OR GENERATORS

When attempting to measure the noise performance of a receiver one needs a stable reference against which to compare it. The most common and convenient form of stable reference used in practice is in fact itself a noise source (9). Four main types are in common use, these being the temperature limited diode, gas discharge sources, avalanche diode sources and thermal sources.

4.1   The Temperature Limited Diode

This forms a convenient source of noise power up to UHF. The figure shows the equivalent circuit of the diode with its source resistance R driving a load resistance R_L. From equation (6), assuming $R_L = R$ we have for the shot noise contribution from the diode

$$P_s = \tfrac{1}{2} e I_D BR C(f) \tag{16}$$

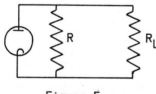

Figure 5

The source resistor will also contribute a thermal noise power proportional to its ambient temperature

$$P_R = kT_AB \tag{17}$$

Thus the total output noise power is

$$P_D = kT_AB + \tfrac{1}{2} e I_D BR C(f) \tag{18}$$

Alternatively, using the equivalent noise temperature concept

$$P_D = kT_DB = kT_AB + \tfrac{1}{2} e I_D BR C(f) \tag{19}$$

$$T_D = T_A + \frac{e I_DR C(f)}{2k} \tag{20}$$

Now e/2k is approximately equal to 5800, and if we express the diode current in milliamperes

$$T_D = T_A + 5.8 \ I_D \ R \ C(f) \tag{21}$$

and for a 50 ohm system

$$T_D = T_A + 290 \ I_D \ C(f) \tag{22}$$

For frequencies up to about 300 MHz $C(f) = 1$, and in that case we see that each milliampere increase in current will produce an additional 290K of noise temperature. For currents up to 30 mA the range of T_D will thus be from ambient up to about 9000K.

4.2 The Gas Discharge Noise Source (9)

These devices may be used from a few hundred MHz up to 100 GHz. A gas discharge is set up in a glass tube. The electrons within the tube have high kinetic energy. When a collision occurs between an electron and an ion or neutral molecule the electron decelerates, the loss in its kinetic energy emerging as a photon of radiation whose frequency is proportional to the kinetic energy change. Since the electron velocities are random, the overall effect is to produce a noise-like emission of energy from the gas discharge. The spectral density of the noise depends upon the mean electron energy, the order of magnitude of the noise temperature produced being 10^4K. Pure rare gases at pressures of a few tens of torr are employed since they possess stable discharges and do not produce ions which might attack the electrodes. Argon, neon and xenon are the most commonly used, with the first being predominant. The gas discharge tubes are coupled into coaxial line structures and waveguides in order to produce noise sources usable at RF and microwave frequencies. In general the mounts are terminated in the characteristic impedance of the transmission line so that they appear as a matched source in the off conditions. These sources are in general not calculable and as such require calibration before use.

4.3 Avalanche Diodes (10)

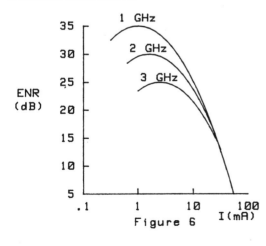

Figure 6

When a semiconductor p-n junction is reverse biassed a breakdown occurs wherein carriers (holes or electrons) absorb energy from the electric field sufficient to ionize atoms with which they collide. More carriers are produced which themselves cause further ionization and an avalanche process develops. When connected across a resistive load a wideband noise spectrum is generated. The spectrum is substantially flat up to a frequency which is

proportional to current density. At any selected frequency the output
increases with current up to a maximum and then decreases with further
current increase.

Figure 6 shows this dependence on current together with the effect
of frequency increase for a particular type of diode. Above a certain
current the ENR is virtually frequency independent. Avalanche diodes
thus lend themselves to application as wideband noise sources, and,
moreover, have in general higher ENRs than other types of noise source
in common use. This, coupled with their small size and weight, low
power consumption, ease of pulsed operation, high reliability and relatively low temperature coefficient make them ideal for many applications. As with the gas discharge sources it is difficult to calculate
their noise output and calibration is required.

4.4 Thermal Sources

We have already defined thermal noise, and thermal noise sources
(resistive elements which may be either heated or cooled) form very
convenient sources of noise. They are of course calculable, and as
such may and very often do form the basis of primary noise standards.
To satisfy these conditions they must be extensively measured before and
during assembly, and monitored during operation.

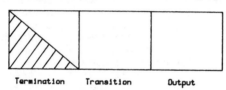

Termination   Transition   Output

Figure 7

The figure shows the basic
construction of a thermal
source. The transition
forms a high thermal
resistance but low elec-
trical resistance between
the termination (heated or
cooled) and the output at
ambient temperature. At
the present time these
sources have been constructed for operation from 4K (liquid helium
cooled) up to around 1300K. Sources cooled by liquid nitrogen (77K)
are very commonly used in assessing the noise performance of low noise
amplifiers.

5   CALIBRATION OF NOISE SOURCES

As with other types of signal generator a noise source must have a
known output level in order to be useful in any noise measurements.
Noise sources must thus be calibrated against a standard.

The UK operates a National Noise Standards calibration service at
RSRE Malvern (11, 12) which serves both BCS approved laboratories as well
as individual customers in some cases. Calibration accuracies of the
order of 1% are typical.

5.1 Methods of Calibration

Noise sources are calibrated using equipments referred to under the
general heading of radiometers. There are many varieties of radiometer,
and only two types will be mentioned here, viz the 'Total Power Radiometer' and the 'Switching (Dicke) Radiometer'. The total power radiometer is shown diagrammatically in figure 8. Two noise sources, a
standard T_s or

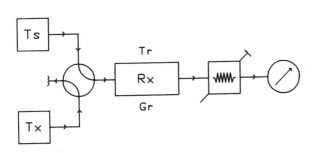

unknown T_x may be
connected via a
switch to a
receiver with gain
G_r and equivalent
noise temperature
T_r. The output
power is monitored
on a meter. We
have, using the
abbreviated term-
inology previously
mentioned,

Figure 8

T_s connected to input $\quad (T_s + T_r)G_r = p$ $\hfill$ (23)

T_x connected to input $\quad (T_x + T_r)G_r = q$ $\hfill$ (24)

In general, the attenuator in the system is used to make the meter reading the same in each case. This has the advantage that the effect of nonlinearities in the metering circuit is eliminated. If the two attenuator values are represented by A_1 and A_2, we have using (23) and (24)

$$(T_s + T_r)G_rA_1 = (T_x + T_r)G_rA_2 \qquad (25)$$

whence

$$\frac{(T_x + T_r)}{(T_s + T_r)} = \frac{A_1}{A_2} \qquad (26)$$

The ratio A_1/A_2 is very often referred to as the 'Y-factor'.

$$Y = \frac{A_1}{A_2} \qquad (27)$$

Combining (26) and (27) and rearranging gives

$$T_x = YT_s + T_r (Y-1) \qquad (28)$$

The system sensitivity is limited by random fluctuations of the receiver output (13, 14), these being the low frequency noise of the signal. The fluctuations have an equivalent RMS value at the input of the receiver of

$$\Delta T = \frac{a \ T_{op}}{\sqrt{B\tau}} \qquad (29)$$

Here, B is the receiver input bandwidth and τ the post-detector time constant, T_{op} having been defined in equation (15) is either $(T_s + T_r)$ or $(T_x + T_r)$; 'a' is a constant which in this case is equal to unity.

Equation (28) is derived on the assumption that the receiver gain remains constant during the measurement period. In all noise measurements we are dealing with small input powers (eg thermal noise from a source at 290K is approximately -204 dBW per unit bandwidth) and thus high receiver gains are required (normally in excess of 100 dB) to provide measurable levels at the output. Significant gain variations are thus possible. Considering the case of calibration of a low temperature source around 77K, a gain change of 0.1 dB can give an error in T_x of around 4K (ie about 5%). Thus for accurate inter-comparisons on noise sources alternatives to the total power radiometer are desirable.

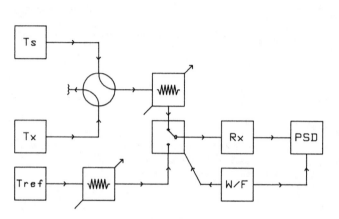

Figure 9

The Dicke (15) or switching radiometer overcomes the problem of gain variation. It is shown in simple form in Figure 9. Here, the manual switch is retained for connecting either T_s or T_x to the system input, but interposed between it and the receiver is a high speed switch which connects either of the input sources or a stable reference source T_{ref} to the receiver input on alternate half cycles of the switching waveform. If the noise powers in each half cycle are not equal, then the input to the receiver will have an amplitude modulation at the switching rate. This modulation may be detected using a phase sensitive detector (PSD). In practice a nulling technique is again employed to remove nonlinearity effects in the receiver. T_s is connected first and the attenuator in the input arm adjusted to give a zero output from the PSD. T_x is now connected and the attenuator re-adjusted for a zero output. This gives the following relationship

$$\frac{T_s - T_a}{T_x - T_a} = \frac{A_2}{A_1} \qquad (30)$$

where T_a is the ambient temperature of the attenuator.

Letting $\qquad (A_1/A_2) = Y$

$$T_x = YT_s - T_a(Y-1) = (T_s - T_a)Y + T_a \qquad (31)$$

Note that compared with equation (28) T_x is now independent of receiver noise temperature but does depend upon T_a, the ambient temperature of the variable attenuator. The system sensitivity is still defined by equation (29), but the value of 'a' is a function of the switch modulation waveform and its correlation in the PSD. Normally 'a' will lie between 2 and 3 for many systems. Since the system is now gain-independent, the value of τ may be increased to any practical limit, and thus the increase in ΔT may be partly offset. The output is normally fed on to a strip chart recorder for a better assessment of the output zero when τ is large (about 20 secs or greater).

In both the total power and switching radiometers the variable attenuator is a very important part of the system. It must be calibrated and must have the necessary resolution. In waveguide systems it is usually a rotary vane attenuator. In coaxial systems only a piston attenuator really possesses the required stability and resolution. However, there are other difficulties associated with the use of a piston attenuator, and as one alternative a temperature limited diode could be used as a directly variable source in the reference arm of a switching radiometer (16).

6   MEASUREMENT OF RECEIVER NOISE TEMPERATURE

General laboratory methods of receiver noise temperature measurements are all based on the total power radiometer concept described in 5.1. The difference is that the two input sources are now chosen to be dissimilar. Let us call these source T_h and T_c, which represent sources with noise temperatures 'hotter' or 'colder' than the receiver respectively. Equation (26) may then be re-written

$$\frac{T_h + T_r}{T_c + T_r} = \frac{A_1}{A_2} \qquad (32)$$

Letting $(A_1/A_2) = Y$ and re-arranging

$$T_r = \frac{T_h - YT_c}{(Y-1)} \qquad (33)$$

The choice of T_h and T_c is important for keeping uncertainties on the measurement low. The graph of figure 10 shows the percentage uncertainty in T_r for a fixed uncertainty in Y for a range of values of T_r. From considerations of this nature it is possible to arrive at some simple rules for the choice of T_h and T_c in order to minimise uncertainties, viz

$$T_r = \sqrt{T_h T_c}; \qquad \frac{T_h}{T_c} < 4; \qquad \text{and preferably} \qquad \frac{T_h}{T_c} > 10 \qquad (34)$$

The noise sources in common use have already been reviewed, and the above rules may be applied to choose suitable sources for various types of receiver. It is common practice, however, to use a thermal noise source at ambient temperature as one of the sources as a matter of convenience. The choice of the other source is then a function of whether or not the receiver noise temperature being measured is likely to be above or below ambient temperature. Table 1 lists some types of receiver amplifiers with typical values of noise temperature and noise factor.

Care must be taken when measuring the noise factor of systems containing crystal mixers. In the Y-factor measurement technique noise from both signal and image frequencies can mix with the local oscillator

to produce an IF output. However, the definition of noise factor
clearly states that the ratio is the total output noise power to that

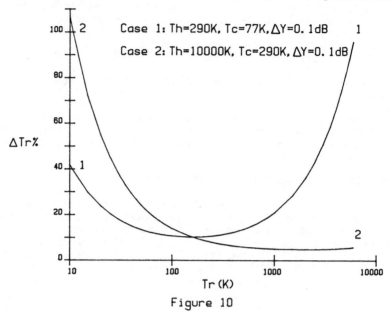

Figure 10

engendered at the input by a standard termination at 290K by the principal input frequency transformation. In the case of a radar receiver,
therefore, a noise factor measured by the Y-factor method should be
increased by 3 dB for mixers with identical responses for signal and
image frequencies (17).

TYPE OF AMPLIFIER	NOISE TEMP.	NOISE FACTOR
Crystal Mixer + IF Amp.	2600K	10dB
Travelling Wave Tube Amp.	290K	3dB
Parametric Amp.	100K	1.3dB
Maser Amp.	14K	0.2dB

TABLE 1

6.1   Cascaded Noise Figures

If we consider a receiver which consists of a number of amplifiers
having different gains and noise figures, it is very often necessary to
calculate the overall noise figure of the assembly. If we have 'n' such

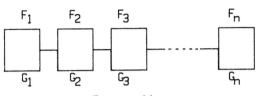

Figure 11

amplifiers then the
total noise figure
is given by the
expression

$$F_t = F_1 + \frac{F_2-1}{G_1} + \frac{F_3-1}{G_1G_2} + \ldots + \frac{F_n-1}{G_1G_2 \ldots G_{n-1}} \qquad (35)$$

or in terms of noise temperatures,

$$T_t = T_1 + \frac{T_2}{G_1} + \frac{T_3}{G_1G_2} + \ldots + \frac{T_n}{G_1G_2 \ldots G_{n-1}} \qquad (36)$$

If it is required to design a system with the lowest overall noise factor from a given assembly of amplifiers, then a more useful parameter to calculate for each unit is the <u>noise measure</u> (18) given by

$$M = \frac{F-1}{1 - \frac{1}{G}} \qquad (37)$$

where F and G are the noise factor and available power gain of the unit. One should then select the amplifier with lowest noise measure to be first in the cascade.

6.2 Noise Contributions from Lossy Networks

In most practical situations noise measurement systems consist of combinations of noise sources, amplifiers and lossy networks (eg attenuators, isolators, lossy waveguide runs). The lossy networks, usually at ambient temperature, generate thermal noise, and it is necessary to calculate how much noise is contributed to the system by such sources.

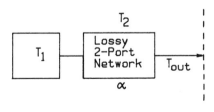

Figure 12

Consider the case shown in Figure 12 where we have a termination at temperature T_1 followed by a lossy two-port network at physical temperature T_2 with available power transmission factor α (see Appendix 1). Then the total output noise temperature is given by the expression

$$T_{out} = T_1\alpha + (1-\alpha)T_2 \qquad (38)$$

A similar expression may be derived for any lossy multi-port network (19).

Applying this to the case illustrated in Figure 13, where we have a lossy network at the standard temperature T_o interposed between a

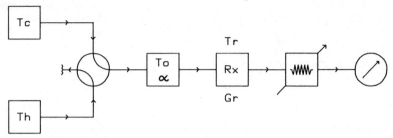

Figure 13

receiver and the sources being used to measure it, we may use equations (25) and (38) to write down

$$\{T_c\alpha + (1-\alpha)T_o + T_r\}G_rA_1 = \{T_h\alpha + (1-\alpha)T_o + T_r\}G_rA_2$$

Letting $\dfrac{A_1}{A_2}$ = Y and re-arranging

$$T_r = \frac{\alpha(T_h-YT_c)}{(Y-1)} - (1-\alpha)T_o \qquad (39)$$

Considering the lossy network and receiver as a single network with overall noise temperature T_r'

Thus
$$T_r' = \frac{T_h-YT_c}{(Y-1)} \qquad (40)$$

Using
$$F = \frac{T_r}{T_o} + 1 \qquad \text{we obtain}$$

from (39)
$$F = \frac{\alpha(T_h-YT_c)}{T_o(Y-1)} + \alpha \qquad (41)$$

from (40)
$$F' = \frac{(T_h-YT_c)}{T_o(Y-1)} + 1 \qquad (42)$$

from (41) and (42)
$$F = \alpha F' \qquad (43)$$

or since $\alpha < 1$
$$F'(\text{dB}) = F(\text{dB}) + \alpha(\text{dB}) \qquad (44)$$

Thus any losses (in dB) immediately preceding the receiver if not accounted for will add directly to the required receiver noise factor in dB. It should be noted that this result is only true for a lossy network at the standard temperature T_0 (290K). If the lossy network is at any other temperature a new expression for F in equation (41) must be derived using the same procedure.

In many practical situations we have the case just considered, eg an aerial coupled to a receiver with lossy waveguide or coaxial components between. Thus we must be certain that our analysis will yield the parameter we have set out to measure. Only a careful breakdown of the system to include contribution from all lossy networks will ensure that this occurs.

7   EFFECTS OF MISMATCH

In considering all the previous situations, one very important assumption has been made, viz that all systems are perfectly matched at all points throughout the assembly. In a practical situation this is rarely, if ever, true.

Noise transmission systems suffer from two effects when mismatch is present. Firstly, in common with all other signal transmission systems, the noise is subject to mismatch loss (20); secondly, the noise temperature of a receiver varies with the source impedance it sees. All the measurement techniques we have considered use a receiver to amplify the noise power and thus the variation of receiver noise temperature brought about by variation in the source match will affect the total power recorded on our output indicator.

7.1 Measurement of Noise Sources

When we are measuring noise sources then the derivations of equations (28) and (31) require that the receiver noise temperature remains constant as one switches between the two sources being compared. Thus it becomes necessary only to modify our systems in such a way as to maintain a constant source immittance during the measurements. A common way of approaching close to this condition in practice is to include an isolator or circulator in front of the receiver. Thus the total power radiometer of figure 8 may be modified as shown in figure 14.

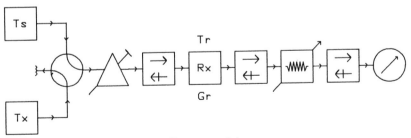

Figure 14

The tuner is used to cancel residual reflections in the isolator and switch such that both sources see a perfectly matched input. The isolators on each side of the attenuator serve to make the mismatch error on the isolator/attenuation/isolator assembly constant. Since all our equations involve a ratio of two attenuations (ie a difference between two dB settings), then the mismatch error cancels. In order to exploit this situation the attenuator must be calibrated with the isolators in position.

We have

T_s connected: $\{T_s\,\alpha_s + (1-\alpha_s)T_a + T_r\}A_1G_r = P$ (45)

T_x connected: $\{T_x\,\alpha_x + (1-\alpha_x)T_a + T_r\}A_2G_r = P$ (46)

Here α_s and α_x are the available power transmission factors (Appendix I) of the switch/tuner/isolator assembly as we switch to T_s and T_x respectively, and T_a is the ambient temperature of the assembly.

Letting $A_1/A_2 = Y$ and re-arranging

$$T_x = \frac{Y(T_s-T_a)\alpha_s}{\alpha_x} + \frac{(T_a+T_r)(Y-1)}{\alpha_x} = T_a$$ (47)

From Appendix I we may write

$$\alpha_s = \frac{(1-|\Gamma_s|^2)\,|S_{21s}|^2}{(1-|\Gamma_{2s}|^2)|1-S_{11s}\,_s|^2}$$ (48)

$$\alpha_x = \frac{(1-|\Gamma_x|^2)\,|S_{21x}|^2}{(1-|\Gamma_{2x}|^2)|1-S_{11x}\Gamma_x|^2}$$ (49)

In these expressions Γ represents a complex reflection coefficient and S a scattering parameter identified by the common suffices.
    If we assume that both sources have a reflection coefficient magnitude < 0.1 (20 dB return loss), that the isolator has an isolation of 30 dB minimum and that two switch paths are symmetrical, then with negligible uncertainty we may write

$$S_{11s} = S_{11x} = 0 \; ; \; |S_{21s}|^2 = |S_{21x}|^2 \; ; \; \Gamma_{2s} = \Gamma_{2x} = S_{22}.$$

Equation (47) then becomes

$$T_x = \frac{Y(T_s-T_a)(1-|\Gamma_s|^2)}{(1-|\Gamma_x|^2)} + \frac{(T_a+T_r)(Y-1)}{\alpha_x} + T_a$$ (50)

    The switching radiometer of figure 9 may be modified as shown in figure 15 in order to reduce mismatch effects. In this case equation (30) is modified by replacing the values of attenuation by the

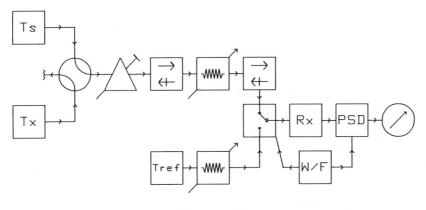

Figure 15

appropriate available power transmission factors and we obtain

$$\frac{T_s - T_a}{T_x - T_a} = \frac{\alpha_x}{\alpha_s}$$

Assuming the same conditions as for figure 14 we may again write

$$\Gamma_{2s} = \Gamma_{2x} = S_{22} \quad \text{and} \quad S_{11s} = S_{11x} = 0$$

Then

$$\frac{T_s - T_a}{T_x - T_a} = \frac{A_2(1-|\Gamma_x|^2)}{A_1(1-|\Gamma_s|^2)} \tag{52}$$

Making the usual substitution $A_1/A_2 = Y$ and rearranging

$$T_x = (T_s - T_a)Y \frac{(1-|\Gamma_s|^2)}{(1-|\Gamma_x|^2)} + T_a \tag{53}$$

7.2 Measurement of Receivers

When measuring receivers the situation is much more complicated and we must either:

a. state a measured noise temperature (noise figure) for a specified source immittance; or

b. provide information on the receiver noise parameters which will enable its noise temperature to be calculated for any source immittance.

The majority of measurements are performed under the assumption that case a. applies, where the specified source immittance is the characteristic impedance of the transmission line. This condition is also the easiest to reproduce if one is concerned with performing measurements on a receiver at different locations using different systems.

When approach b. is being considered then the practical and theoretical aspects differ depending upon the frequency range concerned. Up to a few hundred megahertz the approach based on the classical theory proposed by Rothe and Dahlke (21) is perhaps the best basis. This considers noisy two-ports as networks in which the noise properties are defined in terms of currents, voltages and immittances. At microwave frequencies, however, the wave approach (22) gives a better practical insight into the problem. The noisy two-ports are then defined in terms of noise waves and transmission line parameters such as voltage reflection coefficient.

7.2a The Network Approach

From Rothe and Dahlke (21) we obtain the appropriate equivalent circuit which is shown in figure 16.

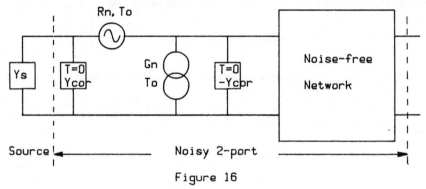

Figure 16

This leads to an expression for the noise figure given by

$$F = 1 + \frac{G_n}{G_s} + \frac{R_n}{G_s} |Y_s + Y_{cor}|^2 \qquad (54)$$

A more familiar expression in the literature is (20)

$$F = F_{min} + \frac{R_n}{G_s} |Y_s - Y_{opt}|^2 \qquad (55)$$

where Y_{opt} is the source admittance required to yield the minimum noise figure F_{min}. A similar expression for noise temperature follows when one substitutes from equation (13).

$$T_r = T_{rmin} + \frac{T_o R_n}{G_s} |Y_s - Y_{opt}|^2 \qquad (56)$$

From equation (56) we may write

$$T_r = T_{rmin} + \frac{T_o R_n}{G_s} \{(G_s - G_{opt})^2 + (B_s - B_{opt})^2\} \qquad (57)$$

In order to define the noise properties of the network fully we need to find T_{rmin}, R_n, G_{opt} and B_{opt}. This may be accomplished using the following procedure (8).

i. With the source conductance G_s constant, measure T_r (or F) for several different values of source susceptance B_s. Plot the results so obtained. From equation (57) we have

$$\frac{\partial T_r}{\partial B_s} = \frac{T_o R_n}{G_s} (- 2B_{opt} + 2B_s) \qquad (58)$$

At the minimum of the curve $\partial T_r / \partial B_s = 0$ and hence

$$B_s = B_{opt} \qquad (59)$$

The results will follow the pattern shown in figure 17a.

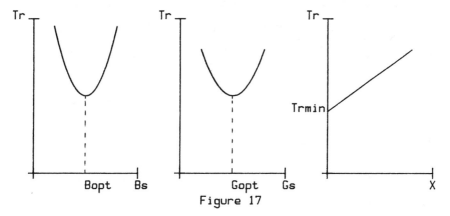

Figure 17

ii. With the source susceptance maintained at the value found from (i), measure T_e for several values of G_s. Plot the results once again. As before from equation (57) we have

$$\frac{\partial T_r}{\partial G_s} = T_o R_n \left(\frac{G_s^2 - G_{opt}^2}{G_s^2} \right) \qquad (60)$$

At the minimum of the curve $\partial T_r / \partial G_s = 0$ and hence

$$G_s = G_{opt} \qquad (61)$$

The pattern of the results is shown in figure 17b.

iii. From the previous data plot T_r against the quantity

$$x = \frac{T_o |Y_s - Y_{opt}|^2}{G_s}$$

The data should lie on a straight line of the form $T_r = T_{rmin} + R_n x$. The slope of the line is R_n and the intercept on the ordinate T_{rmin} as shown in figure 17c.

This practical method is susceptible to the usual measurement uncertainties, and it is difficult to obtain meaningful figures in some circumstances. As an alternative, computational methods may be used since in theory only four different values of source admittance are required to solve equation (57) for the noise parameters. Such techniques using a larger number of source admittance values have been suggested (24, 25, 26) and least squares fitting used to obtain a more representative result. Practical methods at microwave frequencies using similar theoretical models have been proposed by other authors (27, 28).

7.2b The Wave Approach

The wave approach has been discussed by a number of authors (22, 29, 30). Meys' approach (30) gives a simpler solution together with a practical measurement technique. The equivalent circuit as suggested by Meys is shown in figure 18.

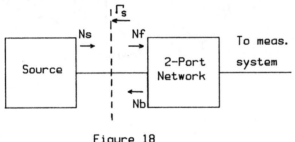

Figure 18

N_f and N_b are the forward and reverse noise waves representing the noise characteristics of the two-port, and N_s is the noise wave from the source which has a reflection coefficient Γ_s. If a noiseless matched load is substituted for the two-port input, the total noise wave incident upon it would be

$$N_t = N_f + \Gamma_s N_b + N_s \tag{62}$$

Assuming no correlation between the source and the two-port noise, but allowing for that between N_f and N_b we have

$$\overline{|N_t|^2} = \overline{|N_f|^2} + |\Gamma_s|^2 \overline{|N_b|^2} + 2\text{Re}\overline{(\Gamma_s N_f^* N_b)} + \overline{|N_s|^2} \tag{63}$$

where $\overline{|N_t|^2}$ represents the mean square value of N_t and N_f^* is the complex conjugate of N_f. If we represent the terms of equation (63) by equivalent Nyquist noise generators then,

$$\overline{|N_t|^2} = kT_tB \; ; \quad \overline{|N_f|^2} = kT_fB \; ; \quad \overline{|N_b|^2} = kT_bB \; ;$$

$$\overline{N_f^*N_b} = kT_cB \, e^{j\phi_c}$$

Further we let

$$\overline{|N_s|^2} = (1-|\Gamma_s|^2)kT_sB \quad \text{and} \quad \Gamma_s = |\Gamma_s|e^{j\phi_s}$$

Thus equation (63) becomes

$$T_t = T_f + |\Gamma_s|^2 T_b + 2T_c|\Gamma_s| \; \cos(\phi_c+\phi_s) + T_s(1-|\Gamma_s|^2) \qquad (64)$$

The practical determination of the four parameters T_f, T_b, T_c and ϕ_c is achieved as follows:

i.   Connect a sliding short circuit to the input (ie $|\Gamma_s| = 1$ and ϕ_s is variable) and equation (64) is modified as shown in equation (65). Plot T_t' for various positions of the short circuit making sure that the total movement of the short circuit exceeds $\lambda_g/2$, where λg is the guide wavelength. The form of T_t' is shown in figure 19.

$$T_t' = T_f + T_b + 2T_c \, \cos(\phi_c+\phi_s) \qquad (65)$$

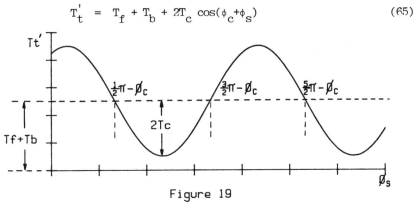

Figure 19

We may thus obtain T_c, ϕ_c and $T_f + T_b$ directly from the graph.

ii.   Connect a matched load of known temperature to the input and measure T_t. Then $T_t = T_f + T_s$, from which T_f may be deduced. We now have all the parameters necessary to define the two-port for any source reflection coefficient.

7.3   Automated Noise Measurements

As with other branches of metrology, automation has made its impact on the noise measurement scene. Systems which have been described in the literature vary depending upon the required noise parameter to be measured. For example most systems applications require that the noise

performance of a receiver be evaluated in a nominally matched transmission line situation. Manufacturers of low noise amplifying
devices, however, may be more interested in the determination of the
complex noise parameters discussed in the previous paragraphs in order
to produce low-noise amplifiers. Noise source calibrations again
impose different constraints on the required measurement system.

7.3a Automatic Receiver Noise Measurements

Instruments classified under the heading "automatic noise figure
meters" have been available for more than 20 years. They range from
the early instruments which simply gave an automatic readout of
Y-factor on a meter suitably scaled to indicate noise figure or noise
temperature directly, to the present day microprocessor controlled
instrumentation. The latter have the advantage that corrections for
source variation with frequency, second stage noise contributions,
detector nonlinearity etc may all be performed before a result is
presented. None of the existing commercially available instruments
make any attempt to include procedures to evaluate complex device
noise parameters. However, they can be used in more sophisticated
systems in which the hardware and software has been enhanced to give
the additional information.

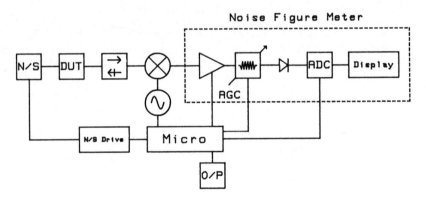

Figure 20

Figure 20 shows the basic elements in an automated noise measurement set-up. The noise source is switched on and off at some controlled
rate to provide the two known sources at the device under test (DUT)
input port. The noise figure meter usually has limited frequency
coverage, and it is therefore often necessary to provide an external
mixer and local oscillator. In some cases the microprocessor is
included as part of the noise figure meter (31). It is generally true
that the additional hardware required for producing the output indication introduces additional uncertainty. Thus automatic methods are in
general less accurate than manual methods. However, it is possible by
careful system refinement to produce a system with uncertainties very
close to those obtainable using manual methods (32).

In order to measure the DUT complex noise parameters it is necessary to provide an automatically variable source impedance. This is normally implemented by the inclusion of an isolator and a programmable tuner between the noise source and DUT. The basic procedure is that previously mentioned in 7.2a, where four known terminations are connected to the input in turn and the DUT noise figure measured in each case. In practice workers have used more than four source impedances and used least squares fitting to enhance accuracy (33, 34). The source impedances used have normally been set up and measured previously using an automatic network analyser. The technique could be improved by incorporation of the impedance measurement into the automated system as indicated in figure 21. Here a six-port reflectometer may be switched into the circuit for source impedance measurements. This would require a switch with good repeatability and symmetrical properties. Any departure from this ideal situation could be established by previous measurements, and the effects included in the results as an additional uncertainty term.

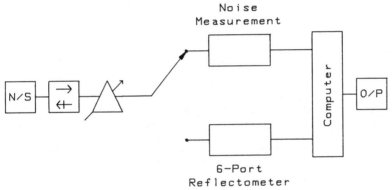

Figure 21

7.3b Automatic Noise Source Measurements

In principle the techniques used for automatic receiver noise measurements could be applied equally well to automatic calibration of noise sources. It is generally true to say, however, that commercially available instrumentation is not orientated towards this application and would require software modifications. Few manufacturers are willing to undertake these modifications for the sake of the relatively small market outlet. Automation of noise source calibration has therefore tended to concentrate in the National Standards laboratories throughout the world, and the UK is no exception here. At RSRE a whole range of automated calibration systems exist covering the frequency range from 1 to 40 GHz at the present time (12, 35). Many of the systems are based on the Dicke type radiometer as depicted in figure 22. The actual measuring process has been automated by replacing two of the main components, viz the input microwave switch and the precision variable attenuator, with automatically controlled items. The microwave switch replacement is a commercially available motor driven switch. The precision variable attenuator has been specially developed for the purpose. It has an anti-backlash gearbox driven by a DC servo motor and a series of absolute shaft encoders giving a digital output related to the attenuation setting (36). The servo motor is driven from the PSD output voltage, thus providing an automatic nulling system. All

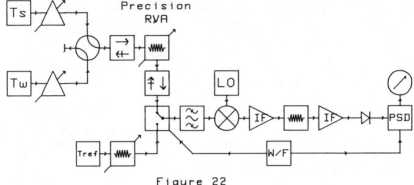

Figure 22

necessary data is collected automatically and a mini-computer controls the measurement sequence and data collection and finally calculates the value of the unknown noise source and the associated uncertainty. Full automation has not yet been achieved since the systems require that the sources are tuned at each frequency of measurement. This is at present a manual exercise carried out using a tuned reflectometer which is built into the radiometer system (not shown in figure 22). It is anticipated that full automation can be achieved by incorporation of a six-port reflectometer in the radiometer. Thus no tuning will then be necessary as measurement of the complex source impedances will allow mismatch effects to be accounted for by calculation.

REFERENCES

1.   Johnson, J.B., July 1928, 'Thermal Agitation of Electricity in Conductors' Phys. Rev. 32, pp 97–109.

2.   Nyquist, H., July 1928, 'Thermal Agitation of Electric Charge in Conductors' Phys. Rev. 32, pp 110–113.

3.   'Joint Service Review and Recommendations on Noise Generators' Joint Service Specification REMC/30/FR, June 1972.

4.   Schottky, W., 1918, 'Spontaneous Current Fluctuations in Various Conductors' (German) Annalen der Physik 57, pp 541–567.

5.   Harris, I.A., Nov 1961, 'The Design of a Noise Generator for Measurements in the Frequency Range 30–1250 MHz'. Proc. IEE Vol. 108, Pt B No 42, pp 651–658.

6.   Van der Ziel, A., "Noise: Sources, Characterization, Measurement'. Prentice-Hall.

7.   Haus, H.A. et al., March 1963, 'Description of the Noise Performance of Amplifiers and Receiving Systems', Proc. IRE Vol. 51, No. 3, pp 436–442

8. Haus, H.A. et al., 'IRE Standards on Methods of Measuring Noise in Linear Two ports,

9. Hart, P.A.H., July 1962, 'Standard Noise Sources', Philips Tech. Rev., Vol. 23, No 10, pp 293-309

10. Haitz, R.H. and Völtmer, F.W., June 1968, 'Noise of a self-sustaining Avalanche Discharge in Silicon: Studies at Microwave Frequencies'. Jnl. Appl. Phys. Vol. 39, No 7, pp 3379-84

11. Blundell, D.J., Houghton, E.W. and Sinclair, M.W., November 1972, 'Microwave Noise Standards in the United Kingdom'. Proc. IEEE Trans. I and M, Vol. I-M21, No 4, pp 484-488.

12. Sinclair, M.W., March 1982, 'A Review of the UK National Noise Standard Facilities', IEE Colloquium on "Electrical Noise Standards and Noise Measurements", Digest No 1982/30, Paper No 1, pp 1/1-1/16.

13. Kelly, E.J., Lyons, D.H. and Root, W.L., May 1958, 'The Theory of the Radiometer', MIT Lincoln Lab., Report No 47.16.

14. Tiuri, M.E., July-October 1964, 'Radio Astronomy Receivers', IEEE Trans. MIL-8, pp 264-272.

15. Dicke, R.H., July 1946, 'The Measurement of Thermal Radiation at Microwave Frequencies', Rev. Sci. Instr. 17, pp 268-275,

16. Halford, G.J. and Robus, E.G., April 1968, 'Noise Source Calibrations in the Decimetre Band'. Radio and Electronic Engineer, Vol. 35, No 1.

17. Pastori, W.E., May 1983, 'Image and Second-Stage Corrections Resolve Noise Figure Measurement Confusion'. Microwave Systems News, pp 67-86.

18. Haus, H.A. and Adler, R.B., 1959, 'Circuit Theory of Linear Noisy Networks', New York, Wiley.

19. Wait, D.F., Sept 1968, 'Thermal Noise from a Passive Linear Multiport'. IEEE Trans. on MTT Vol. MTT-16, No 9, pp 687-691.

20. Kerns, D.M. and Beatty, R.W., 'Basic Theory of Waveguide Junctions and Introductory Microwave Network Analysis'. Pergamon Press.

21. Rothe, H. and Dahlke, W., June 1956, 'Theory of Noisy Fourpoles', Proc. IRE Vol. 44, pp 811-818.

22. Penfield, P. Jnr., March 1962, 'Wave Representation of Amplifier Noise', IRE Trans. on Circuit Theory, pp 84-86.

23. Haus, H.A. et al., January 1960, IRE Sub-committee 7.9 on Noise: 'Representation of Noise in Linear Two-ports'. Proc. IRE, pp 69-74.

24.   Lane, R.Q., August 1969, 'The Determination of Device Noise Para-
      meters', Proc. IEEE, pp 1461-62.

25.   Hartmann, K., October 1976, 'Noise Characterization of Linear
      Circuits', IEEE Trans. CAS, Vol. CAS-23, No 10, pp 581-90.

26.   Caruso, G. and Sannine, M., September 1978, 'Computer-Aided
      Determination of Microwave Two-Port Noise Parameters'. IEEE
      Trans. MTT. Vol. MTT-26, No 9, pp 639-642.

27.   Vlaardingerbroek, M.T., Knol, K.S. and Hart, P.A.H., 1957,
      'Measurements on Noisy Fourpoles at Microwave Frequencies'.
      Philips Res. Rep. 12, pp 324-332.

28.   Lange, J., June 1967, 'Noise Characterization of Linear Two-ports
      in Terms of Invariant Parameters'. IEEE Jnl. of Solid State
      Circuits, Vol. SC-2, No 2, pp 37-40.

29.   Otoshi, T.Y., 'The Effect of Mismatched Components on Microwave
      Noise-Temperature Calibrations'. September 1968. IEEE Trans.
      MTT, Vol. MTT-16, No 9, pp 675-686.

30.   Meys, R.P., January 1978, 'A Wave Approach to the Noise Proper-
      ties of Linear Microwave Devices'. IEEE Trans. MTT, Vol. MTT-26,
      No 1, pp 34-37.

31.   Swain, H.L. and Cox, R.M., April 1983, 'Noise Figure Meter Sets
      Record for Accuracy, Repeatability and Convenience'. Hewlett
      Packard Journal, pp 23-34.

32.   Meys, R.P., March 1982, 'A Microprocessor-Controlled Automatic
      Noise-Temperature Meter'. IEEE Trans. I&M, Vol. IM-31, No·1,
      pp 6-8.

33.   Lane, R.Q., August 1978, 'Derive Noise and Gain Parameters in
      10 Seconds', Microwaves, pp 53-57.

34.   Abbott, D.A. and Shurmer, H.V., October 1982, 'Automatic Noise
      Figure Measurements with Computer Control and Correction'. The
      Radio and Electronic Engineer, Vol. 52, No 10, pp 468-474.

35.   Sinclair, M.W. and Lappage, R., 1980, 'An Automated Radiometer for
      Measurement of Above-Ambient Noise Sources'. CPEM Digest,
      pp 478-482.

36.   Warner, F.L., Watton, D.O., Herman, P and Cummings, P., Dec 1976,
      'Automatic Calibration of Rotary Vane Attenuators on a Modulated
      Sub-Carrier System'. IEEE Trans. on I&M, Vol. IM-25, No 4,
      pp 409-413.

APPENDIX I

1.   Summary of Useful Two-Port Network Parameters (20)

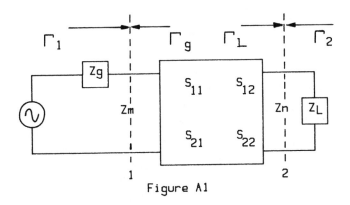

Figure A1

The mismatch factor at each measurement plane is defined as

$$\text{Mismatch Factor} \ = \ \frac{\text{Power Delivered to Input}}{\text{Power Available from Source}}$$

At plane 1

$$\frac{P_{1d}}{P_{1a}} \ = \ M \ = \ \frac{(1-|\Gamma_g|^2)\,(1-|\Gamma_1|^2)}{|1-\Gamma_g\Gamma_1|^2} \qquad\qquad (\text{AI.1})$$

At plane 2

$$\frac{P_{2d}}{P_{2a}} \ = \ N \ = \ \frac{(1-|\Gamma_2|^2)\,(1-|\Gamma_L|^2)}{|1-\Gamma_2\Gamma_L|^2} \qquad\qquad (\text{AI.2})$$

The efficiency of a two-port network is defined as

$$\text{Network Efficiency} \ = \ \frac{\text{Power Delivered to Load}}{\text{Power Delivered to Input}}$$

$$\frac{P_{2d}}{P_{1d}} \ = \ \frac{(Z_m/Z_n)\,|S_{21}|^2\,(1-|\Gamma_L|^2)}{(1-|\Gamma_1|^2)\,|1-S_{22}\Gamma_L|^2} \qquad\qquad (\text{AI.3})$$

Note that efficiency is independent of the source impedance.

The available power transmission factor of a two-port network is defined as

Network Available Power Transmission $=$ $\dfrac{\text{Power Available at Output}}{\text{Power Available from Source}}$
Factor

$$\frac{P_{2a}}{P_{1a}} = \alpha = \frac{(Z_m/Z_n)\ |S_{21}|^2\ (1-|\Gamma_g|^2)}{(1-|\Gamma_2|^2)\ |1-S_{11}\Gamma_g|^2} \qquad (AI.4)$$

Note that α is independent of the load impedance.

In general in a coaxial or waveguide system we have $Z_m = Z_n = Z_0$ (characteristic impedance), and the preceding expressions may be simplified accordingly.

We also have that

Power delivered to Input = Power Available from Source x
Mismatch Factor

$$P_{1d} = MP_{1a} \qquad (AI.5)$$

Power delivered to Load = Power Available at Output x
Mismatch Factor

$$P_{2d} = NP_{2a} \qquad (AI.6)$$

now

$$\eta = \frac{P_{2d}}{P_{1d}} \qquad (AI.7)$$

Hence from (AI.5), (AI.6) and (AI.7)

$$\frac{P_{2d}}{P_{1d}} = \frac{N}{M}\frac{P_{2a}}{P_{1a}} = \eta \qquad (AI.8)$$

but

$$\frac{P_{2a}}{P_{1a}} = \alpha$$

$$\therefore \eta = \frac{N}{M}\alpha \qquad (AI.9)$$

Chapter 11

Sampling principles

P. Cochrane

1   PROLOGUE

Applying a sampling technique to the solution of any
problem is often an admission of a fundamental inability
to encompass all the data involved. In effect sampling is
specifically used to reduce information to its essentials
- or at least to a more manageable level. At microwave
frequencies it is often the only practicable technique
available, and in this forum time domain sampling may be
classified into three broad functional categories:-

    i.   Switching between two or more reference/measure
       paths

    ii. Part of an Analogue to Digital conversion
       process

    iii. Overcoming fundamental hardware speed
       limitations precluding real-time operation.

In turn each of these categories involve processes
that may be perceived to be Sub-Nyquist, Nyquist or Super
Sampled. It is therefore essential that we examine the
action of sampling in some depth before we go on to
consider specific microwave measurement applications.
This chapter is thus solely concerned with explaining the
action and limitations of sampling in the time domain -
but with due reference to the frequency domain
implications. Some reference is also made to the
practical difficulties encountered and the engineering
solutions available. In the chapter that follows we
concentrate on specific examples of instruments and
measurement techniques employing sampling and explain
their actions, applications and limitations. An initial
consideration is also given to the design and realisation
of practical time samplers.

1.1   Mathematical Symbols, Notation and Definitions

i = an integer

n = an integer

T = a period of time

τ = a period of time

f_c = carrier frequency

f_m = modulation frequency

f_s = sampling frequency

B = a finite frequency band

β = modulation index

J_n = Bessel function of the first kind

$g(t)$ = generalised function of time

$G(f)$ = generalised function of frequency

<=> signifies a Fourier transformation

* signifies a convolution integral

$\displaystyle\sum_n$ = a summation over all 'n'

$\displaystyle\sum_n'$ = a summation over all 'n' except n=o

$s(t) = \text{comb}_T [\ ] = \displaystyle\sum_i \delta(t-iT)$ = the sampling function
(at T intervals)

$\text{Rep}_{\frac{1}{T}} [\ ]$ = the replication function (at $\frac{1}{T}$ intervals)

$\text{rect}(t) = u(t+\tfrac{1}{2}) - u(t-\tfrac{1}{2})$ = the rectangular function

$\text{sinc}(t) = \dfrac{\sin(\pi f t)}{\pi f t}$

$\delta(t) \triangleq \displaystyle\int_{-\infty}^{+\infty} \delta(t)dt = 1; \ \delta(t) = 0, \ t \neq 0$ = the Dirac delta function

2   INTRODUCTION

The derivation of the sampling theorem is often attributed to Nyquist [1], Shannon [2], and less commonly to Gabor [3], but in fact it was known to Cauchy [4] much earlier. In many respects it was rediscovered and refined by Whittaker [5,6] before it emerged in its more popular guise via communication theory [7]. Briefly this theory states:-

If a signal is band-limited and if the time interval is divided into equal parts forming subintervals such that each subdivision comprises an interval T seconds long where T is less than half the period of the highest significant frequency (B) component of the signal; and if one instantaneous sample is taken from each subinterval in any manner; then a knowledge of the instantaneous magnitude of each sample plus a knowledge of the instants within each subinterval at which the sample is taken contains all the information of the original signal.

There are at least 5 generalised sampling theorems [8] that spring from this definition - and many more [9,10] related to the relaxation of the bandlimiting constraint. Fortunately we are only concerned with two relatively straightforward interpretations pertaining to systematic and random sampling of bandlimited signals.

3   UNIFORM SAMPLING

3.1   The Nyquist Case.

Let us consider a bandlimited signal $g(t)$ with a frequency spectrum $G(f)$, sampled by evenly spaced Dirac elements [11-15] as follows:-

$$g(t) \cdot s(t) = \text{comb}_T \, [g(t)] \equiv \sum_i \delta(t-iT) \cdot g(t) \qquad \cdots \cdots (1)$$

$$\begin{array}{c} \wedge \\ \| \\ \vee \end{array} \qquad\qquad \begin{array}{c} \wedge \\ \| \\ \vee \end{array}$$

$$\frac{1}{T} \text{Rep}_{\frac{1}{T}} [G(f)] \equiv \frac{1}{T} \sum_n G(f-\tfrac{n}{T}) \qquad \cdots \cdots (2)$$

Alternatively, an amplitude modulation view of this process may be derived:-

$$\frac{1}{T} \sum_n G(f-\tfrac{n}{T}) = \frac{1}{T} \sum_n \delta(f-\tfrac{n}{T}) * G(f) \qquad \cdots \cdots (3)$$

$$= \frac{G(f)}{T} + \frac{1}{T} \sum_n' \delta(f-\tfrac{n}{T}) * G(f) \qquad \cdots \cdots (4)$$

Therefore $g(t) \cdot s(t) = \dfrac{g(t)}{T} \left\{ 1 + 2 \sum_{i=1}^{\infty} \cos 2\pi i \tfrac{t}{T} \right\} \qquad \cdots \cdots (5)$

Pictorially the sampling of a bandlimited signal may thus be viewed as an amplitude modulated harmonic series as indicated in Fig 1.

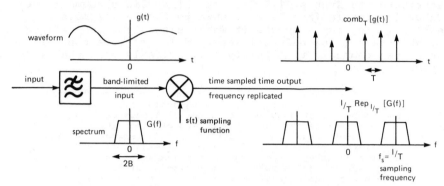

Fig 1   Sampling a band-limited signal

Clearly the limiting condition for separability - or recovery of the original information undistorted, is when $f_s \geqslant 2B$ - this is commonly referred to as the Nyquist condition [1] and 2B called the Nyquist frequency.

3.2   Sub-Sampling.

If the sampling frequency is not at least twice that of the highest component present, or if the signal is not strictly band-limited, aliasing distortion [16] occurs as the frequency bands overlap.

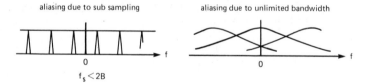

Fig 2   Alias distortion in the frequency domain

This phenomenon of non-ideal-separability may also be conveniently viewed in the time domain as depicted in Fig 3.

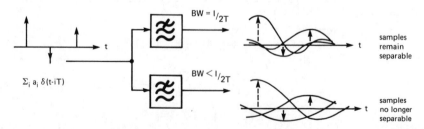

Fig 3   Alias distortion in the time domain

In the case of continuous signals that are not strictly band-limited this effect has to be minimised to prevent serious distortion. The effects of aliasing are often considered as noise-like processes and a power criterion is commonly used to restrict its effect. Typically this might be < 1% of the total recovered sample power depending upon the application [17]. For periodic signals this problem is generally less serious because the frequency spectrum is discontinuous and the spectra can be interleaved by the sampling process. The signal thus remains perfectly recoverable provided the sampling frequency is adjusted to achieve spectral interleaving.

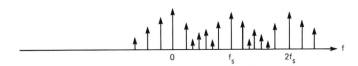

Fig 4   Interleaving of sample spectra

This form of sub-sampling is predominant in instrumentation where sinusoidal or repetitive waves of limited harmonic content are to be processed. For a sinusoidal wave, in particular, the sampling action may be thought of as a "down conversion" modulation:-

$$\text{comb}_T \left[\cos 2\pi f_c t \right] <=> \frac{1}{2T} \text{Rep}_{\frac{1}{T}} \left[\delta(f-f_c) + \delta(f + f_c) \right] \qquad \dots \dots (6)$$

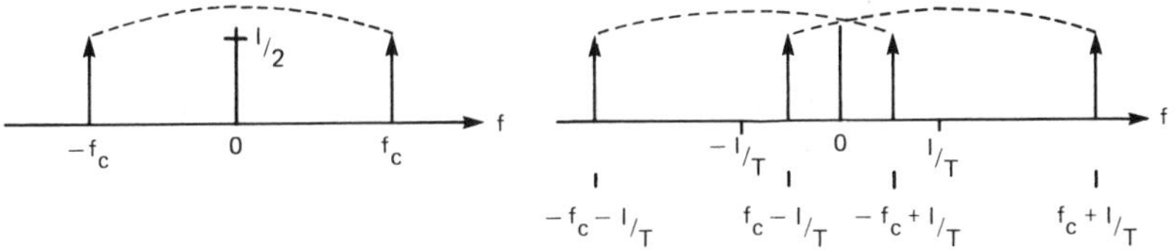

Fig 5   Down conversion via sampling

3.3   Super Sampling.

Strictly speaking any sampling process that uses $f_s > 2B$ falls in the super sampling category. In fact the title is more commonly reserved for cases where $f_s \gg 2B$, which implies a high degree of separability making possible simple filtering. When defining super sampling we have to take care in terms of our definition of the original signal. Consider the two cases of a single carrier and a carrier with bandlimited modulation as shown in Fig 6.

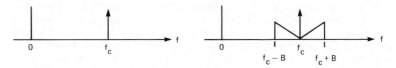

Fig 6   Example spectra

On the one hand to recover the carrier we might expect to sample at $f_s > 2f_c$. On the other, to retrieve the modulation information we have to sample at $f_s > 2B$ - not $f_s > 2(f_c + B)$!! How is this so? Consider the action of the sampler:-

signal $g(t) = [1 + m(t)] \cos\omega_c t$ ..... (7)

sampled signal $h(t) = comb_T [\cos\omega_c t] + comb_T [m(t) \cos\omega_c t]$ ..... (8)

$$\wedge \\ \| \\ \vee$$

$$H(f) = \frac{1}{T} Rep_{\frac{1}{T}} \left\{ [\delta(f) + M(f)] * \frac{1}{2} [\delta(f-f_c) + \delta(f+f_c)] \right\}$$

                                                                      ..... (9)

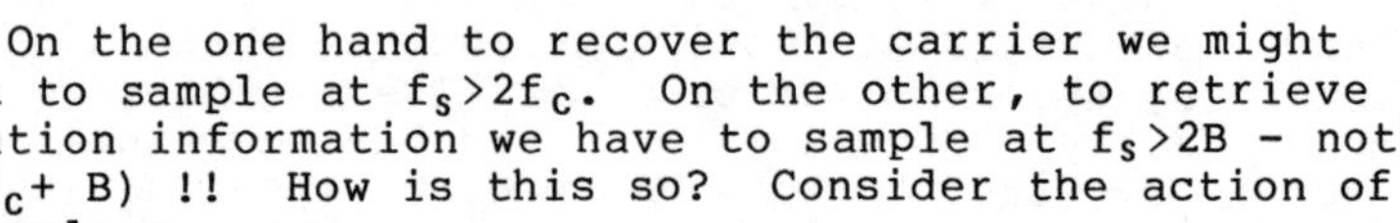

Fig 7   Sampler output spectrum for a carrier plus modulated carrier

Qualitatively we can argue that a carrier alone contains very little information (only amplitude and frequency) and occupies zero bandwidth, therefore requiring a sampling rate $\ll f_c$. Similarly the information bearing energy of the sidebands lies in a bandwidth B, and a sampling rate of $f_s \geqslant 2B$ is sufficient. In short, the process of down conversion can be achieved by a sampler of rate $\geqslant 2B$.

It is clearly necessary to take some care to define sampling systems in terms of the information to be recovered; for a repetitive wave containing fixed amplitude and frequency information we may in **principle** sample as slowly as we wish.

4`   NON-UNIFORM SAMPLING

4.1   Statement of Constraints.

The sampling theorem tells us that we may recover all of the original information provided that samples are

taken in T spaced slots (satisfying the greater than twice
frequency criterion) and that both the amplitude and time
information is recorded [8]. This condition is satisfied
for random and deterministic distributions of the sample
times provided they do not stray beyond the boundary given
by $\pm$ T/2 on the uniform case [18,19].

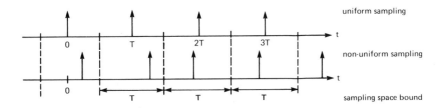

Fig 8   Non-uniform sampling

4.2   Practical Occurrence.

The randomisation of the sampling process can arise
via three principal mechanisms:-

   i.   Uncertainty in the signal arrival time - such as
        pulse from a non-stationary source.

   ii. Jitter in the sampler action.

   iii. Deliberate randomisation, or deterministic
        perturbation, introduced to avoid harmonic or
        sub-harmonic beats between the signal and
        sampling frequencies.

Providing the sampling time shifts are constrained
and both amplitude and position information is retained
each of these categories may be catered for in practice
[21]. However, in the field of nuclear physics the above
conditions are further compounded by amplitude and shape
variations from sample to sample. Fortunately this lies
outside our sphere of interest!

4.3   Deterministic Perturbation Analysis.

As any well behaved deterministic distribution of
the sample instants may be described by a suitable Fourier
series, we consider the special case of a sinusoidal
variation, about the uniform points. Extension to an
arbitrary variation is then a straightforward matter of
including similarly disposed components [22,23]. For
manipulative convenience we therefore apply a sinusoidal
modulation to the sampling instants via a cosine series
description:-

$$s(t) = \frac{1}{T} \left\{ 1 + 2 \sum_i \cos 2\pi f_s it \right\} \qquad \ldots\ldots \text{(10)}$$

With the introduction of a sinusoidal sampling deviation this takes on the deterministic form:-

$$s_d(t) = \frac{1}{T} \left\{ 1 + 2 \sum_i \cos(\omega_s it + \beta \sin \omega_m t) \right\} \qquad \ldots\ldots \text{(11)}$$

$$= \frac{1}{T} \left\{ 1 + 2 \sum_i \sum_n J_n(\beta) \cos(i\omega_s + n\omega_m)t \right\} \qquad \ldots\ldots \text{(12)}$$

Where $J_n(\beta)$ are Bessel functions of the first kind

β is the modulation index

ω_m is the modulation frequency

Sampling a bandlimited signal with this wave gives:-

$$s_d(t).g(t) \Longleftrightarrow \frac{1}{T} \left\{ \delta(f) + \sum_{i,n} J_n(\beta)[\, \delta(f-if_s-nf_m)+\delta(f+if_s+nf_m)\,] \right\} * G(f)$$

$$\ldots\ldots \text{(13)}$$

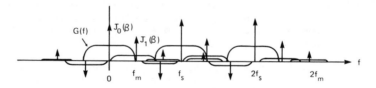

Fig 9   Sampling spectra with deterministic
         perturbation

Hence provided f_s is made sufficiently large, the deviation βf_m is constrained, and we have a knowledge of f_s (ie when the samples were taken) we can recover $g(t)$ intact. When $f_s \gg \hat{f}_G$ - the highest frequency in the bandlimited signal - we need only consider the $i = 0$ component in isolation:-

$$s_d(f)*G(f)\big|_{i=0} = \frac{G(f)}{T} + \sum_n \frac{J_n(\beta)}{T} [\, G(f) - nf_m) + G(f + nf_m)]$$

$$\ldots\ldots \text{(14)}$$

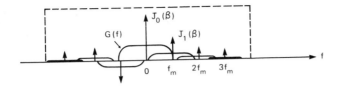

Fig 10   The fundamental sampling spectrum with a
          deterministic perturbation

A further separability condition is now apparent.
When $2 \hat{f}_G < f_m$; $G(f)$ may be recovered by filtering out the
baseband component about $f = 0$.

When the above conditions for separability do not
apply our only recourse is to deconvolve [24] or log both
amplitude and time information of each sample [25,27] –
simple recovery by filtering is not generally possible.

4.4   Random Perturbation Analysis.

Unfortunately there is no concise closed form
analysis available for this case [28] and we thus adopt a
series solution to illustrate the process and its
similarities with deterministic perturbation. Stipulating
that the random modulation introduced by $\phi(t)$ obeys the
condition $|\phi(t)| \ll 1$, we can assume the following
approximations:-

$$s_R(t) = \frac{1}{T} \left\{ 1 + 2 \sum_i \left[\cos\omega_s it - \phi(t)\sin\omega_s it \right] \right\} \qquad \ldots\ldots (15)$$

$$S_R(f) = \frac{1}{T} \delta(f) + \sum_i \left[\delta(f-if_s) + \delta(f+if_s) \right] + j \sum \left[\Phi(f-if_s) - \Phi(f+if_s) \right]$$

$$\ldots\ldots (16)$$

$$S_R(f)*G(f) = \frac{G(f)}{T} + \frac{1}{T}\sum_i [G(f-if_s)+G(f+if_s)] + \frac{j}{T}\sum [\Phi(f-if_s)-\Phi(f+if_s)]*G(f)$$

$$\ldots\ldots (17)$$

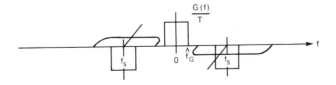

Fig 11   Recoverable spectra with constrained
          random perturbation

Again separability is possible provided; $f_s \gg \hat{f}_G$ and $f_s \gg (\hat{f}_G + \hat{\Phi})$, otherwise a record of all sample amplitude and times has to be made as per the deterministic case.

5   PRACTICAL LIMITATIONS

5.1   Aperture Distortion.

The Dirac δ function is physically unrealizable and in practice we have to content ourselves with more modest functions [29,30]. These are not only limited in their amplitude and width, but also in their shape. Although a rectangular pulse is unrealiseable - as is a rectangular filter in the frequency domain, we use this ideal as a convenient approximation to demonstrate the practical limitations imposed by finite amplitude and width sampling pulses.

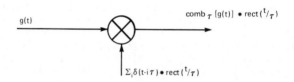

Fig 12   Aperture distortion model

With this imperfect sampling pulse the output is both amplitude scaled and time domain smeared by the convolution process, which leads to a reduction of effective bandwidth in the frequency domain:-

$$\text{comb}_T \left[g(t) \right] \;*\; \text{rect} \left(\frac{t}{\tau} \right) <=> \frac{1}{T} \text{Rep}_{\frac{1}{T}} \left[G(f) \right] . \tau \; \text{sinc} \; (f\tau) \qquad \ldots . . \; (18)$$

| perfect sampling | smearing function | perfect spectrum | bandwidth reduction |

The sampler output thus suffers a frequency domain droop dependent upon the particular shape of the sampling pulse. For all practical pulses of interest the resulting distortion is bounded - best to worst - by sinc and sinc^2, ie the pulse shape lies between the rectangular and triangular. The resulting amplitude distortion introduced by these functions, as well as that for a Gaussian pulse are given in Fig 13.

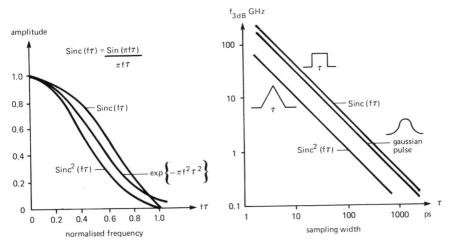

Fig 13   Aperture distortion with pulse width and shape

For some applications this limitation can be partly off-set by introducing an amount of high frequency compensation prior to the sampler as indicated in Fig 14. However, this is not a popular technique at microwave frequencies as it often introduces other complications related to phase distortion which is also compounded by the variability of practical samplers - generally speaking engineering solutions tend towards producing the best sampler possible.

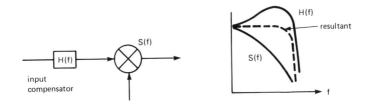

Fig 14   Sampler frequency compensation

For many applications the concept of a bandwidth reduction penalty is sufficient - in others it is somewhat imprecise.   Consider the examples given in Fig 15; clearly for the smooth sinusoid the lack of bandwidth is not critically important - it only results in an amplitude scaling.   For step or pulse type signals however the penalty can be more severe.   In these cases an engineering rule of thumb says; the sampling pulse width should not exceed 1/10 the transient period.

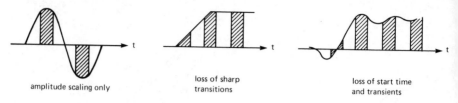

amplitude scaling only

loss of sharp
transitions

loss of start time
and transients

Fig 15   The importance of sampling width

5.2   Synchronised Sampling.

When dealing with periodic signals the sampling process works provided the sampling frequency and signal frequency are not directly (or closely) related. To illustrate this feature let us consider the case when:-

$$g(t) = \cos\left[\frac{2\pi t}{nT}\right]$$ ..... (19)

where n is an integer and $f_s = \frac{1}{T}$

now $\text{comb}_T\left[\cos\frac{2\pi t}{nT}\right] <=> \frac{1}{2T}\text{Rep}_{\frac{1}{T}}\left[\delta(f - \frac{1}{nT}) + \delta(f + \frac{1}{nT})\right]$ ..... (20)

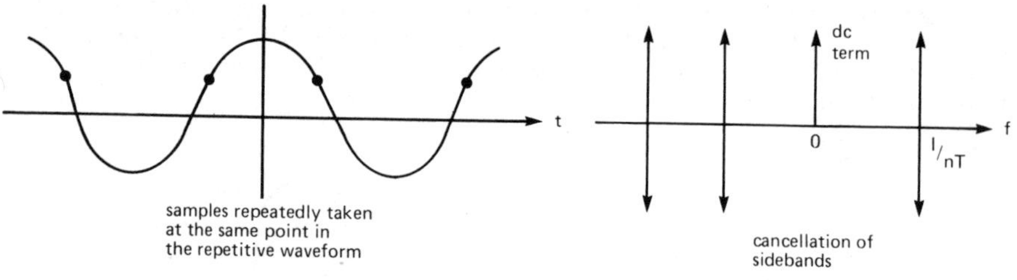

samples repeatedly taken
at the same point in
the repetitive waveform

dc
term

0

$^1/_{nT}$

cancellation of
sidebands

Fig 16   Synchronised sampling of a sinusoid

This clearly results in a repetitive sampling of the same value in the wave which manifests itself as sideband cancellation in the frequency domain. Recovery of the original wave is thus impossible and hence the sampling theorem requirement $f_s > 2B$. For the case of closely related sampling and period time - ie 'n' is 'not quite' an integer - a slow roll or progression becomes evident.

$$\text{comb}_T\left[\cos\frac{2\pi t}{(n+\Delta)T}\right] <=> \frac{1}{2T}\text{Rep}_{\frac{1}{T}}\left[\delta(f - \frac{1}{(n+\Delta)T} + \delta(f + \frac{1}{(n+\Delta)T})\right]$$ ..... (21)

The resulting sampling process thus suffers from a "beat frequency", which in terms of any measurement or observation is undesirable and should be avoided. In contrast, when the sampling frequency and repetition rate of a periodic wave are related in a sub-harmonic manner ($f_s \ll f_g$), then synchronised sub-sampling is achieved giving an expanded time display of the original wave as shown in Fig 17. At present this is the most commonly used sampling CRO display technique for microwave frequencies, and as per the previous case, precise synchronisation must be maintained or a beat frequency (display roll) is produced.

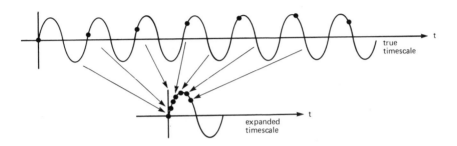

Fig 17   Synchronised sub-sampling of a sinusoid

5.3   The Hold Process.

Because we inherently use narrow sample pulses it is often necessary to stretch the output from the sample gate before it can be further processed [26,29]. Attempting an A/D conversion or trying to display these direct would clearly be impossible. It is thus necessary to hold the sample values so that an A/D converter can converge [31-34] or a CRO tube can be sufficiently illuminated.

The physical realisation of the hold function is dependent upon the particular application, but in general it takes the form of an integrator, which leads naturally to the integrated product description for sampling depicted in Fig 18. In many microwave applications it is necessary to distribute this process using dispersive transmission line, amplification and the sophisticated integration to counteract temperature, voltage and other drift parameters.

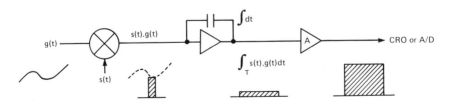

Fig 18   Sample and hold model

5.4   Sampling Jitter and Noise.

All practical sampling schemes suffer from fundamental circuit limitations and signal uncertainties that give rise to both amplitude noise and phase jitter. Given that every care is exercised to minimise these at the design stage, further reduction is possible by signal averaging along both the amplitude and time axes, as well as more sophisticated image processing [31,35]. Broadly speaking averaging techniques achieve performance improvements by the voltage addition of the wanted signal and power addition of the noise/jitter as depicted in Fig 19.

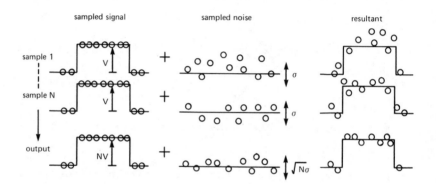

Fig 19   Signal averaging example

$$\text{Original } S/N = \left(\frac{v}{\sigma}\right)^2 \qquad \qquad \dots \dots (22)$$

$$\text{New } S/N = \frac{(nv)^2}{n\sigma^2} \qquad \qquad \dots \dots (23)$$

The improvement attained is thus α 10 log n dB.

For small deviations sampling jitter may be considered to be a noise-like process and, by virtue of phase to amplitude conversion, the above averaging description is applicable. However, amplitude and time axis averaging may be applied independently by operating on the full sample array [31,36-38].

6   A FINAL NOTE ON THE PRINCIPLES

Although we have concentrated on time domain sampling alone, it should be recognised that many of the developments described have a dual role in the frequency

domain. This duality is principally introduced by the
nature of the Fourier Transform Pair. Sampling in time or
frequency produce very similar phenomenon in the other
transform domain. Further analogies may also be drawn
with travelling wave devices which effectively use spatial
sampling to achieve wide bandwidth operation. A good
introductory text explaining some of these aspects, has
been written by Bracewell [13].

7   REFERENCES

1.   Nyquist, H.: 1928, Trans AIEE, 47, 617-644.

2.   Shannon, C.E.: 1949, Proc IRE, 37, 10-21.

3.   Gabor, D.: 1946, Journal IEE, 93, pt 3, 429-457.

4.   Cauchy, A.L.: 1841, Compt Rend Paris, 12, 283-298.

5.   Whittaker, E.T.: 1915, Proc Royal Society Edinburgh,
      35, 181-194.

6.   Whittaker, E.T.: 1935, Cambridge Tracts in
      Mathematical Physics, 33.

7.   Luke, H.D.: 1978, Nachrichtentechnik, 31/4, 271-273.

8.   Yen, J.L.: 1956, IRE Trans on Cct Theory, 251-257.

9.   Kishi, G. & Sakanina, K.: 1981, Electron and Comms
      in Japan, 64-A/3, pp 34-42.

10.   Jerri, A.J.: 1977, Proc IEEE, 65/11, 1565-1596.

11.   Muth, J.M.: 1977, Transform Methods with
      Applications to Engineering and Operations
      Research. Prentice Hall.

12.   Brigham, E.O.: 1974, The Fast Fourier Transform.
      Prentice Hall.

13.   Bracewell, R.: 1965, The Fourier Transform and its
      Applications. McGraw-Hill.

14.   Van der Pol, B. & Bremmer, H.: 1955, Operational
      Calculus Based on the Two-Sided Laplace Integral.
      Cambridge University Press.

15.   Lathi, B.P.: 1968, An Introduction to Random Signals
      and Communication Theory. Intertext Books London.

16.   Wozencraft, J.M. & Jacobs, I.M.: 1965, Principles of
      Communication Engineering. Wiley, New York.

17.   Schwartz, M.: 1970, Information Transmission
      Modulation and Noise. McGraw-Hill, New York.

18.   Papoulis: 1965, Probability Random Variables and
      Stochastic Processes. McGraw-Hill, New York.

19.   Middleton, D.: 1960, Introduction to Statistical
      Communication Theory. McGraw-Hill, New York.

20.   Nahman, N.S.: 1983, IEE Trans IM-32/1, 117-124.

21.   Nahman, N.S.: 1978, Proc IEEE, 66/4, 441-454.

22.   Sarhadi, M. & Aitchison, C.S.: 1980, IEE Elec Lett,
      16/10, 350-352.

23.   Sarhadi, M. & Aitchison, C.S.: 1979, IEE Proc EMC,
      Brighton, England, 345-349.

24.   Nahman, N.S. & Guillaume, M.E.: 1981, Deconvolution
      of Time Domain Waveforms in the Presence of Noise.
      US NBS, Technical Note 1047.

25.   Andrews, J.R.: 1973, Trans IEEE, IM-22/4, 375-381.

26.   Frye, G.J. & Nahman, N.S.: 1964, IEE Trans,
      Vol IM-13/1, 8-13.

27.   Sugarman, R.: 1957, Rev Sci Inst, 28/11, 933-938.

28.   Bendat, J.S. & Piersol, A.G.: 1971, Random Data:
      Analysis and Measurement Procedures. Wiley, New York.

29.   Grove, W.M.: 1966, IEEE Trans, MTT-14/12, 629-635.

30.   Tielert, R.: 1976, IEE Elec Lett, 12/3, 84-85.

31.   Jones, N.B.: 1982, Digital Signal Processing. IEE
      Control Engineering Series 22. Peter Perigrinus.

32.   Isaacson, R. et al: 1974, Electron, 19-33.

33.   Bucciarelli, T. and Picardi, G.: 1975, Alta
      Frequenza, XLIV/8, 454-461.

34.   Kinneman, G. & Wilde, R.: 1978, Nachrichtentechnik
      Elecktronik, 28/8, 337-339.

35.   Reed, R.C.: 1977, IEE Euromeas Conf Pub 152, 101-103.

36.   Haralick, R.M., Shanmucam, K. & Dinstein, I.: 1973,
      IEEE Trans, SME-3/6, 610-621.

37.   Stuller, J.A. & Kurz, R.: 1976, IEEE Trans Comm,
      1148-1195.

38.   Nahman, N.S.: Proc IEEE, 55/6, 855-864.

Chapter 12

Sampling systems

P. Cochrane

1   INTRODUCTION

The use of wideband sampling in microwave
measurements has now become commonplace and has been
developing rapidly since the mid 1950's. Looking back to
its origins, in many respects, it appears to have been
developed principally to overcome the limitations posed
by real time oscillography [1-3]. Here Travelling Wave-
like Cathode Ray Tubes [4-8] have been developed to
display waveforms up to about 7 GHz [5]. These are
extremely expensive devices to produce - as evidenced by
the fact that currently available commercial units have
only in recent years reached 1 GHz bandwidth real time.
In contrast, sampling techniques offer the key attractions
of being both fast and reasonably priced with currently
available commercial units operating up to about 20 GHz.

From these early display applications sampling has
been used increasingly to overcome other measurement
limitations in voltmeters, power meters, counters, timers
and comparison schemes [9-20]. Until recent years the
techniques and instruments commonly in use offered only
equivalent time (multiple occurrence/Sub-Nyquist)
operation. Lately device and circuit developments have
made possible instruments capable of real time (single
occurrence/Nyquist) sampling [21-23]. However, perhaps
the most profound development - for both modes of
operation - has been the development of suitable
digitizers and subsequent linking to the ubiquitous mini-
computer. Next to overcoming fundamental speed
limitations, this one step has increased the measurement
capability and versatility most dramatically.

In this chapter we review the techniques used in
commonly available commercial equipment, and also consider
a few examples of instruments developed for research
applications. Our purpose in this approach is to try to
convey an operational understanding of existing
instrumentation and perhaps pre-empt the arrival of new
instruments in the laboratory. Throughout we utilise the
defined and derived relationships of the previous Chapter

to explain the operation and limitations of specific
schemes considered - which have been classified into two
broad functional groups; visual display and parametric
measurement.

2   PRACTICAL SAMPLERS

As the sampler is the key element in all the
instruments and techniques described in the text to
follow, we briefly consider the realisation of such
devices and their practical performance limitations. The
space constraint on this text only allows us to consider
in detail some commonly used examples. Due reference is
thus made to the many varied techniques developed or
currently being investigated.

2.1   Sample Pulse Generation.

As described in the previous chapter, generating a
suitably short sample pulse is fundamental in attaining
the required bandwidth. For electrically driven samplers
this is sometimes realised using a directional coupler,
but more commonly by a short circuit transmission line or
cavity as depicted in Fig 1.

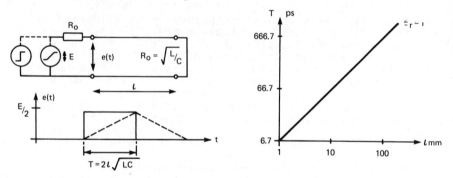

Fig 1   Sample pulse generation using a short
circuit line

The step or ramp drive to the transmission line may
be generated by a variety of devices; Tunnel Diode
(T $\sim$ 20 ps), Avalanche Transistor (T $\sim$ 200 ps), Snap Off or
Step Recovery Diode (T $\sim$ 100 ps) and Mercury Wetted Relay
(T $\sim$ 10 ps). It is often the case that, in turn these have
to be driven or triggered by a high amplitude short
duration ($\tau \sim$ 10T) transition. In practice a cascade of
two or more such circuits may be necessary [24-28].

More recent work with Josephson devices has produced
high repetition pulses/samplers with T $\sim$ 10 ps [28-31],
and laser pulse driven photoconductive devices have
achieved T $\sim$ 100 ps [23]. However, these techniques tend
to introduce considerably more complexity - ie cryogenic
environment - which so far relegates them to specialised

research applications and we therefore neglect them in the remaining discussion. To avoid unwanted transients due to drive circuit operation and/or mismatches between the short circuit transmission line and an "on device", it has often proved necessary to shape or taper the line/cavity [32-35] as per the example shown in Fig 2. However, a number of commercial and experimental instruments now employ a degree of integration, coupled with thin film construction for this function, which overcomes many of the limitations due to transients generated by stray reactive elements and impedance mismatches [36-38]. A typical sample pulse generator circuit used in a commercial CRO is given in Fig 3. This utilises an avalanche breakdown transistor energised by an NPN transistor strobe device operating with a 30 μs period. The avalanche transistor delivers a 40 volt, 5 ns risetime, balanced drive step wave to the snap-off diode, which in turn operates as a current switch delivering a step of about 20 mA (< 40 ps risetime) to the short circuit stub lines.

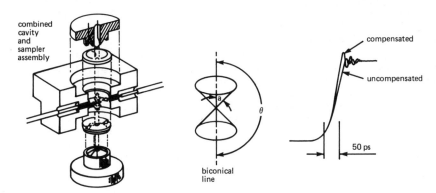

Fig 2   Tapered cavity sampling head

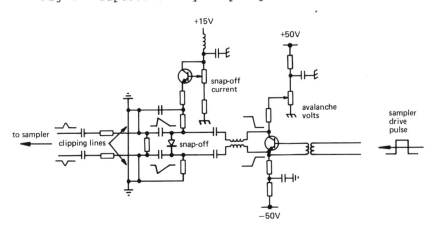

Fig 3   A commercial sample pulse generator

In commercial instrumentation such circuits are generally operated at kHz (Sub-Sampling) rates. However, it is possible to increase the operating rate up to the low MHz region - usually < 30 MHz - before the circuit reaches the power dissipation limit of the avalanche device. Making samplers for real time microwave rates (Nyquist and Super Sampling) can be extremely difficult - and the subsequent signal processing even more so! However, sampling pulse windows of < 200 ps have been achieved at GHz rates using conventional ECL logic drives. Josephson [31] and Optical [22] devices offer the prospect of doing much better in the future.

2.2   Sampling Gate.

This part of the hardware may take many forms; from a single unbalanced diode [39], double balanced ring of four [36], to a series of six floating diodes [37] depending on the specific application. The most commonly used devices include the Tunnel, Schottky and Hot-Carrier diodes giving a sampling bandwidth up to about 40 GHz. Recent work with Josephson devices has given a performance extension to about 70 GHz [28].

Generally, the balanced sampler is preferred as it avoids the "kick out" effect and may be realised in a floating earth, dc coupled, configuration as per the examples shown in Fig 4. The term "kick out" is an American expression coined to describe the leakage of the sampling pulse from the sampler. Typically this phenomenon can be suppressed by > 35 dB, by a balanced circuit, which is necessary to prevent interference to, and reflections from, the circuit under test.

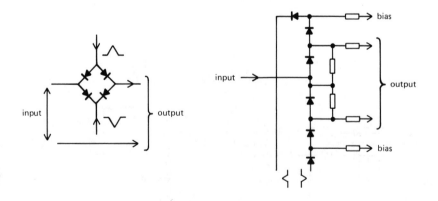

Fig 4   Typical commercial sampling gate
         configurations

2.3   Sampler Circuit Function.

The two diode circuit shown in Fig 5 below is indicative of the types commonly found in commercially available equipment, and serves as a model for further discussion. In this circuit the signal is sampled by balanced strobe pulses applied through the capacitive couplers. Because the input line is floating (ie can be set at an arbitrary dc level) and the signal may be amplitude asymmetric, the diodes can be expected to assume an unequal charge storage state. An output is thus produced that is related to both the amplitude and polarity of the input signal.

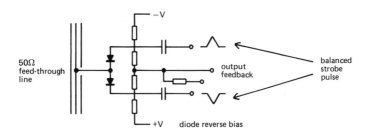

Fig 5   Two-diode sampling gate

A feedback path is provided to reset the reverse bias applied to the diodes after each sampling action so that subsequent samples only detect voltage changes in the input signal. This arrangement increases the dynamic range of the sampler and maintains a near optimum bias.

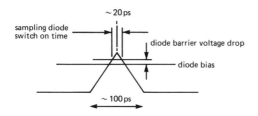

Fig 6   Sampling diode bias

Since the sampler output is proportional to input signal differences, an integrator is required to create an output proportional to the input. In its simplest form the sampler thus appears as shown in Fig 7(a). Taking account of the real-life components a near true equivalent circuit is a good deal more complex as Fig 7(b) indicates.

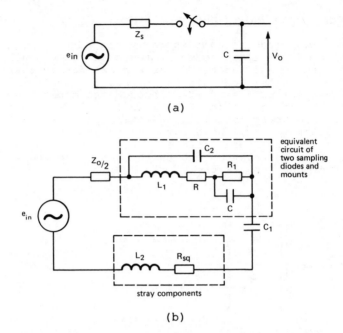

(a)

(b)

Fig 7   (a) Simple equivalent sampler circuit
        (b) A real component equivalent circuit

This is an unbalanced equivalent circuit of the two
diode balanced array. As demonstrated by Grove [32], this
model may be subsequently reduced by assuming $C_1 \gg C_2$ and
that C_2 is anyway masked by embedding the diodes in the
dielectric of the transmission line. The equivalent
sampling circuit is thus reduced to:-

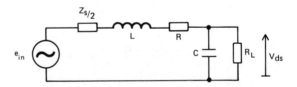

Fig 8   Simplified equivalent circuit for a diode
        sampling gate

Grove derived a standard form complex frequency
domain description for a step input signal:-

$$V_{ds} = \frac{e_{in}\left[\dfrac{1}{LC}\right]}{s^2 + s\left[\dfrac{R + Z_o/2}{L}\right] + \dfrac{1}{LC}} = \frac{e_{in}\,\omega_n^2}{s^2 + 2\delta\omega_n s + \omega_n^2} \quad \ldots\ldots (1)$$

Using this relationship it is possible to optimise the sampler to give maximum flatness - minimum overshoot for whatever application. As a matter of interest Grove [40] selected the following design values for an optimised 15 GHz design:-

$$\delta = \frac{1}{\sqrt{2}} \qquad L = 250 \text{ pH}$$

$$C = 0.2 \text{ pF}$$

$$Z_o = 50 \ \Omega \qquad R = 10 \ \Omega$$

these give an indication of the difficulties of mechanical construction experienced prior to the introduction of integration and thin film technology.

2.4   Sampler Time Domain Analysis.

Whilst the analysis by Grove [32,40] et al [35,41] provides both an adequate design description and switching model, an alternative approach by Nahman [42] provides (in the author's view) a somewhat better picture of the overall operation. Because Nahman's description has only appeared in lecture notes, we briefly reproduce his analysis as follows:-

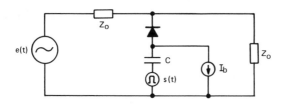

Fig 9   Nahman's single ended model of a balanced sampling gate

The scheme we wish to analyse consists of a uniform transmission line terminated in its characteristic impedance with a sampling gate connected at some point across the line as per Section 2.3. This may be reduced to the equivalent (Thevenin's) signal source and sampling pulse generator circuit shown in Fig 10.

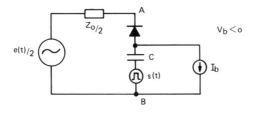

Fig 10   Thevenin equivalent sampler circuit

Here the sampling process requires that e(t), s(t)
and observation of the voltage developed across the
capacitor C be synchronised. The current source I_b
develops a negative potential to keep the diode cut-off,
and as a result C is charged to that bias level V_b (which
is set by the reverse resistance of the diode). For
reasons previously discussed, the sampling pulse s(t) is
very narrow and in this application exceeds the magnitude
of V_b with an excess magnitude greater than the peak to
peak signal e(t). Consequently, the sampling diode is
only turned on by the presence of the sampling pulse.
When s(t) is present the diode is rapidly switched "on-
off" and behaves as a time dependent resistor r(t) over
the time interval $t_1 \leqslant t \leqslant t_2$ (for a single event).

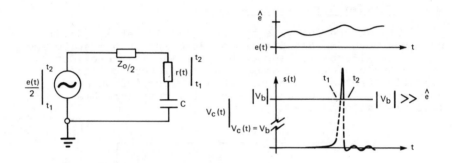

Fig 11   Sample gate switching action

When the sampling pulse amplitude exceeds $|V_b|$, the
diode resistance becomes very small (compared to the
reverse resistance) and the resulting current flow
$(t_1 \rightarrow t_2)$ develops a voltage across C. This current is
governed by the differential equation:-

$$\frac{e(t)}{2} = \frac{Z_o}{2} i(t) + r(t)\ i(t) + \frac{1}{C} \int i(t)\ dt \qquad \ldots\ldots (2)$$

The sampler output is the capacitor voltage change:-

$$y(\tau) = \frac{1}{C} \int_{t_1}^{t_2} i(t)\ dt \qquad \ldots\ldots (3)$$

Because this changes very little during the sampling
period, it may be considered essentially constant and
almost equal to $V_c(t_1) = V_b$. Equation (2) may thus be
approximated by:-

$$e(t) - 2V_b = \left[Z_o + 2r(t)\right] i(t) \qquad \ldots\ldots (4)$$

$$i(t) = \frac{e(t) - 2V_b}{Z_o + 2r(t)} = \left[e(t) - 2V_b\right] g(t) \qquad \ldots\ldots (5)$$

Where $g(t)$ is the gating function:-

$$g(t) = \left[Z_o + 2r(t)\right]^{-1} \qquad \ldots\ldots (6)$$

Entering (5) in (3) now gives for the one sample sample $(t_1 \rightarrow t_2)$:-

$$y(\tau) = \frac{1}{C} \int_{t_1}^{t_2} \left[e(t) - 2V_b\right] g(t - \tau)dt \qquad \ldots\ldots (7)$$

Where $g(t)$ is synchronised to $e(t)$ but delayed by a selectable time τ.

$$\therefore y(\tau) = \frac{1}{C} \int_{t_1}^{t_2} e(t)g(t - \tau) + \frac{2V_b}{C} \int_{t_1}^{t_2} g(t - \tau)dt \qquad \ldots\ldots (8)$$

Since the last term yields a constant:-

$$y(\tau) = K + \frac{1}{C} \int_{t_1}^{t_2} e(t) \, g(t - \tau)dt \qquad \ldots\ldots (9)$$

This sampling function is thus of the integrated product type and in practice would be of a balanced form (ie two diodes at least). The constant term 'K' of equation (9) would thus be eliminated by the contribution of a second diode, capacitor and sample pulse as indicated in Fig 12. Sampling pulse leakage (kick-out) is also clearly balanced out by this circuit configuration.

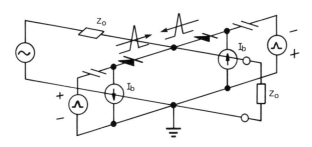

Fig 12   A two diode balanced sampling gate
         configuration

2.5   Sampler Probes.

A major problem in microwave measurements arises
when we try to co-locate test equipment and the system
under test. Obviously test leads and interconnections
must be kept as short as possible to avoid waveform
distortion due to cable/waveguide attenuation, phase and
impedance mismatches. Sampling probes offer a novel means
of achieving this objective.

Because the sampler may be constructed in a very
small size, complete with pulse generator, it is ideally
suited to probe type applications. Here the sampling
diodes can be located at the back of a connector or probe
tip (solder-in or touch) and present a well defined high
frequency interface [36]. The output fed to the main body
of the test equipment is then predominantly low frequency
and has no influence on the measurement path.

3   VISUAL DISPLAY TECHNIQUES

Under this generic heading we can group not only
sampling oscilloscopes, but correlators, time domain
network analysers, sampling (down conversion) network
analysers, sampling spectrum analysers, amplitude
detectors and transient recorders [22]. Fortunately, most
of these methods are being considered separately in other
chapters of this book - whilst others only tenuously fall
within the sampling system category. In this chapter we
therefore concentrate on the single and multiple sampler
oscilloscope.

3.1   The Sequential Sampling Oscilloscope.

In its conventional single sampler form this
instrument synchronously sub-samples a waveform in a
similar manner to that described in the previous
Chapter, and more closely to that outlined by Yen [43]
(Sampling with a Single Gap in an Otherwise Uniform
Distribution). The physical realisation of this process
requires a trigger signal [44-46] (generated external to
the main CRO hardware and separate from the signal - or
picked off within the CRO from the signal) to initiate
regularly spaced samples relative to some fixed point in
the signal. Each successive sample is delayed by a fixed
amount Δt as depicted in Fig 13.

In practice there is normally a delay between the
trigger signal and the samples being taken - this is due
to the effective path length differences between signal
and trigger. Whilst this is not usually a problem for
waveforms of the kind shown it does sometimes create
difficulties for pulse display, as will be explained
later. The problem can be effectively overcome by the use
of a delay line in the input signal path or by random
sampling, as described in Section 3.2. Most commercial
sampling CROs use samplers at a rate ~ 100 kHz with a

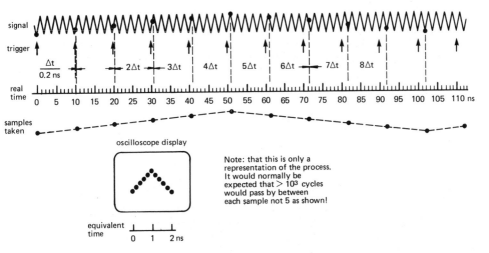

Fig 13   Sequential sampling and display of a
periodic waveform

display of 1-3 k dots giving the illusion of a continuum -
not as shown in the example. At this rate of sampling it
can take ~1/50 of a second to build up the display - just
fast enough to avoid flicker.

Much of the sampling CRO hardware and its operation
remains the same as its conventional real time counter -
part, but as the sampling process is with reference to a
particular point in the waveform, it is also capable of
displaying repetitive but irregularly spaced pulses as
shown in Fig 14.

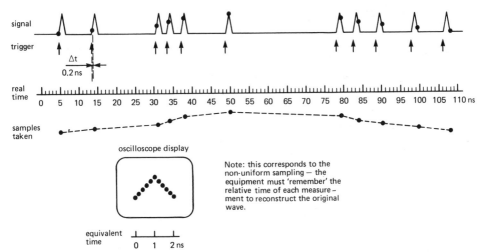

Fig 14   Sequential sampling and display of a
repetitive but non periodic waveform

An indication of the hardware configuration and associated control waveforms of a commercial sampling CRO is given in Fig 15. The optional delay line in the input path to the sampler (shown dotted) overcomes the problem of initiating the trigger sample command prior to the arrival of a pulse at the sampler. However, unless this line is super-cooled [48,49] its bandwidth is unlikely to be more than 1 GHz or so! This is a severe limitation of the sequential sampling technique for pulse type applications, which fortunately, may be overcome without cryogenic plant as described in the next section.

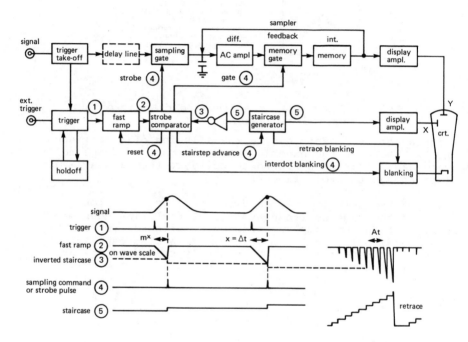

Fig 15   Schematic of a commercial sampling CRO

3.2   The Random (Display) Sampling Oscilloscope.

In the sequential case just described, the samples are taken in an orderly fashion one after another and displayed as a uniformly spaced succession of finely focussed spots. The distance between each spot thus represents a fixed interval in the original wave. For low repetition rate, or random arrival time pulses, this approach is inadequate as the first sample taken will miss the leading edge. This limitation is introduced by the different signal and trigger path delays within the CRO [45,46]. If the use of a delay line in the signal path cannot be tolerated, and if there is no external pre-trigger available, then we have to resort to a random sampling scheme.

The reader should note that; in most commercial equipments the word random does not imply truly random sampling or random pulse arrival times – indeed we are generally restricted to repetitive waveforms. The term arises from the randomisation of the final display due to processing uncertainties.

The principle behind this scheme is to trigger on one pulse and then delay the sampling action to just before the arrival of the next – thereby capturing the leading edge as well as the rest of the pulse using the sequential sampling routine previously described. The elements of a random sampling CRO are given in Fig 16.

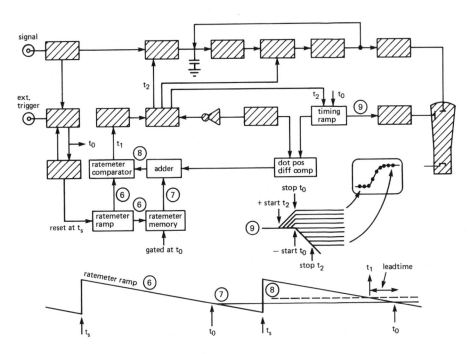

Fig 16   Schematic of a commercial "random sampling" CRO

Here key additional elements, over and above those shown in Fig 15, are shown unshaded. It should also be noticed that a number of interconnections have now been omitted and/or introduced between the new and existing blocks. The key element in this new configuration is the ratemeter which forms an estimate of the pulse repetition rate – and so positions the sampling window one (or more) pulse position/s on from the trigger point. This is achieved by initialising the slow ramp ⑥ shortly after the trigger pulse (time t_s) and then stopping the ramp at the next t_0 (ie the same amplitude point in the pulse). The ratemeter memory then stores the amplitude ⑦ and a DC component is added ⑧ to create a sampling lead-time, allowing the whole of the pulse to be displayed and positioned mid-screen.

In principle this is a simple and effective scheme, but in practice there are some severe snags; any incidental noise, instability or FM on a signal continually changes the difference between successive pulse arrival times (t_0) and thus the estimated sampling time (t_1) also varies. In addition the ratemeter ramp can be extremely slow - and thus flat - thereby introducing switching point uncertainty at the comparator for ⑥ and ⑧ . The sampling lead time can thus be subject to large uncertainties that would, if not corrected for, lead to an incoherent display. This limitation is overcome by the timing ramp which introduces an x-position correction ⑨ based on the difference between the sample position and trigger time. Display dots thus arrive on the screen in the correct position but not necessarily equispaced or in the correct sequence - leading to the description "random sampling".

3.3   The (True) Random Sampling Oscilloscope.

This scheme was devised [41] to record pulses of the same shape arriving in a random manner - specifically the condition of no synchronisation between signal and sampling [50]. For it to do this it has to record both amplitude and position information of each sample taken - see previous Chapter Sec 4. A schematic of the necessary hardware is given in Fig 17. Here it can be seen that the input pulse ① and trigger ③ arrive simultaneously which triggers the X time ramp ⑦ at time t_0. A sampling pulse ② arrives at some arbitrary time t_1 and a sample is taken ④ and passed to the Y memory ⑤. The sampling pulse ② passes through the delay ⑥ (allowing for the non-zero start up of the ramp generator) and stops the ramp ⑦ at time $t_1 + \tau$. The two memories now contain the initial information of sample amplitude and position which can be passed on to the CRT for ~ 3 μs of display, the memories are then reset for the next sample.

Although regularly spaced sample pulses ② are shown, this is not necessary - random sampling pulses and random signal or periodic signal and random sampling can equally well be accommodated by this scheme. However, the principal mode of operation is likely to be preferred as this gives a higher likelihood of signal and sampling pulse being coincident. The arrival of the spots on the CRT are very definitely random for this instrument as illustrated by the examples of Fig 18 for N successive samples recorded.

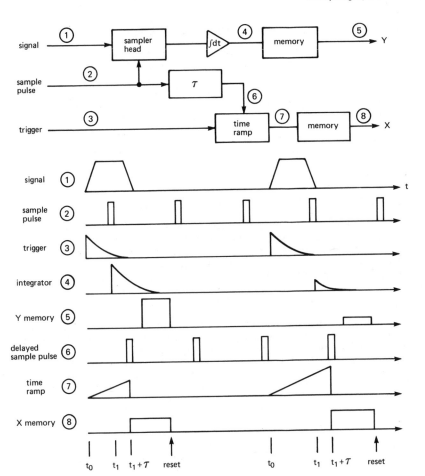

Fig 17   A true random sampling scheme

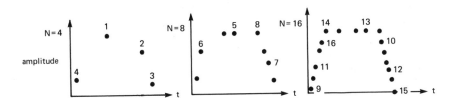

Fig 18   Random sample display assembly

3.4   The Digital Processing Oscilloscope (DPO).

A most powerful extension of the standard sampling CRO is now commercially available under the general title of the DPO [51]. This type of instrument is also reported in the literature under the heading of Time Domain Network Analyser [51]. Such equipments utilise A/D conversion of the CRO X and Y outputs (the X is now very often realised via a programmable delay as opposed to a time base ramp) to feed a signal processing computer. Here facilities such as Convolution, Deconvolution, FFT and IFT [54] may be used to manipulate recorded signals as described in previous chapters.

3.5   The Multiple Sampler Oscilloscope.

Unlike the CROs described so far, which function by time stretching - operating in equivalent time - the schemes we are now to consider operate in real time. Capturing a fast single shot event is extremely difficult at microwave frequencies, but has been achieved with a single sampler using a super-cooled recycling delay line and by using a multi-head sampling process [21,22,38].

In the recycling line case a pulse is injected into a super-cooled 'racetrack' [55] passing by a standard sub-sampler a number of times. At each pass a successively delayed sample is taken and a complete picture of the pulse recorded. Because the sampler has to operate at μs rates a long delay line is required that has to be super-cooled to avoid serious dispersion effects. This technique does not appear to have proved popular and is only scantly referenced.

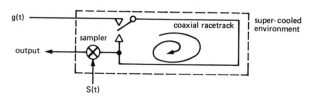

Fig 19   Racetrack multiple sampler

A more attractive arrangement, also scantly referenced is the multi-sampler solution [56,57] shown in Fig 20. Here tens, and sometimes hundreds of samplers, are arranged along a relatively short line; a 10 ps sample spacing dictates ~ 3 mm between each head, whilst 100 ps requires ~ 3 cm. Thus for 100 samplers the line only need be ~ 0.3-3 m long. These present a wide real time sampling window to a pulse which may be captured at a single shot.

A/D conversion or direct readout and display may follow the multi-head sampling operation - as per the more popular single sampler CRO previously described.

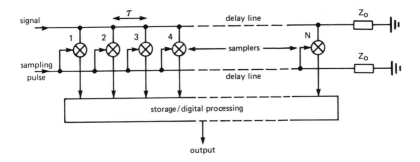

Fig 20   Multi-sample gate solution

4   PARAMETRIC MEASUREMENT TECHNIQUES

There are a very large number of techniques that
fall under the general heading of this section. These use
random or synchronised sub and super-sampling to perform
time domain processing and frequency domain down
conversion [36]. We therefore consider some of the more
common, and mention some not so common, examples of these
types of processes.

4.1   Random Sampling Voltmeter.

Producing detectors (ac to dc converters) for
microwave applications is extremely difficult if wide
dynamic range and flat frequency response are required.
Generally, semiconductor diodes change their mode of
operation with signal amplitude and frequency [58]. For
example; the exponential characteristic of a diode
dictates that it tends to be a good peak detector for
sinusoidal signals of $\geqslant 500$ mV amplitude and gives an rms
value for $\leqslant 30$ mV. In addition, the stray L, C associated
with both the diodes and their mounts also render them
frequency sensitive. Such limitations obviously become
further compounded for non-sinusoidal and modulated
signals. These factors coupled with the temperature
sensitivity and inherently noisy characteristics of such
detectors precludes their general use in microwave
measurements. Fortunately all of these limitations can
(largely) be overcome by the use of a sub-sampling
voltmeter [59]. The schematic of a typical instrument is
given in Fig 21(a). Instruments of this type rely on a
large number of samples taken at random in the wave to
allow such parameters as the; peak, mean, rms, power and
in some cases the pdf to be recorded. Randomisation of
the sampling process is necessary for the reasons stated
in the previous Chapter, and in this particular scheme is
achieved by frequency modulating the sampling rate. A
typical sampling frequency would be 10-20 kHz modulated at
10 Hz as depicted in Fig 21(b). Alternatively a noise
source of pseudo-random modulation sequence may equally
well be employed.

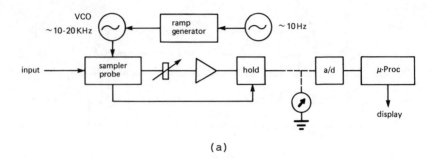

(a)

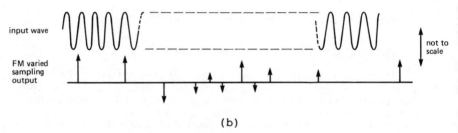

(b)

Fig 21   (a) Random sampling voltmeter
         (b) Randomised sampling

Instruments based on this scheme are principally
limited by the type of sampler used, but can be expected
to yield:-

Sensitivity	$< 50~\mu v$
Resolution	$< 20~\mu v$
Accuracy	$\pm~3\%$
Convergence	< 2 seconds
(after autoranging)	

4.2   Phase Locked Sampling Vector Voltmeter.

In this type of instrument the wideband sampler is
effectively used to down convert a signal to a pre-defined
Intermediate Frequency (IF) [60]. This is achieved by
synchronous sub-sampling controlled by a Phase Locked Loop
(PLL) circuit as indicated in Fig 22.

Because the samplers produce a comb-like spectrum
over a very wide range, the incoming signal is convolved
onto the spectral lines to create a low frequency
component that lies within the pass-band of the IF
filters. When a signal is initially applied to reference
channel 'A' and the PLL is not locked, the modulation
components may not fall within the pass-band centred on
f_s. The PLL search generator (Fig 23) produces a ramp
which sweeps the sampling VCO; when the modulation
component moves into the IF filter range the PLL locks and
the search sweep is inhibited. Under this condition the
difference between the Kth harmonic of f_s and f_a is f_i.

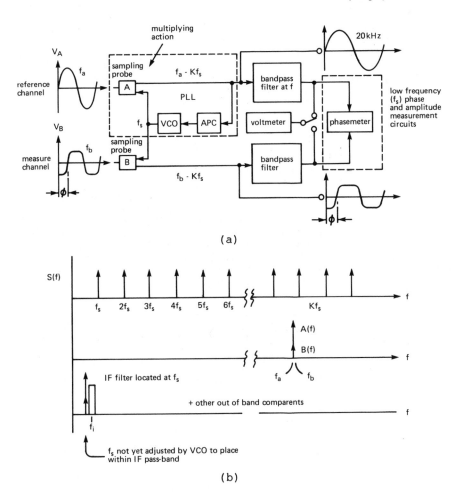

(a)

(b)

Fig 22   Phase locked vector voltmeter

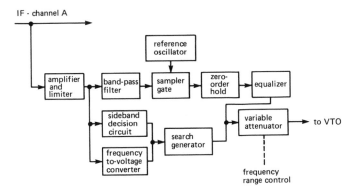

Fig 23   Phase lock search generator

Commercial instruments [61] typically use ~ 20 kHz sampling rates with IF bandwidth ~ 1 kHz and exhibit >90 dB dynamic range. This is mainly afforded by the noise reduction performance of the narrow-band IF, which also allows precision amplitude and phase measurements to be made with relatively modest hardware following the IF filters. Typical performance figures are 0.1 dB and 0.1° resolution.

4.3   Phase Locked Sampling Frequency Meter.

As described in Section 3.2 of the previous Chapter, any sub-sampling system inherently implies a frequency down conversion process [60]. Provided sufficient information about the signal being measured is available, then it follows that its frequency may be recorded using a low frequency counter. However, such information - necessary to perform measurements on the correct spectral line - is not generally available without some sophisticated pre and post-sampler processing [60-63]. One scheme that gives the correct frequency scaling - or spectral line selection - automatically [64] is shown in Fig 24.

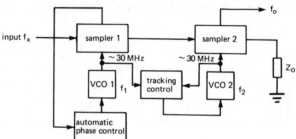

Fig 24   Automatic phase locked sampling frequency meter

The two feed-through samplers (as described in Section 2) are operated at different sampling rates, f_1 and f_2, which are known with precision. This is achieved by the frequency offset PLL arrangement of Fig 25.

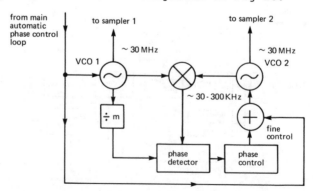

Fig 25   Offset PLL sampling clock generator

In turn the two samplers are phase locked to the unknown frequency such that:-

$$f_i = nf_1 \qquad\qquad \dots\dots (10)$$

and
$$f_o = n(f_1 - f_2) = f_n \left(1 - \frac{f_1}{f_2}\right) \qquad\qquad \dots\dots (11)$$

but
$$\frac{f_1}{m} = f_2 - f_1 \qquad\qquad \dots\dots (12)$$

∴
$$\frac{f_2}{f_1} = \frac{m+1}{m} \qquad\qquad \dots\dots (13)$$

so
$$f_o = \frac{f_x}{m} \qquad\qquad \dots\dots (14)$$

The value of 'm' is usually 10^2-10^4 and counting the frequency is therefore straightforward - even at 20-40 GHz. This approach overcomes many of the limitations and spectral line uncertainties of prescalers using mixers and IF amplifiers [65]. However, with no front-end protection for the first sampler the PLL design becomes more difficult as this has to cope with a range of input levels - which can be adversely affected. The two main requirements of the input PLL are:-

i.   An automatic search capability - usually achieved by a ramp sweep of the VCO - disabled on acquisition.

ii. A wide gain margin to maintain loop stability over a wide range of harmonic numbers; n = 10-500.

A good description of this design problem is given by Allen [64] and a more complex alternative using triple parallel sampling is described by Aitchison [63].

5   A FINAL NOTE ON SYSTEMS

In this Chapter we have briefly looked at the design, realisation and operation of samplers, and their application in a few instruments concerned with display and parametric measurement. Although we have cited numerous references and made comment on these, it should not be supposed that the coverage is comprehensive - for it is not! We have merely taken a snap-shot (a sample!) of what is available in the published literature - when this contribution was researched more than 400 pertinent references were found - with relative ease!

Where next - what does the future hold? Remarkable
as it might seem we can expect even faster commercial
samplers based on quantum and optical devices that should
break the 1 ps barrier. Even more powerful and compact
computers will also become available in the near future to
give higher degrees of sophistication in the related
signal processing and measurement control.

6   ACKNOWLEDGEMENTS

Many of the diagrams and illustrations given in this
chapter are based on previously published works by the
staff members of the Tektronix and Hewlett Packard
Companies. I am indebted to both organisations for
permission to reproduce this material and, in particular,
wish to extend my thanks to Brian Leake of HP, John
Schmid, Peter Derby and Brian Curant of Tek for their kind
assistance.

During the various stages of draft preparation my
proof readers for these companion chapters were;
Fred Westall (BTRL), Roy Borge (NELP), Derek Ingram
(Cambridge), Dr Richard Collier (Kent) and
Dr Norris Nahman (NBS Colorado). I thank these gentlemen
for their diligence in vetting the manuscript and for
providing helpful comments and additional references.

Acknowledgement is also made to the Director of
British Telecom Research Laboratories for permission to
publish this contribution.

7   REFERENCES

1.   Janssen, J.M.: 1950, Philips Tech Rev, 12, 52-81.

2.   Germeshausken, K.J., Golderberg, S. & Mcdonald, D.F.:
       1957, Trans IRE, ED-4, 152-158.

3.   Frye, G.J., Bruce, J.D. &.Nahman, N.S.: 1961, Trans
       IRE, 1/10, 85-89.

4.   Bradley, D.J. & New, G.H.C.: 1974, Proc IEE, 62,
       313-345.

5.   Clement, G. & Loty, C.: 1973, Electron, 16/1, 102-111.

6.   Prosser, R.D.: 1972, Proc IEEE, 60, 645-646.

7.   Stonebaker, D.M.: 1968, NBS Jnl, 72/2.

8.   Andrews, J.R. & Nahman, N.S.: 1971, IEEE 11th Symp on
       Electron, Ion and Laser Beam Technology, San
       Francisco Press Inc, 141-146.

9.   Bologlu, A. & Barber, V.A.: 1978, HP Jnl, 2-16.

10. Schneider, R.F., Felenstein, R.E. & Offermann, R.W.: 1979, HP Jnl, 41-47.

11. Cole, R.H.: 1982, IEEE Conf on Precision Electromagnetic Meas C/18-20.

12. Faulkner, N.D. & Vilar, E.: 1982, ibid M/10-11.

13. Sampling: 1982, A key to the future in hyperfrequency instrumentation. Toute Electron, 470, 23-5.

14. Malherbe, J.C. & Kirilenko, J.F.: 1981, Electron Indust, 8, 45-7.

15. Perrett, J.J.: 1981, ibid 19, 53-4.

16. Anderson, J.M.: 1980, Rev of Sci Inst, 52/1, 145-146.

17. Cittins, D.R.: 1979, Comm Int, 6/1, 48-50.

18. Fujisawa, K., Yamhmoto, Y., Ito, T. & Iwai, T.: 1978, 8th European Microwave Conf, 508-512.

19. Gaddy, O.L.: 1960, IRE Trans Inst 1/9, 326-333.

20. Honnold, G.H. & Nahman, N.S.: 1964, IEEE Trans Inst and Meas, 123-128.

21. Nahman, N.S.: 1967, Proc IEEE, 55/6, 855-864.

22. Nahman, N.S.: 1978, Proc IEEE, 66/4, 441-454.

23. Nahman, N.S.: 1983, IEEE Trans, IM-32/1, 117-124.

24. Andrews, J.R.: 1973, Trans IEEE, IM-22/4, 375-381.

25. Tieler, R.: 1976, IEE Elec Lett, 12/3, 84-85.

26. Pulse and waveform generation with step recovery diodes. HP Appl Note 918.

27. Andrews, J.R. & Baldwin, E.E.: 1978, NBS Tech Note 888.

28. Tuckerman, D.B.: 1980, App Phys Lett, 36/12, 1008-1010.

29. Weker, N.K. & Bedard, F.D.: 1977, URSI Conf on Meas in Telecomm, Lannion, 155-158.

30. Faris, S.M.: 1980, App Phys Lett, 36/12, 1005-1007.

31. Hohkawa, K., et al: 1983, Elect Lett, 19/8, 291-292.

32. Grove, W.M.: 1966, IEEE Trans, MTT-14/12, 629-635.

33. Sternes, K.J.: 1972, IEEE Trans, IM-21/3, 209-214.

34. Ramo, S., Whinner, J.R. & Van Du Zer, T.: 1965
Fields and Waves in Communication Electronics.
John Wiley, 463.

35. Riad, S.M.: 1982, IEEE Trans, IM-31/2, 110-115.

36. Sayed, M.M.: 1980, HP Jnl, 31/4.

37. Mulvey, J.: 1970 Sampling oscilloscopes. Tektronix,
Beaverton, Oregon.

38. Tektronix: S-6 sampling head. ibid.

39. Sugarman, R.: 1957, Rev of Sci Inst, 28/11, 933-938.

40. Grove, W.M.: 1966 12.4 GHz Feedthrough Sampler
HP Jnl.

41. Nahman, N.S. & Riad, S.M.: 1978, IEEE Microwave
Symposium Digest, Ottowa, 267-269.

42. Nahman, N.S.: The Characteristics of Sampling.
Lecture notes for a joint University of Colorado/NBS
course on "Time Domain Measurements" Spring Semester
1976 and also reproduced for Ecole d'Ete Tegor, CNET,
Lannion 1978.

43. Yen, J.L.: 1956, IRE Trans on Cct Theory, 251-257.

44. Best, A.I., Howard, D.L. & Umphrey, J.M.: 1966, HP Jnl
27/3.

45. Tektronix: 7S11 Sampling Unit Manual. Tektronix,
Beaverton, Oregon.

46. Tektronix: 7T11 Sampling Sweep Unit. ibid.

47. Lawton, R.A. & Andrews, J.R.: 1976, IEEE Trans,
25/1, 56-60.

48. Nahman, N.S.: 1973, Proc IEEE, 61/1, 76-79.

49. Andrews, J.R.: 1974, IEEE Trans, IM-23/4, 468-472.

50. Frye, G.J. & Nahman, N.S.: 1964, IEEE Trans IM-13/1,
8-13.

51. Rousseau, T. & Cox, B.: 1980, Tekscope, 12/3, 3-9.

52. Bancroft, J. & Johnston, R.: 1973, Proc IEEE, 20/1,
472-473.

53. Andrews, J.R.: 1978, Proc IEEE, 66, 414-423.

54. Gans, W.L.: 1976, <u>IEEE Trans</u> <u>IM-25</u>, 384-388.

55. Cummings, A.J. & Wilson, A.R.: 1964, <u>Proc IEEE</u>, <u>52</u>, 1749.

56. Schwarte, R.: 1972, <u>Elec Lett</u>, <u>8/4</u>, 95-96.

57. Davies, T.J. & Nelson, M.A.: 1976, <u>Applied Optics</u>, <u>15/6</u>, 1404-1410.

58. Detwiller, W.L.: 1979, <u>Communications</u>, <u>16/2</u>, 42-48.

59. Boatwright, J.T.: 1966, <u>HP Jnl</u>, <u>17/11</u>, 2-8.

60. Yen, C.S.: 1965, <u>IEEE Trans</u>, <u>IM14/1</u>, 64-68.

61. Carlson, R. & Weinert, F.K.: 1966, <u>HP Jnl</u>.

62. De Bella, G.B.: 1968, <u>HP Jnl</u>.

63. Underhill, M.J., Sarhadi, M. & Aitchison, C.S.: 1978, <u>IEE Elec Lett</u>, <u>14/12</u>, 366-367.

64. Allen, R.L.: 1967, <u>HP Jnl</u>.

65. Chappel, R.: 1978, <u>Electronic Eng</u>, 39-43.

Chapter 13

Spectrum analysis

S. J. Gledhill

1 SIGNAL ANALYSIS

The analysis of electrical signals is vital to all
engineers, of equal importance is the type of analyzer
which is used. An electrical signal, such as a
sinewave, can be characterized in three ways:-

- Amplitude
- Time
- Frequency

Fig. 1 shows a sinewave with a second harmonic
represented on a three dimensional graph with the
amplitude, time and frequency axes being mutually
perpendicular.

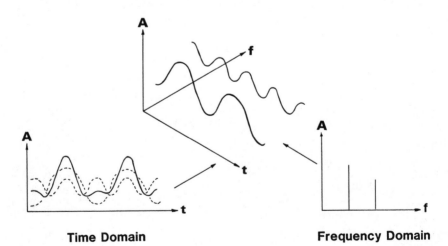

Time Domain   **Frequency Domain**

Fig. 1 Signal analysis

Three dimensional displays are not used in practice so we have to choose which of the two dimensions are to be displayed and what type of instrument is used.

We are all familiar with one type of signal analyzer, the oscilloscope, which displays a signal in the form of amplitude against time. When examining a distorted sinewave the oscilloscope would show a waveform as depicted in Fig. 2.

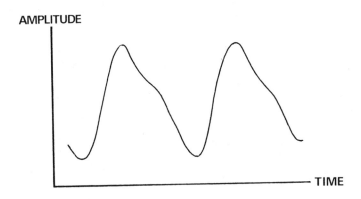

Fig. 2   Oscilloscope display of a
distorted sinewave

An oscilloscope can show timing and phase differences between signals and can indicate transients but it is not very good at showing many signals; further problems are that oscilloscopes do not readily cover microwave frequencies and they can not measure low level signals.

A spectrum analyzer does not suffer from such problems. It displays amplitude against frequency, the same distorted sinewave would be displayed as in Fig. 3.

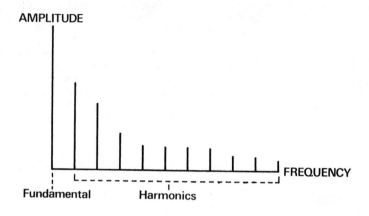

Fig. 3 Spectrum analyzer display of a distorted
sinewave

Much more information is given. The amplitudes and
frequencies of each harmonic are clearly shown.

2   WHAT IS A SPECTRUM ANALYZER?

A spectrum analyzer is basically a swept-tuned
superheterodyne radio receiver with a visual display as
can be seen from the simplified block diagram in Fig. 4.
The ramp generator controls the voltage controlled
oscillator as well as the spot on the display so that
the spot moves across the screen in synchronism with the
tuned frequency of the analyzer.

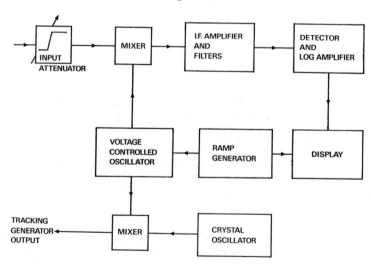

Fig. 4   Simplified block diagram of a spectrum
analyzer

Such a simple block diagram only shows the concept, in practice there are many frequency conversion changes to ensure high selectivity and freedom from spurious responses. Most of the blocks are variable, the sweep centre frequency, sweep width and IF filter bandwidth (resolution bandwidth) can be changed so that the instrument can be adjusted to either view a wide part of the spectrum using a wide IF filter, or a narrow part of the spectrum using a narrow IF filter.

3   FURTHER CONSIDERATIONS

In the 1960's it was rare to find spectrum analyzers outside research and development laboratories. Since then the situation has changed, spectrum analyzers are easier to operate and their specifications have been improved, opening up many new applications.

Ease of use has been the major innovation because with the early spectrum analyzers, and many less sophisticated modern instruments, it is very easy to obtain totally incorrect readings. The problems arise from the fact that when sensitivity and resolution are changed then other parameters need to be changed or optimised, and these parameters are interdependent.

3.1 Sensitivity

Consider what appears to be the most simple operation, increasing the sensitivity. This can be carried out either by reducing the attenuation before the instrument's input mixer or by increasing the gain at a later stage in the IF chain.   Which is correct? An unknowing operator could easily change the wrong control. If the attenuation at the input is reduced then this may overload the input mixer and produce intermodulation and harmonic distortion; a measurement of a signal may then be incorrect because the instrument itself may be creating the distortion which is seen on the display. To be safe the operator may leave the RF attenuation alone and increase the IF gain, the problem here is that this may introduce an unacceptably high level of noise which may mask the lower level signals which are to be analyzed.

This is quite a dilemma in itself, but the problem is further compounded because the optimum conditions change according to the input signal level and the required resolution bandwidth. A narrower bandwidth causes the noise level to fall thus posing a new intermodulation and noise trade-off condition. These problems are now solved in sophisticated instruments, they are internally programmed to select the optimum operating conditions without introducing errors.

3.2   Resolution

Once the sensitivity has been changed an operator will then invariably want to change the tuned frequency and then reduce the scan width so that selected signals can be viewed with narrow spans. As the scan width is reduced then narrower filters need to be selected, but a narrow filter may appear to broaden and the amplitude will be reduced if it is swept too fast; the sweep speed thus must also be optimised. This is further considered in a later section. On older instruments this was a problem but it is no longer a problem since sweep speeds are selected automatically by the instrument.

If the sweep speed is too slow, and sweeps as long as 100 seconds are not uncommon, then a storage medium is required, for this purpose digital storage of displays is now used. Previous generation instruments used storage tubes which required skilful, time-consuming, manipulation.

When one considers that as the resolution bandwidth is reduced the noise/intermodulation ratio may change it is easy to appreciate why the innovation of coupled pre-programmed controls was such a breakthrough.

3.3   Additional Features

The features which contribute to ease of use explain why modern spectrum analyzers have become much more widely used. We also need to look at innovations and improvements in the specification to consider what else has contributed to increased acceptance. Horizontal resolution is one aspect that is vitally important, especially if one wishes to examine a carrier signal close in to determine noise and line-related sidebands. Very few instruments can carry out this measurement, but it is the acid test of any quality spectrum analyzer. The specification points to look for are the narrowest resolution bandwidth filter, the shape factor of the filter and the stability of the local oscillators used in the instrument.

The shape factor of a filter is the ratio of the 3 dB bandwidth to the 60 dB bandwidth, a good spectrum analyzer has filters with a shape factor of 11:1, further improvement of shape factor is not readily practicable, it is better to concentrate on achieving a narrower bandwidth. A 10 Hz filter with a shape factor of 11:1 can be used to measure 50 Hz sidebands approximately 60 dB down. A 3 Hz filter can typically resolve close-in noise and hum down to less than 85 dB. The local oscillator stability is obviously also important here, crystal-derived local oscillators are used to ensure that noise which is measured is from the signal being analyzed and not imparted onto it by the local oscillator.

Vertical Resolution is yet another key area, a switchable vertical scale is necessary so that typically 10 dB/division and 1 dB/division can be selected. More modern instruments tend to be even better, 0.5 db/division scales are now more widely available to assist in resolving small changes in signal level.

Lower frequency spectrum analyzers become even more useful instruments if they are used with a tracking generator. A tracking generator sweeps in synchronism with the input tuned frequency so that the frequency responses of devices, particularly filters, can be measured over a wide dynamic range. The usefulness of a tracking generator however depends on its ability to accurately track the input tuned frequency of the spectrum analyzer, this is absolutely vital when narrow filters are used in the analysis. This can still be a short-coming of some instruments and it can only readily be overcome by having an integral tracking generator which is closely locked to the input tuned frequency of the instrument.

It is important to realise why a tracking generator is so significant. Many swept measurements are made with sweep generators, they are fine for measuring flatness and small amounts of attenuation, but they cannot measure filters such as crystal filters or SAW filters with selectivities to 100 dB or more. The reason is that a sweeper uses a broadband detector which not only detects the stimulus signal it also detects harmonics and noise of the stimulus. A spectrum analyzer/tracking generator combination on the otherhand uses a detector which is synchronously tuned only to the frequency of the stimulus and is thus not detecting unwanted products and will therefore provide a much wider measurement dynamic range.

4 SOME THEORY

Before proceeding further with some practical spectrum analyzer measurements we shall consider some basic theory.

Mathematically it can be shown that for a repetitive waveform the individual sinusoidal components are related by the Fourier series:

$$F(x) = \frac{ao}{2} + (a_1 \cos x + b_1 \sin x) + (a_2 \cos 2x + b_2 \sin 2x) +$$

$$(a_n \cos nx + b_n \sin nx) \dots\dots\dots\dots (1)$$

$F(x)$ = the original function.

The lowest frequency terms cos x and sin x represent the fundamental frequency whilst the other terms, which are integral multiples of this fundamental frequency, are the harmonics of the fundamental. The above equation can be represented in an alternative form.

$$F(x) = \frac{ao}{2} + \sum_{n=1} (a_n \cos nx + b_n \sin nx \ldots (2)$$

where the angle x is in radians.

However, the functions normally encountered will have an arbitrary period T rather than 2π , the series still applies except that all functions have to be scaled by 2π and the expression then becomes:

$$F(t) = \frac{ao}{2} + \sum_{n=1} (a_n \cos \frac{2n\pi x}{T} + b_n \sin \frac{2n\pi x}{T}) \quad \ldots (3)$$

The coefficients a_n, b_n, are the amplitudes of the frequency components and are given by:

$$a_n = \frac{2}{T} \int_{-\frac{T}{2}}^{+\frac{T}{2}} F(x) \cos \frac{2n\pi x}{T} \ dx$$

$$b_n = \frac{2}{T} \int_{-\frac{T}{2}}^{+\frac{T}{2}} F(x) \sin \frac{2n\pi x}{T} \ dx$$

$$\ldots \ldots (4)$$

When displayed on a spectrum analyzer the two terms combine to display a single amplitude factor

$$C_n = a_n^2 + b_n^2$$

Comprehensive tables of Fourier coefficients for various waveforms are available, and provide a powerful tool for the mathematical analysis of waveforms. From the result of this type of analysis we can calculate what should appear in the frequency domain for a given waveform.

5   SPECTRA OF COMMON SIGNALS

5.1   Sinewaves

A pure sinewave has energy at only one frequency. The spectrum will appear to be a single line of energy at a particular frequency. However, a pure sinewave is very rare and some distortion is usually present. We can use the theory to predict that energy will appear at harmonics of the fundamental frequency. By measuring the amplitude ratio of these harmonics to the fundamental frequency component we can determine the individual harmonics. The total distortion present in the signal can also be calculated.

5.2 Amplitude modulation

When an RF signal of frequency Fc is amplitude modulated by a pure tone Fm, two additional signals are produced, these are termed the upper and lower sidebands which are at a frequency of Fc ±Fm. With pure amplitude modulation the amplitude of these sidebands will be equal, but the level will vary dependent upon the depth of modulation, the carrier amplitude will however remain constant. At 100% modulation depth half the total power will reside in the sidebands, therefore each sideband will contain one quarter of the power, or will be 6 dB below the carrier level. From these facts we arrive at the expressions:

$$\text{Percentage modulation depth} = \frac{2 \ (\text{individual sideband voltage})}{\text{carrier voltage}} \times 100$$

$$\dots\dots\dots\dots\dots\dots \quad (5)$$

The modulation depth can thus be measured by examining the spectrum of the modulated signal. If distortion of the modulated signal exists, this will appear in the form of additional sidebands harmonically related to the modulating frequency and can be measured as for distortion. Furthermore if frequency modulation is present as well as the AM then sideband assymetry will occur.

Fig. 5 compares time domain (oscilloscope type) with frequency domain (spectrum analyzer type) displays for an amplitude modulated carrier.

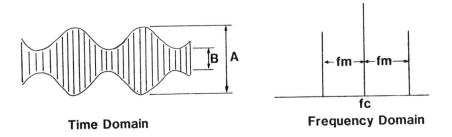

Time Domain   **Frequency Domain**

Fig. 5 Amplitude modulation

5.3   Frequency modulation

A frequency modulated signal may be defined by four main parameters.

Fc   - carrier frequency

Fd   - frequency deviation

Fm   - modulating frequency

M   - modulation index = $\dfrac{Fd}{Fm}$

Theoretically frequency modulation produces an infinite number of sidebands which are spaced symmetrically around the carrier and separated by a frequency equal to Fm. In practice however the amplitude of the sidebands falls quite rapidly outside the peak deviation frequency, and in cases of very low deviation only one pair of significant sidebands are present. The process of frequency modulation is a constant power process, therefore in amplitude terms we can expect that both the sidebands and the carrier amplitudes will vary with changing modulation index. It can be shown that the relative amplitudes of the various components obey the same relationship as the relative amplitudes of Bessel functions of the first kind acording to the expression below.

$Jn\left(\dfrac{Fd}{Fm}\right)$ or Jn (M)

.... (6)

where n = 0 for the carrier, n = 1 for the first sideband etc.

As the modulation index changes, the amplitudes of the various components change, relative to the unmodulated carrier, and are given by:

$A = Jn \left(\dfrac{Fd}{Fm}\right)$

.... (7)

A characteristic of Bessel functions is that the amplitude of both the Jo - carrier component, and the Jn - sidebands, will reduce to zero at specific values of the argument, i.e. the modulation index or $\dfrac{Fd}{Fm}$

As many excellent sets of tables of Bessel functions exist, it is unnecessary to calculate the individual values of modulation index which will produce carrier or sideband zeros or nulls. Fig. 6 graphically shows a wide range of Bessel nulls.

The tables below indicate the modulation indices for some carrier nulls and for sideband nulls.

Null Number	Carrier Nulls Mod. Index	1st Sideband Nulls Mod. Index
1	2.40	3.83
2	5.52	7.02
3	8.65	10.17
4	11.79	13.32
5	14.93	16.47
6	18.07	19.62
7	21.21	22.76
8	24.35	25.90
9	27.49	29.05

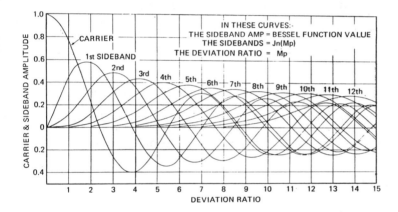

Fig. 6   Bessel functions for the first 12 orders

As deviation is equal to the modulating frequency times the modulation index, the deviation of an FM signal can be accurately set by selecting an appropriate modulating frequency and adjusting the modulator until a carrier zero is indicated on the spectrum analyzer. As a swept frequency analyzer is unable to detect or display phase information, the FM sidebands are all shown as being in a positive sense, however, from an examination of the Bessel series mathematics, it will be noted that the upper and lower odd sidebands are 180 degrees out of phase with each other.

With narrow deviation where the modulation index is less than 1, only two sidebands are produced and the deviation is calculated from the following:

$$\frac{S}{C} = \frac{Fd}{2Fm} \text{ where } \frac{S}{C} = \text{ sideband to carrier voltage ratio}$$

$$\dotsc\dotsc\dotsc\dotsc\dotsc (8)$$

Although when viewed on a swept analyzer the signals will all appear to be in a positive amplitude sense, the two sidebands are in reality of opposite polarity. This alternation is displayed to advantage when an amplitude modulation signal contains some low deviation residual FM. This causes one of the AM sidebands to be increased in level by the amplitude of the FM sideband, whilst the other sideband is reduced by this value. Thus it is possible to calculate both the AM modulation depth and the residual FM deviation from the one combined display so long as the predominant modulation is know.

5.4 Pulse spectra

From the mathematical analysis it can be shown that for a train of rectangular pulses of amplitude V, period T and pulse width t, the component amplitudes Cn can be expressed according to the following equation:

$$\text{The component amplitudes } Cn = \frac{Vt}{T} \sin \left(n\, W\, o\, \frac{t}{2} \right) \bigg/ n\, W\, o\, \frac{t}{2}$$

$$\cdots\cdots\cdots\cdots\cdots \quad (9)$$

If we substitute x for the term $n\, W\, o\, \dfrac{t}{2}$ the equation becomes the familiar form $\dfrac{\sin x}{x}$.

The theoretical spectrum of such a signal is shown in Fig. 7.

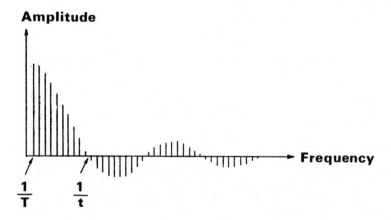

Fig. 7   Theoretical spectrum of a pulse waveform

The spacing of the individual spectral components is $\dfrac{1}{T}$ which is termed the pulse repetition frequency, and the zero crossing occur at intervals of $\dfrac{1}{t}$.

As mentioned previously, a swept analyzer shows no phase information, thus the spectrum appears on the analyzer with the negative sections having been apparently inverted.

5.5 Pulsed RF

When a rectangular pulse waveform is used to amplitude modulate an RF carrier, the observed effect is as if the basic pulse spectrum has been translated from a DC maximum amplitude term, to a spectrum which now has its maximum amplitude term occurring at the carrier frequency. The spectrum will extend both up and down in frequency in the same form as the original modulation spectrum. As considered previously, the amplitudes of upper and lower sidebands are at a given equal offset from the centre frequency and will be equal in amplitude unless asymmetry occurs due to the effect of residual frequency modulation. This factor is extremely useful when examining the output of pulsed radar systems, where the cause of the frequency modulation is frequency pulling of the RF source.

Pulsed RF

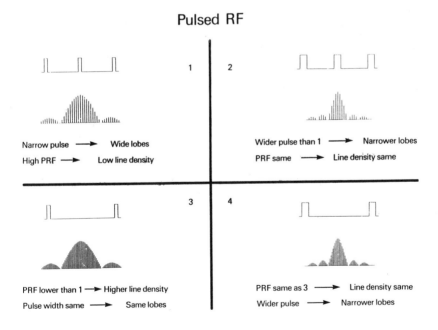

Fig. 8 Four spectrum analyzer displays of pulsed RF showing how the display is changed according to the pulse width and pulse repetition frequency of the modulating signal.

6 MORE PRACTICAL CONSIDERATIONS

6.1 Stability

In achieving the objective of spurious free operation, the practical spectrum analyzer has become a multi-stage superhet with several oscillators either fixed in frequency, variable or swept. It is essential to ensure that the accuracy of the oscillators is maintained, and that the relative drift rates of the individual oscillators do not combine in such a way as the make narrow sweep widths impossible to use due to the drift. Furthermore, as was mentioned earlier, the oscillators must have excellent close-in sideband noise.

6.2 Resolution

This is defined as the ability of an analyzer to distinguish between two adjacent signals, and is a function of the IF resolution filter bandwidth and its shape factor, that is the shape of the filter between the 3 dB and 60 dB bandwidth points. As the local oscillator sweeps across a signal consisting of a single spectral line, the shape mapped out on the display will be that of the IF filter, and therefore, if two signals are closely spaced, with one being large with respect to the other, it is possible that the small signal will be lost in the skirt of the filter. In practice shape factors of around 11:1 are normally used.

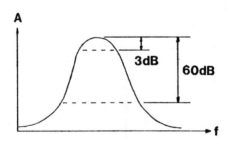

Fig. 9 Filter shape factor

To achieve the greatest resolution, the IF resolution filters should be only a few Hz wide with a very steep skirt. Unfortunately, the narrower a filter is made, the longer it will take to respond to the incoming signal. In the spectrum analyzer the filter is effectively being swept, therefore the narrower the filter the slower it has to be swept otherwise errors will occur as shown in Fig. 9. This shows that there has been an apparent reduction in the amplitude, a shift in the centre frequency and widening of the bandwidth of the filter as a result of incorrectly increasing the sweep speed. In practice high resolution filters are required when using narrow frequency sweeps in order to examine signals which are closely spaced. As the span width is increased wider filters may be used. Thus a modern high resolution spectrum analyzer will offer a series of filters extending from 3 or 10 Hz to 1 or 3 MHz to cover various applications.

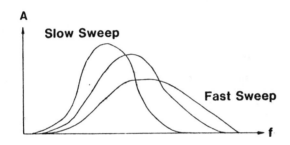

Fig. 10   Effect on response if sweep speed is too fast

6.3   High sensitivity

Sensitivity is a measure of the smallest signal that can be resolved by the analyzer. The limiting factor is the analyzer noise figure which is added to the thermal noise level. In practice this is usually measured by applying a signal which is adjusted to provide an indication on the screen of the analyzer which is 3 dB above the residual noise level. Modern spectrum analyzer can measure signal levels as low as -140 dBm normalized to a 1 Hz bandwidth.

6.4   Dynamic range

The term dynamic range is often quoted but is generally misunderstood. Dynamic range is defined in many ways depending on the performance that is required for a certain type of measurements.

6.4.1 Display dynamic range. This is defined as the maximum difference between the levels of two non-harmonically related sinusoidal signals each of which can be totally displayed and simultaneously measured with a specified accuracy.

Modern quality spectrum analyzers have 100 dB of displayed dynamic range.

6.4.2 Non-harmonic dynamic range. This is defined as the maximum difference between the levels of two non-harmonically related sinusoidal signals simultaneously present at the input which can be measured with a specified accuracy.

6.4.3 Harmonic dynamic range. This is defined as the maximum difference between two harmonically related sinusoidal signals simultaneously present at the input mixer which can be measured with a specified accuracy.

6.5 Intermodulation rejection

Intermodulation occurs when two or more signals mix together in a non-linear system to produce additional frequency components.

If two signals F_1 and F_2 are fed into a non-linear device then harmonic distortion will occur giving rise to signals at $2F_1$, $3F_1$, ... and $2F_2$, $3F_2$.... These components mix together to form intermodulation products such as $2F_1 - F_2$.

The input of a spectrum analyzer must have very good intermodulation performance to allow for the measurement of intermodulation products as low as 80 or 90 dB down.

6.6 Frequency range

Spectrum analyzers are available which cover from 50 Hz to 20 GHz or thereabouts and even higher when external mixers are used. Microwave users will inevitably use such instruments but it is interesting to consider this further.

Many microwave system measurements are made at IF frequencies such as 70 or 140 MHz. Very high resolution is required which is often incompatible with an instrument covering to microwave frequencies, furthermore to achieve such resolution over the full frequency range is extremely expensive.

Manufacturers thus provide high frequency spectrum analyzers with poorer resolution for the RF measurements and less expensive very high resolution instruments for baseband and IF measurements.

7 SPECTRUM ANALYZER USES SUMMARY

7.1 Harmonic analysis

The relative levels of individual harmonics in dBs can be read directly from the screen when operated in a logarithmic mode, whilst the absolute levels may be found by taking into consideration the reference level setting of the analyzer and converting to dBm.

Total harmonic distortion is given by:

$$D = \sqrt{V_2^2 + V_3^2 + V_4^2 + ..V_n^2} \quad .... \quad (10)$$

where V_2 = Voltage ratio between the fundamental and the second harmonic.

V_3 = Relates to the third harmonic etc.

If the relative levels of a particular series of harmonics were as follows:

Harmonic	Level w.r.t. carrier	Voltage ratio
2	−30dB	1/32
3	−38dB	1/79
4	−45dB	1/178

$$\text{Total harmonic distortion} = \sqrt{\frac{1}{32}^2 + \frac{1}{79}^2 + \frac{1}{178}^2} = 0.034 = 3.4\%$$

where $\frac{S}{C}$ = sideband to carrier voltage ratio ....(11)

7.2 Amplitude modulation

The frequency spacing of the sidebands from the carrier is equal to the modulating frequency whilst the percentage modulation depth is given by:

$$\text{Modulation Depth} = 100 \times \frac{2S}{C}$$

Asymmetry of the sidebands results from residual FM in which case the AM depth is calculated from the average level of the two sidebands and the frequency deviation $= \frac{2 \times Fm \times FMS}{C}$

$$.................... (12)$$

where Fm = modulation frequency, and
      FMS = the level of the FM sideband.

Any modulation distortion is calculated as for harmonic analysis.

7.3 Frequency modulation

7.3.1 Narrow deviations. In the case of narrow deviation where the modulation index is less than 1 only two sidebands are significant. The ratio of sideband to carrier level is equal to half of the modulation index.

$$\frac{S}{C} = \frac{Fd}{2Fm} \quad \text{where} \quad \frac{S}{C} = \text{sideband to carrier voltage ratio}$$

$$Fd = \text{deviation} \quad \ldots\ldots\ldots\ldots\ldots\ldots \quad (13)$$

The frequency spacing of the sidebands from the carrier is equal to the modulating frequency Fm.

7.3.2 Wider deviations. For a modulation index greater than 1, the Bessel zero method is used. The ratio Fd/Fm is set to achieve a carrier or sideband null. The sideband spacing equals the modulating frequency Fm. Alternatively more recent spectrum analyzers incorporate FM slope detectors which display frequency deviation against time.

7.4 AM/FM bandwidth.

On spectrum analyzers providing a Maximum Hold or Peak Memory mode of operation, the occupied bandwidth of a signal may be obtained by leaving the analyzer set to this mode for a period of time. Maximum frequency excursions over that period can thus be measured.

7.5 Frequency drift and occupancy.

This is measured in the same way as AM and FM bandwidth using the Maximum Hold mode and measuring the maximum to minimum frequencies excursions.

7.6 Spectrum surveillance

The spectrum analyzer may be connected to an antenna and used to observe a particular frequency band. This may be used to identify a specific or illegal transmission, to find unoccupied portions of the frequency band, for positioning new transmitters, or for monitoring purposes. The high sensitivity of the analyzer, coupled with wide frequency range, and the ability to ˜home˜ onto a signal for more detailed examination makes a spectrum analyzer a very useful instrument for this purpose.

7.7 Pulsed radar performance checks.

Using the spectrum analyzer suitably coupled to the output of a radar transmitter the following tests can be carried out:

Carrier frequency. Determined by measuring the frequency of the main lobe peak.
Modulation pulse width. Computed from the reciprocal of the side lobe frequency width.

Pulse Repetition Frequency (PRF): Determined from the frequency spacing of the spectral sample lines.

Pulse shape: This can be estimated from the ratio of the amplitude of the main lobe to the first side lobe. The ideal rectangular pulse will have a ratio of 13.2 dB.

FM: The presence of FM will cause asymmetry of the upper and lower sidebands a lift in the side lobe amplitudes and a lack of distinct nulls.
Frequency drift. Measured as previously mentioned.

7.8 Radio Frequency Interference (RFI) testing.

Using suitable antennae and search coils, identification and location of interference can be rapidly carried out.

7.9 Frequency response measurements

By using the tracking generator as the input to a circuit, the output frequency response is displayed directly on the analyzer. Responses of over 120 dB can be measured in this way and the use of the frequency counting ability enables the bandwidth of the circuit to be measured rapidly. Use of the store mode of operation, enables easy observation of changes in response due to adjustments and an A-B mode further simplifies comparisons.

7.10 Insertion loss.

Can be measured by first connecting the tracking generator output directly to the spectrum analyzer input and storing the combined response, the device to be tested is then inserted between the tracking generator and analyzer. A Normalize mode available on recent spectrum analyzers stores the response digitally and automatically corrects for the instrument and cable frequency response. The difference in level observed between two traces is the insertion loss of the device.

7.11 Swept return loss

May be measured by using an external RF bridge. The tracking generator is connected to the bridge input, the output of which is connected to the spectrum analyzer. A known high quality low ،VSWR load is connected to one port of the bridge and the level on the analyzer is stored. The unknown impedance is then connected to the remaining port of the bridge, and the difference in dB⁻s between the stored trace and the new trace is the direct reading of the return loss of the device.

7.12 Other measurements

This list of measurements is by no means exhaustive particularly when it is considered that in a modern high quality calibrated spectrum analyzer, there are the attributes of a high quality receiver, frequency counter, modulation meter, frequency selective voltmeter, distortion analyzer and swept amplitude response analyzer. It is also not surprising therefore that spectrum analyzers are now being so widely used, and that the above list can only give a general guide to the uses.

8 ADVANCES IN SPECTRUM ANALYZERS

The design of spectrum analyzers has steadily progresed over the last two or three decades. It is constructive to point out some of the more important milestones.

- Automatically coupled controls to prevent operator error.

- Digital trace storage to facilitate slow sweeps.

- Electronic graticules to remove parallax error.

- Steerable markers to measure levels and frequencies at any point on the display.

- Keyboard entry of control settings with on-screen readout of their values.

- Full programmability through the GPIB (General Purpose Interface Bus).

- Automatic self-calibration using an internal reference source to improve accuracy.

- Improved resolution, resolution bandwidths down to 10 Hz or 3 Hz are more common.

- Improved intermodulation performance.

Perhaps the most important innovations are the RF performance improvements and GPIB control. Instruments can now talk and listen and may be used in a variety of ATE environments. In the simplest from an instrument can talk to a GPIB plotter to give hard copies of screen displays and at the other extreme can be incorporated into a sophisticated ATE.

The middle ground is represented by automated systems which use a desk top controller to carry out routine measurements and compare with specification limits.

REFERENCES

1. Panter, ˝Modulation, Noise and Spectral Analysis˝, McGraw-Hill.

2. Starr, ˝Radio and Radar Technique˝, Pitman.

3. Engelson and Telewski, ˝Spectrum Analyzer Theory and Application˝, Artech House.

4. BS 6329, ˝Expression of the Properties of Spectrum Analyzers˝

Chapter 14

Swept and stepped frequency methods

R. N. Clarke

1.   INTRODUCTION

For nearly 20 years, until the late 1970's, a microwave
metrologist who wished to characterise a component quickly
and conveniently over a broad band of frequencies would be
most likely to turn to a swept frequency measurement system
in order to carry out the task. At some cost to accuracy,
when compared with fixed frequency measurements, he would
thereby gain the benefits of speed, convenience and the
ability to display measurement results instantaneously. In
more recent years, automatic measurement systems have to
some extent taken over this role from manual swept frequency
measurements. Self contained, microprocessor-driven
instruments now offer the additional advantages of lower
measurement uncertainty and still further improvement in the
presentation of measured results.

Nevertheless, there are a number of reasons why the
principles of swept frequency measurement should continue to
be part of the education of a microwave engineer. The
automatic network analyser (ANA) and spectrum analyser may
well have been welcomed because they relieve much of the
burden of microwave metrology, but it is all too easy to use
them without being aware of the metrological principles
which have gone into their broadband design. Even where
automatic test equipment (ATE) of this type is being used,
experience has shown that knowledge of the principles of
swept frequency measurements can often save a good deal of
time and effort and can improve the quality of measurements.
Furthermore, it must be recognised that suitable ATE is not
always available to the metrologist. In its absence the
older manual techniques can still be implemented relatively
easily and inexpensively. They have the additional advantage
that they are more amenable to 'hands-on' adjustments - for
instance a change of measurement frequency range, which can
easily be accomplished on a manual swept system, may require
a complete recalibration of an ANA.

The following paragraphs will introduce the principles
of swept frequency measurements. They are discussed in more
detail in Ely, (1), application note, (2) and in Laverghetta,

(3,4). We will be concerned with the following topics:

(1) The properties of components used in broad-band frequency-domain measurement systems.

(2) The limitations of such components and their influence upon measurement uncertainty.

(3) The measurement methods themselves.

(4) Estimation and reduction of uncertainties.

2.   SWEPT AND SYNTHESISED MICROWAVE SOURCES

At the heart of all swept frequency sources is a voltage controlled oscillator (VCO). Common types of microwave VCO are the Backward Wave Oscillator (BWO), and various solid-state sources: FET oscillators, Gunn diodes and Impatt diodes. Many frequency synthesizers also employ VCOs which are phase-locked to a crystal reference.

Swept frequency measurements became popular with the advent of backward wave oscillators (BWOs) (Warner, (5)) during the late 1950's - a time when the only alternative methods were manual discrete frequency techniques which were slow and laborious. BWOs were the first widely available voltage-tuned microwave sources able to cover a full waveguide band. They are still in common use but they have a limited lifetime and in many laboratories they have been replaced by solid-state swept sources.

Gunn diode and FET oscillators may use varactor-diode tuning or else Yttrium Iron Garnet (YIG)-sphere tuning (Dethlefsen, (6), Zensius et al., (7)). These oscillators are preferred because they have lower phase noise than Impatt diodes. However their output powers fall rapidly with frequency and in the millimetre-wave region bias-tuned Impatt diodes are used because they are more powerful (Kramer and Johnson, (8)). BWOs have also continued in use above 40GHz because they do not exhibit the phase noise and limited tuning range of Impatts. Other swept millimetre-wave sources are based upon frequency multipliers fed by lower frequency sweepers (Napier and Cowell, (9))

2.1   Microwave Sweepers

For many years sweepers were manufactured on a modular basis. The mainframe would supply 'services', such as power-supplies and linear voltage ramps, for a set of plug-ins which would each contain a microwave VCO covering a single waveguide band. All of the necessary amplifiers and filters and an internal power-levelling loop would be included in the plug-in (see section 2.3). More recently, sweepers have become available which cover the band 10MHz to 40GHz without the need to change plug-ins - a development which has been associated with the introduction of a new

2.92mm coaxial connector (the K Connector) which does not overmode at frequencies up to 45GHz. The following points require special consideration when using sweepers:

2.1.1   Output Power. A general-purpose broad-band sweeper may supply 10 mW of levelled power up to 18GHz but only 4mW at 40GHz. High power sweepers are available which can offer 40mW up to 18GHz, but millimetre wave sweepers may give only 1 or 2 mW. For some measurements on active devices (such as signal compression in detectors and mixers) it is convenient to sweep or step the output power as well as the frequency. Some sweepers allow this to be done in steps of 0.1dB or 0.01dB under automatic control.

2.1.2   Harmonics and Spurious Output. These are typically specified to be at most 20dB and 50dB below carrier respectively. The presence of harmonics and especially of broad-band noise must not be forgotten in measurement systems which use broad-band detection (Clarke and Granville-George, (10)) - they can certainly give rise to spurious results, particularly if the detected power at the carrier frequency is low. Harmonics can also interfere with the proper operation of levelling loops (section 2.3).

2.1.3   Frequency Accuracy and Linearity of Sweep. These can be checked with a counter by switching to the manual sweep mode and simultaneously recording the VCO voltage. Note, however, that YIG-tuned sweepers exhibit hysteresis (6, 7). It is important to check the frequency at the lower end of a broad-band plug-in, as a small offset in the voltage supplied to the VCO may result in a discrepancy of 10% or more in the output frequency.

2.1.4   Source Line-Width. This is important in narrow-band swept-frequency determinations of the properties of resonators or filters. Microwave resonators with high Q-factors may have resonance line-widths which are up to an order of magnitude smaller than that of the source. It is as well to check the validity of such measurements by inspecting the signal source with a spectrum analyser.

2.1.5   Additional Sweeper Facilities. For metrological purposes the following features are of value: a sweep analogue output; internal levelling; an input for external levelling; internally-generated amplitude modulation (AM); external sweep-triggering and amplitude modulation. Internal AM is usually 100% modulation at 1kHz and many proprietary detector systems require this facility to be present. An audio output synchronised to the 1kHz is always useful.

2.2   Microwave Synthesisers.

The deployment of computer-controlled ATE during the 1970's promoted what was effectively a movement back to discrete frequency measurements. The most accurate ATE systems in fact use stepped frequency sources whose frequency is required to vary discretely with time rather than smoothly as in swept measurements. Such systems are designed to be broad-band, so they have much in common with earlier swept measurement systems, but they are able to gain from the benefits of a new class of microwave source – the microwave synthesiser (11). Synthesisers are nowadays able to change frequencies very rapidly, within 50 ms with phase-lock, but some can also be used in a fast swept mode; such instruments are called 'Synthesised Sweepers'.

All points made in section 2.1 with regard to sweepers apply equally to synthesisers except that the problems of achieving frequency accuracy mentioned in 2.1.3. are absent when synthesisers are used in a phase-locked mode. Note that great care should be taken to ensure that spurious signals are not affecting measurements.

2.3   Levelling of Swept Sources.

Power output from a swept oscillator can vary greatly across a band (by over 10dB in the case of BWOs). This variation can be very inconvenient for measurements since it might be impossible to distinguish the properties of the device-under-test (D.U.T.) from those of the swept source. The technique of Automatic Level Control (ALC) uses a feed-back loop (Fig. 1) to overcome this problem by reducing peak powers to the same level as the minimum power available in the band. Most sweepers have optional internal ALC, generally also used as a continuously variable output power control for the sweeper. Typically the power can be levelled internally to within +1dB across a band at the lower microwave frequencies but this may degrade to +2dB or more at millimetre-wave frequencies.

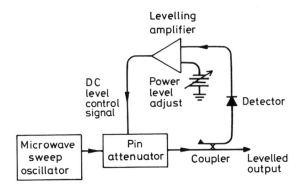

Fig. 1   Automatic Level Control (ALC)

External levelling is to be preferred to internal levelling because it brings the reference point at which levelling occurs closer to the measurement plane. It also compensates for components situated between the source and the measurement port if they are placed within the levelling

loop. A flatness of +0.5dB can be achieved across
a band in this way. This is illustrated in Fig. 2, which
shows a measurement in which the sweeper output power has
been boosted by means of an amplifier. A low-pass filter is
incorporated since the amplifier and source may both produce
harmonics. It is also placed within the ALC loop to prevent
the harmonics from reaching the broad-band levelling
detector. The detector may be a diode or a power meter (see
section 3). The levelled output power is set by the
reference input of the low-frequency levelling amplifier, as
shown in Fig. 1. The amplifier gain can be increased until
an optimum degree of levelling is achieved without
instability (2). The output from the amplifier is fed to the
control attenuator which is usually a PIN diode.

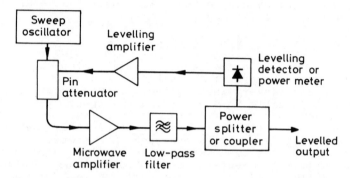

Fig. 2   An ALC loop containing an amplifier and filter

The functions of the power-splitting component in the
ALC loop are (i) to ensure that the levelling detector sees
a signal which tracks the transmitted signal at all
frequencies, and (ii) to ensure that the ALC loop presents a
good equivalent source match. A directional coupler with
high directivity and good output match can perform these
tasks, but a symmetrical two-resistor power splitter
can also be used (section 4.2). To explain point (ii), it
should be noted that the equivalent source reflection
coefficient of a levelled source can be made to be largely
independent of that of the source alone. Assuming infinite
loop-amplifier gain, and taking port 1 of the power splitter
to be the input, port 2 the output, and port 3 the side-arm,
the equivalent source reflection coefficient is given
((1),(2), or Johnson (12)) by:

$$\Gamma_e = S_{22} - S_{21} \times (S_{32}/S_{31}) \quad \dotsc\dotsc\dotsc\dotsc\dotsc\dotsc (1)$$

$$= \Gamma_c - TD$$

where Γ_c is the output reflection coefficient of the power

splitter itself, T is its forward transmission coefficient
and D ($= S_{32}/S_{31}$) is its directivity.

3. DETECTORS, MIXERS AND POWER METERS

3.1 Detector Diodes

Broad band detector diodes are widely available, but one should note the following characteristics:

(a) Frequency response (flatness). +0.2dB per octave is possible up to 8GHz but this may degrade to +0.6dB above 18GHz.

(b) Match (reflection coefficient). This can vary across the band and with power level. In manufacturers' specifications VSWRs vary from approx. 1.15 below 4GHz to as high as 2.2 in the 18 - 26 GHz range.

(c) Harmonic generation. Being inherently non-linear devices, diodes can reflect harmonics back into a broad-band system. They may pass relatively unhindered through a device such as a directional coupler and may be detected as a spurious signal in other detectors in the system.

(d) Signal compression at high input powers. Diode detectors are usually used in the square-law (low input power) region. This region is extended to above -10dBm by resistive loading (2). Checks should be made to determine the law of the diode if higher powers are used.

(e) Sensitivity. In the square law region this is quoted in mV/µW but the figure will vary with temperature. A typical value of 0.5mV/µW can be obtained from Schottky Barrier diodes. This figure should be taken into account when estimating the effects of noise upon a measurement.

3.2 Mixer Diodes

These can be used with a fixed local-oscillator (LO) frequency or else with the LO phase-locked to the signal to produce a fixed intermediate frequency (IF). In the former case the IF produced will be swept over the same bandwidth as the source. Such a frequency-changing technique may be useful if the IF produced is in a more convenient part of the spectrum. The second technique is used in heterodyne network analysers (section 5.3). Mixers provide better signal-to-noise ratio and linearity than detector diodes but they share with them problems due to lack of flatness across a band. Preference for their use in swept measurements may often be a matter of convenience.

3.3 Power Meters

Thermistor power meters (13) possess a number of advantages over diode detectors:

(a) They are usually flatter across a waveguide band.

(b) Thermistor power meter heads vary less in their properties from one to another than do diodes.

(c) Thermistor power meters based on RF/DC substitution produce only a zero-drift with changing temperature, whereas detector diodes also exhibit a change in <u>sensitivity</u>.

(d) Input impedance is independent of input power.

(e) Thermistors are more linear than other detectors and do not produce harmonics.

The major disadvantage of thermal power meters is their slow response, which necessitates slow sweep speeds and which may induce oscillation in levelling loops.

3.4   Frequency Response Test Sets

These instruments offer a conveniently packaged detector and display facility for swept measurements (Fig. 3). They provide diode detectors which are coupled to linearising circuitry, to amplifiers, and to optional logarithmic convertors. One diode is used as a reference detector allowing the instrument to be used in a ratio mode to eliminate source amplitude variations. The sweep-analogue-output from the source drives the horizontal deflection. Scalar network analysers are similar instruments which incorporate the sweeper source.

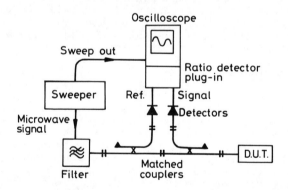

Fig. 3   Reflectometry using the ratio detector in a frequency response test-set plug-in.

4.   PASSIVE COMPONENTS

4.1   Directional Couplers

Key parameters to be considered are:

(a) Frequency range
(b) Coupling coefficient
(c) Coupler match (VSWR)
(d) Directivity
(e) Frequency Sensitivity (coupling flatness)
(f) Transmission coefficient
(g) Tracking between pairs of couplers

The first two parameters will be determined by the task at hand but (c) to (g) all influence measurement uncertainty (see section 5.1). There is an engineering trade-off between bandwidth and the other parameters and hence it may be necessary to decide whether to cover a broad band of frequencies with one coupler and accept larger measurement uncertainties or else to suffer the inconvenience of using two or more couplers to cover the frequency range.

Directivities of 40dB across a band with a VSWR of 1.05 are typical for waveguide but equivalent figures for coaxial couplers are 25dB and 1.2 respectively. In discrete frequency measurements, directivity can be improved by the use of tuners but this is not normally possible in swept measurements, which are consequently less accurate. Tuners with automatic control have been developed, however, to allow this method to be used in computer-controlled frequency-stepped measurements. Many systems use matched pairs of couplers which are required to track each other well over a band. Some manufacturers produce couplers specifically optimised for reflectometer use: one coaxial model covering 7 to 12.4GHz has 10dB coupling with 35dB directivity, 1.035 VSWR and a flatness of $\pm$1.2dB with 0.3 dB tracking between a matched pair.

Fig. 4 shows a method for improving the flatness of coupling across a band. If the two couplers are matched the transmission coefficient of the coupler on the side-arm will compensate to some extent for the coupling coefficient variations of the other. Tracking of the through and coupled arms may be improved to 0.2dB if this technique is used. If such a combination is used in a levelling-loop, the degree of levelling is improved (2).

4.2   Power Splitters

Resistive power splitters employing either 2 or 3 resistors (Fig. 5) are used to separate power equally into two channels in coaxial systems. Output tracking between arms can be better than 0.25dB up to 18GHz, which is better than directional-coupler performance, but this is achieved

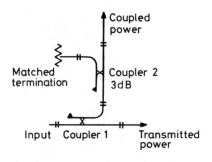

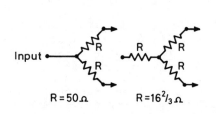

Fig. 4   A technique for
improving coupler
tracking.

Fig. 5   Two- and three-
resistor power
splitters.

at the expense of approximately 6dB insertion loss in each
arm. Three-resistor splitters give a good match at all
three ports in a matched system but they should not be
used as the coupling element in a levelling loop, since the
action of the loop produces a very poor equivalent source
match, as can be seen by inserting relevant figures into
equation (1) (1,2,12). Conversely a two-resistor
splitter which normally gives a poor equivalent source match
gives a good match when used in a levelling loop (2, 12).
The same considerations apply to power splitters used in
ratio systems (see section 5.1). Two-resistor splitters are
marketed specifically for both of these applications.

4.3   Reflectometer Bridges

Swept coaxial reflectometry based on the use of
directional-couplers has limited accuracy because of their
relatively poor directivities (typically 25dB). However,
directivities of better than 40dB up to 18GHz can be
achieved with resistive bridges, as shown in Figs. 6 and 7.
If the bridge is constructed with well-matched RF resistive
elements, the signal presented to the detector is given by:

$$E_{det} \quad \alpha \quad \frac{Z_x - R_0}{Z_x + R_0} \quad \dots\dots\dots\dots(2)$$

Which is just the reflection coefficient of Z_x
referenced to impedance R_0.

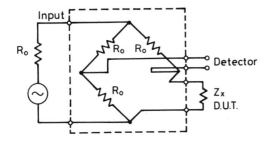

Fig. 6   A coaxial reflectometer bridge

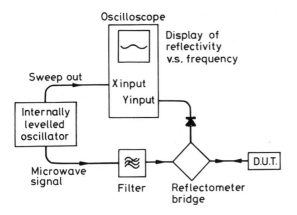

Fig. 7   Swept reflectometry with a bridge

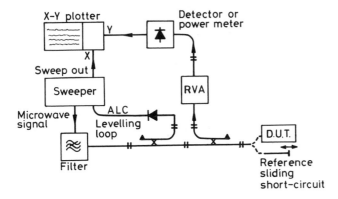

Fig. 8   Waveguide reflectometry using a rotary-vane
attenuator to plot a calibration grid

5.   MEASUREMENT TECHNIQUES AND ERROR REDUCTION

The systems described below have been chosen to illustrate the basic principles of swept and stepped measurements. There is not room to mention all types of proprietary swept frequency systems - they range from the most sophisticated of vector heterodyne ANAs (Donecker, 14) to simple frequency-response test set plug-ins (Fig. 3). In the calibration laboratory it is worth noting that even six-port reflectometers, previously thought to be too slow, have been developed for use as fast swept-frequency systems (Luff et al., (15), Kaliouby and Bosisio, (16)). Millimetre-wave swept measurements are still relatively novel and new techniques are constantly under development (Paul, (17), Gianfortune, (18), Application Note, (19)).

5.1   Basic Swept Measurements

Figs. 3 and 7 to 10 illustrate how measurement systems can be assembled from the components discussed above.

All microwave measurements involve errors caused by component imperfections, but swept measurements are beset by an additional source of error - the lack of uniformity of response of devices across the frequency band. Indeed, errors are exacerbated by the fact that engineering compromises are necessary to allow devices to be used over a broad band at all. Component imperfections such as mismatch and finite directivity of couplers are always worse for broadband components. A number of hardware techniques have been developed to correct for these errors:

(a) Swept sources can be levelled using ALC, as discussed in section 2.3 and illustrated in most of the figures. Alternatively, or additionally, a ratio detector can be used which has much the same effect (Fig. 3).

(b) Pairs of matched components should be used to compensate for lack of flatness. Figs. 3 and 8 show this approach applied to reflectivity measurement. One matched coupler monitors the source and provides the reference signal for the ALC or ratio detector. The two detectors also form a matched pair. Fig. 9 shows how matched directional couplers can be used in transmission measurements. Rather than taking the signal directly from the D.U.T. to the detector, the second coupler is interposed to compensate for the lack of tracking between main-arm and side-arm of the coupler in the levelling loop. Fig.4 demonstrates another form of compensation, as discussed in section 4.1.

(c) Substitution techniques should be used to compare the properties of the D.U.T. with those of a known device or standard. Fig. 8 illustrates the use of a rotary-vane attenuator (RVA), X-Y plotter and sliding short-circuit to calibrate a reflectometer across a band. A calibration grid is plotted by adjusting the RVA to a number of fixed settings with the reflectometer terminated by the short-circuit. If the frequency is swept slowly and the position of the short circuit is swept rapidly during this process, the effects of directivity are revealed in the envelope of the trace (2). Subsequently, the RVA is set to minimum loss, and the short-circuit is replaced by the D.U.T. to obtain the trace of its reflectivity on the calibraton grid. The RF substitution method in Fig. 9 uses a similar approach for transmission measurements.

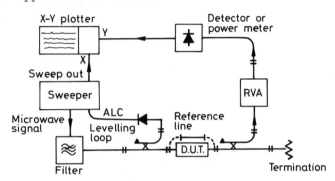

Fig. 9   A substitution transmission measurement using matched compensating couplers.

(d) Display devices such as ratiometers and storage normalisers can be used to clean-up the presentation of data. However, one should not forget that the detected power variation across a band during system calibration may not all be due to source-power or detector-sensitivity changes. For example, display devices may even increase measurement errors if the variations are caused by directivity errors.

(e) Coupler directivity errors in swept reflectometry are usually dominant if the D.U.T. is a near-match. Coaxial measurements should ideally be performed on a high-directivity bridge (section 4.3 and Figs. 6 and 7). Alternatively, or additionally, one can use a technique which allows one to distinguish clearly between the coupled and leakage signals. This is achieved, for example, in the long-line reflectometer (section 5.2.1).

(f) In reflectometry, a mismatched source produces multiple reflections which cause large errors if the D.U.T. is also badly matched. In transmission measurements, mismatches likewise produce large errors when measuring low loss components. The problem can be reduced by using an ALC loop to improve source match (section 2.3). If sufficient power is available, well-matched resistive attenuators can be used to buffer mismatched components. Broad-band isolators should not be used because they are usually not well matched themselves. If attenuators are used, however, their own frequency response should be checked.

Exact analyses of the effects of directivity, tracking and matching errors are to be found in the literature (1,2).

5.2   Other Swept and Stepped Techniques

A number of swept techniques have been developed which have no strict discrete frequency analogue. The long-line reflectometer (section 5.2.1), still makes a valuable contribution to exact metrology but other methods mentioned below may have been rendered obsolete by the advance of ATE. They are nevertheless included because of their ingenuity.

5.2.1   The Long-Line Reflectometer (Hollway and Somlo, 20) uses a long transmission line (waveguide, coaxial or microstrip) as an impedance reference. It is particularly useful for impedance matching (Fig. 10). In this application a small discontinuity is deliberately placed at the end of the line nearest to the reflectometer (in waveguide a small ball-bearing can be used, held in place by a magnet). The tuner and the mismatched D.U.T. are connected to the other end of the line. Reflections from the discontinuity and the tuner/D.U.T. combine to produce rapid oscillations of detected signal as the source is swept. Their magnitude is reduced to zero when the tuner correctly compensates for the reflection from the mismatched device. If the line is sufficiently long the source frequency excursions need not depart far from the centre frequency, where match is to be achieved. Long-line reflectometers are also used for reflectometry across a full band. Improved accuracy is obtained because the long air-line separates the reflections from the D.U.T., which are changing rapidly in phase, from the leakage signals originating in the reflectometer (Lacy and Andres, 21, Lacy and Oldfield 22).

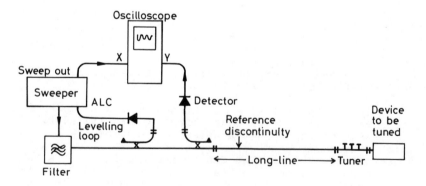

Fig. 10   A long-line reflectometer configured for impedance matching.

5.2.2   Slotted Lines can be used with swept frequency sources to provide a direct display of VSWR against frequency (1, 2). This approach can provide high levels of accuracy (Staufer et al., 23).

5.2.3   The Locating Reflectometer was developed to locate accurately the position of discontinuities in waveguide systems (Somlo, 24). The usual method for doing this - time domain reflectometry (TDR) - does not have very high resolution in waveguide because of wave dispersion. The reflectometer operates by using microwave components to carry out Fourier transformations!

5.2.4   The Locating Absorptometer (Fantom, 25) has a similar motivation but possesses the capability of distinguishing between resistive and reactive discontinuities. As TDR cannot always do this very well, the technique is also useful in coaxial line.

5.3   Automated Stepped Frequency Measurements

Automated measurements allow the errors discussed in 5.1. to be greatly reduced by the use of software error correction. The effects of mismatch, poor directivity, tracking etc. can all be compensated for by correction parameters which are determined for each frequency step in a pre-calibration procedure; but the following residual sources of error should be born in mind:

(a) In amplitude-only measurements full error correction is not possible, unless a six-port analysis is being used (15, 16). This is not usually possible with proprietary equipment such as scalar network analysers.

(b) Detector non-linearities may not have been taken into account.

(c) Repeatability errors cannot be eliminated. Lack of repeatability of the voltage-controlled local oscillators in some heterodyne network analysers, for instance, gives rise to errors as high as 0.25dB.

(d) Pre-calibration is, at best, only as good as one's knowledge of the standards used. At lower radio frequencies, it is sometimes difficult to get three known impedance standards to calibrate a network analyser.

(e) In stepped measurements it is possible to lose an important feature of the response of the D.U.T. between frequency increments if they are too sparsely distributed.

REFERENCES

1.   Ely, P.C., 1967, 'Swept-Frequency Techniques',
     Proc. IEEE, 55, No.6, 991-1001.

2.   Hewlett Packard Application Note No. 183, 1978, 'High
     Frequency Swept Measurements'.

3.   Laverghetta, T.S., 1981, 'Microwave Measurements and
     Techniques', Artech House, Washington.

4.   Laverghetta, T.S., 1981, 'Handbook of Microwave
     Testing', Artech House, Washington.

5.   Warner. F.L., 1977, 'Microwave Attenuation
     Measurement', IEE Monograph Series No. 19, Ch. 10,
     Peter Peregrinus Ltd., Stevenage.

6.   Dethlefsen, A., 1982, 'An Overview of Microwave
     Design Considerations for Swept Sources', Proceedings
     of the Hewlett Packard RF & Microwave Measurement
     Symposium, London.

7.   Zensius, D.P., et al., 1983, 'GaAs FET YIG Oscillator
     Tunes from 26 To 40 GHz', Microwaves and RF
     22, 129-139, No. 10, Oct.

8.   Kramer, N.B. and Johnson, R.A., 1984, 'Generating
     power at mm-wave frequencies', Microwaves and
     RF, 23, 243-249, No. 5, Dec.

9.   Napier, R.S. and Cowell, J., 1982, 'Extending Standard
     Sources up to Millimeter Waves', Microwaves and
     RF, 21, 74-109, No. 13. Dec.

10. Clarke, R.N. and Granville-George, D.A., 1982,
     'Effects of broadband noise in radio-frequency
     six-port scattering-parameter measurements',
     Electron. Lett., 18, 1110-2, 9 Dec.

11. Special Issue, 1984, 'Technology Closeup -
     Synthesizers', Microwaves and RF, 23
     No. 6, June.

12. Johnson, R.A., 1975, 'Understanding Microwave Power
     Splitters', Microwave Journal, 18, 49-56,
     No.12.

13. Hewlett Packard Application Note No. 64-1, 1977,
     'Fundamentals of RF and Microwave Power
     Measurements'.

14. Donecker, B., 1984, 'Accuracy Predictions for a New Generation Network Analyzer', Microwave Journal, 27, 127-141, No.6, June.

15. Luff G.F., et al., 1984, 'Real-time six-port reflectometer', IEE Proc., 131, Pt. H. 186-190, June.

16. Kaliouby, L. and Bosisio, R.G., 1984, 'A New Method for Six-Port Swept Frequency Automatic Network Analysis', IEEE Trans., MTT-32, 1678-82, Dec.

17. Paul, J.A., 1983, 'Wideband Millimeter-Wave Impedance Measurements', Microwave Journal, 26 95-102, No. 4, Apr.

18. Gianfortune, P.A., 1984, 'Create a millimetre-wave vector-measurement system', Microwaves and RF 23, 150-159, No. 6, June.

19. Hewlett Packard Application Note No. 8510-1, 1984, 'Millimeter-wave vector measurments usinf the 8510 Network Analyzer'.

20. Hollway, D.L. and Somlo, P.I., 1969, , 'A High-resolution Swept-Frequency Reflectometer', IEEE Trans., MTT-17, No. 4 April.

21. Lacy, P.D. and Andres, 1982, I., Wiltron Technical Review, November.

22. Lacy, P.D. and Oldfield, W., 1980, 'Calculable Physical Impedance References in Automated Precision Reflection Measurement', IEEE Trans. IM-29, 390-395, No.4, Dec.

23. Stauffer, G.H., et al., 1974, 'Direct-Reading Swept-Frequency Slotted-Line System with Slope Correction', IEEE Trans., IM-23, 394-399, No.4, Dec.

24. Somlo, P.I., 1967, 'The Locating Reflectometer', IEEE Trans., MTT-15, 250-259, Apr.

25. Fantom A.E., 1979, 'The Location of Reflections in two-port microwave components', NPL Report No. DES 61, NPL, Nov.

Chapter 15

Analogue system measurements

D. A. Williams

1. INTRODUCTION

The measurement theories and techniques described in the following relate to microwave transmitting and receiving systems or combined systems such as RADAR transceivers.

2. TRANSMITTING SYSTEM MEASUREMENTS

The function of any transmitter is primarily to generate a "carrier" signal. Data may be imposed upon this signal by the process of "modulation".

2.1. Carrier Measurements

The two basic parameters used to characterise a carrier signal are:-

        a)   Frequency
        b)   Power

These measurements generally involve the use of digital frequency meters and absolute power meters.

It is impossible to generate an absolutely pure and stable carrier signal, time variant fluctuations in frequency and power are always present to some degree and may be quantified as:-

        c)   The long term stability of the signal
        d)   The short term stability of the signal.

The difference between these two parameters lies in the time interval between frequency measurements. The long term stability of a signal generator may be measured by the use of a frequency counter with a long sampling interval. Measurement of the short term stability involves the use of modulation theory since the variations in frequency may be very rapid and give rise to modulation "sidebands" either side of the carrier.

2.2.   Modulation measurements (Ref. 1)

The term modulation relates to the process of imposing data onto a carrier signal by time variation of the carrier amplitude, frequency or phase.

2.2.1.   Amplitude modulation

The process of varying the amplitude of a radio frequency carrier by a modulating voltage is known as amplitude modulation. This process is depicted in figure (2.1)

FIGURE 1   DEPICTING AMPLITUDE MODULATION

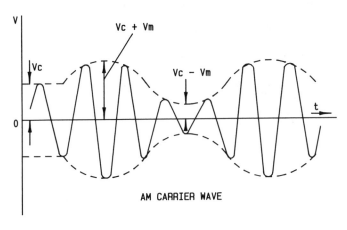

MODULATING SIGNAL

AM CARRIER WAVE

The amplitude of the modulated carrier varies sinusoidally between the values $V_c + V_m$ and $V_c - V_m$.

If $V_m/V_c = m$ is the <u>modulation</u> <u>factor</u> then $V_m = mV_c$. The expression for the modulated carrier is:-

$$v_c = (V_c + V_m \text{Sin} \, \omega_m t) \, \text{Sin} \, \omega_c t$$

or $\quad v_c = V_c \text{Sin} \, \omega_c t + mV \text{Sin} \, \omega_c t \times \text{Sin} \, \omega_m t$.

Now $\quad \text{Sin} \, \omega_c t \, \text{Sin} \, \omega_m t = \frac{1}{2} \left[\text{Cos} \, (\omega_c - \omega_m)t - \text{Cos}(\omega_c + \omega_m)t \right]$

Hence $\quad v_c = V_c \, \text{Sin} \, \omega_c t + \dfrac{(mV_c)}{2} \cos \, (\omega_c - \omega_m)t - \dfrac{(mV_c)}{2} \cos \, (\omega_c + \omega_m) \, t$

Thus if the amplitude modulated signal is displayed on a spectrum analyser, the display will resemble figure (2)

Insert fig. 2.2

FIGURE 2   THE AM SPECTRUM

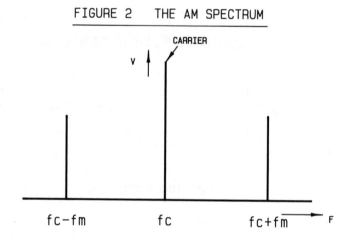

The expression above shows that an AM carrier wave contains three frequency components.

The first term represents the carrier, the second and third terms represent the lower and upper sidebands respectively.

The modulation index (m) may be measured using an oscilloscope for low frequency carriers or a spectrum analyser for microwave carriers.

Since $\quad = \quad P_c \quad \propto \quad (V_c)^2$

Power in each sideband $P_{SB} \quad \propto \quad \left(\dfrac{mV_c}{2} \right)^2$

Sideband to carrier ratio (dB)

$\quad = \quad 10\log_{10} \quad \dfrac{m^2}{4}$

2.2.2. Frequency Modulation

The process of varying the frequency of a carrier wave in proportion to the modulating signal is known as frequency modulation.

This process is depicted in figure (2.3)

FIGURE 3   DEPICTING FREQUENCY MODULATION

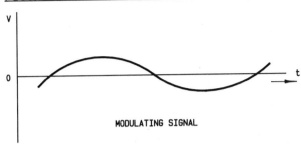

MODULATING SIGNAL

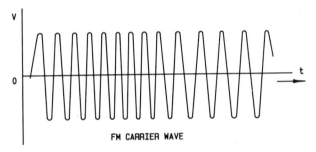

FM CARRIER WAVE

To obtain an expression for an FM wave, let the instantaneous carrier wave be represented by

$$v_c = V_c \sin \omega_i t = V_c \sin 2\pi f_i t$$

where f_i is the instantaneous frequency. For a positive increase in frequency, we have

$$f_i = f_c + \Delta f_c \sin \omega_m t$$

where f_c is the carrier frequency and Δf_c is the frequency deviation of the carrier wave, due to the modulating signal of frequency f_m.

If the instantaneous carrier phase is φ_i then

$$\tfrac{1}{2} \frac{d\varphi}{dt} = f_i = f_c + \Delta f_c \sin \omega_m t$$

or

$$\frac{d\varphi_i}{dt} = 2\pi f_i = \omega_c + 2\pi \Delta f_c \sin \omega_m t$$

By integration and a correct choice of the phase angle, we obtain

$$\varphi_i = \omega_c t - \frac{\Delta f_c}{f_m} \cos \omega_m t$$

or $\qquad \varphi_i = \omega_c t - m_f \cos \omega_m t$

where $m_f = \Delta f_c / f_m$ is called the modulation index.

Since $v_c = V_c \sin \varphi_i$ we obtain

$$v_c = V_c \sin \left[\omega_c t - m_f \cos \omega_m t \right]$$

which represents an FM carrier wave.

2.2.3. The FM spectrum

Expanding v_c yields

$$v_c = V_c \left[\sin \omega_c t . \cos(m_f \cos \omega_m t) - \cos \omega_c t . \sin(m_f \cos \omega_m t) \right]$$

Now

$$\cos(m_f \cos \omega_m t) = J_0(m_f) - 2J_2(m_f)\cos 2\omega_m t + 2J_4(m_f)\cos 4\omega_m t ...$$

$$\sin(m_f \cos \omega_m t) = 2J_1(m_f)\cos \omega_m t - 2J_3(m_f)\cos 3\omega_m t + ...$$

The coefficients $J_n(m_f)$ are Bessel functions of the first kind and order n. They are generally tabulated and a typical plot is shown in Fig. 4 Substituting into v_c yields the result

$$v_c = V_c \left[J_0(m_f)\sin \omega_c t - J_1(m_f) \{ \cos(\omega_c + \omega_m)t + \cos(\omega_c - \omega_m)t \} \right.$$
$$\left. - J_2(m_f) \{ \sin(\omega_c + 2\omega_m)t + \sin(\omega_c - 2\omega_m)t \} + . \right]$$

which reveals an infinite set of sidebands whose amplitudes are determined by the Bessel functions $J_0(m_f)$, $J_1(m_f)$etc. See figure 2.4

FIGURE 4

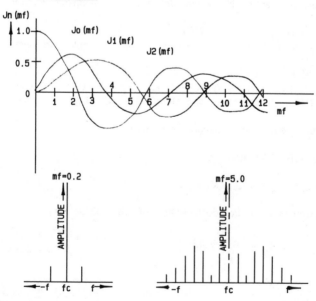

Typical spectra for m_f = 0.2 and m_f = 5 are shown in figure 2.4
The modulation index of an FM signal may be measured by the use of a spectrum analyser and noting the amplitude of particular sidebands. This technique is only feasible for a single modulating frequency and in the case where the modulating signal consists of white noise it is necessary to construct a frequency discriminator to analyse the amplitude of modulation sidebands at particular modulating frequencies (Ref 2,3)

3. RECEIVING SYSTEM MEASUREMENTS

The function of any receiver is primarily to detect small signals, amplify them to a usable level (this often involves processes of frequency translation) and then remove any information that may be present on the carrier (de-modulation).

3.1. Noise figure

In a receiver, the noise figure is usually of prime importance since it is this factor that determines the sensitivity of the receiver. Refs. (4) and (5).
The noise figure (F) of any two port system (amplifier, mixer, receiver etc.) is a measure of the "noisiness" of the system and relates two signal-to-noise ratios, one at the input end and the other at the output end of the system.
If S_i/N_i is the ratio of signal power to noise power at the input of the system and S_o/N_o is the ratio of signal power to noise power at the output of the system then:-

$$F + \frac{S_i/N_i}{S_o/N_o}$$

Noise figure is often quoted in terms of decibels ie.

$$F_{dB} = 10 \, Log_{10} \; \frac{S_i/N_i}{S_o/N_o}$$

There are two methods in common use for the measurement of Noise Figure.
a) Y Factor method

This method of Noise Figure measurement involves injecting two different known levels of noise power into the system under test and measuring the ratio of the available output powers under the two condions of available input power.
Probably the most accurate and straightforward Y factor method for amplifies having noise figures less than 4dB involves the use of a "hot and cold" load.
This device consists of a high quality matched load ($50\,\Omega$)

mounted in a chamber which may be either cooled by liquid nitrogen or heated by a thermostatically controlled electric heater.

Nyquist's noise formula for the thermal noise power associated with any resistor shows that:-

$$Pn = KTB \text{ watts}$$

where

Pn = noise

K = Boltzmanns' Constant (1.38×10^{-23} Joules/$^{\circ}$K)

T = absolute temperature of the resistor $^{\circ}$K

B = Bandwidth over which the noise power is measured.

Thus by setting the temperature of the load to two accurately known values then two different accurately known noise powers will be generated.

A diagram of a "hot and cold" noise figure measurement set is shown in fig. (3;5)

FIGURE 5   HOT AND COLD
NOISE FIGURE MEASUREMENT SYSTEM

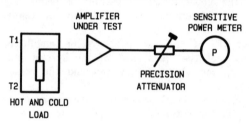

The amplifier under test is connected to the noise source and feeds a precision attenuator and sensitive power meter. The power meter is only used as a indicator, and a superhetrodyne receiver system is often used for this application.

The noise power at the input of the amplifier is P_{ni} = KTB.

The noise power at the output of the amplifier can be calculated from the definition of noise factor

$$F = \frac{S_i/P_{ni}}{S_o/P_{no}}$$

The output signal is simply the input signal multiplied by the amplifier gain.

i.e.

$$F = \frac{S_i KTB}{G\, S_i/P_{no}}$$

$$F = \frac{S_i}{KTB} \times \frac{P_{no}}{G\, S_i}$$

and $\underline{P_{no} = FGKT_1 B}$ $= P_1$

If now the temperature of the load is changed to T_2 the output power will change by an amount:

G.K $(T_2 - T_1)$ B and the output power will become

$P_2 = FGKT_1 B + GK(T_2 - T_1)B$

Since the Y factor is the ratio of the output powers under two conditions of temperature.

$$Y = \frac{P_2}{P_1} = \frac{FGKT_1 B + GK(T_2 - T_1) B}{FGKT_1 B}$$

$$\therefore \quad Y = \frac{FT_1 + (T_2 - T_1)}{FT_1}$$

$$Y = 1 + \frac{(T_2 - T_1)}{FT_1}$$

$$\therefore \quad (Y-1) = \frac{1}{F} \left[\frac{T2}{T1} - 1 \right]$$

$$\therefore \quad F = \frac{\frac{T_2}{T_1} - 1}{(Y - 1)}$$

The measurement procedure is as follows.

Cool the load with liquid nitrogen i.e.

$T_1 = 77.3\,^{\circ}K$

Set the precision attenuator to a convenient value A_1 so that a usable deflection on the indicator is achieved, this deflection is noted as in A_1.

The load is now heated to some high temperature (often $100\,^{\circ}C$ or $373.2\,^{\circ}K$) and the attenuator is adjusted until the same deflection is obtained.

The value of attenuation A_2 is noted.

The difference in attenuation gives the Y factor

i.e.   10 log Y   = $(A_2 - A_1)$

The Noise Figure of the amplifier may then be calculated using

$$F = \left(\frac{\frac{373.2}{77.3} - 1}{(Y - 1)} \right)$$

For these two temperatures $T_1 = 77.3\,^{\circ}\text{K}$

$T_2 = 373.2\,^{\circ}\text{K}$

$$F_{dB} = 10 \, \log_{10} \quad \boxed{\frac{3.83}{(Y-1)}}$$

b)   Switched Noise Source Method

In the analysis of the Y factor measurement method, it was shown that for an amplifier whose input is terminated in a resistance

$$P_{no} = FGKT_1 B$$

Expansion of this expression yields –

$$P_{NO1} = G \, KT_1 B + (F-1) \, KT_1 B \qquad (3.1)$$

A circuit diagram of a network representing this equation is shown in figure 3.6a Noise Source A is the input termination. In addition, noise source B is an imaginary source which represents the amplifiers contribution to the noise output, referred to the input.

Figure 6a can be modified as shown in figure 3.6b (3.6 a & b)

FIGURE 6   EQUIVALENT NOISE REPRESENTATION
OF A NOISY NETWORK

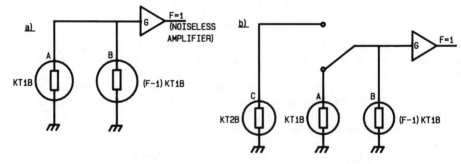

A switch is added, so that the original input termination (A) can be disconnected and the input terminated in a second source (C) of a temperature T_2.

This temperature is usually that of the gas-discharge noise generator.

The noise output power using source C is

$$P_{NO2} = G\ KT_2B + (F-1)\ KT_1B \qquad (3.2)$$

Dividing by equation $\qquad\qquad (3.1)$

$$\frac{P_{NO_2}}{P_{NO_1}} = \frac{T_2 + T_1\ (F-1)}{T_1F}$$

$$\therefore\quad \frac{P_{NO_2}}{P_{NO_1}} = \frac{T_2 + T_1F - T_1}{T_1F}$$

$$= \frac{T_2}{T_1F} + 1 - \frac{1}{F}$$

$$= \frac{1}{F}\left[\frac{T_2}{T_1} - 1\right] + 1$$

$$\therefore\quad \frac{P_{NO_2}}{P_{NO_1}} - 1 = \frac{1}{F}\left[\frac{T_2}{T_1} - 1\right]$$

$$\therefore\quad F = \frac{\dfrac{T_2}{T_1} - 1}{\dfrac{P_{NO_2}}{P_{NO_1}} - 1}$$

The term $\left[\dfrac{T_2}{T_1} - 1\right]$ is called "excess noise ratio" and is usually expressed as ENR.

It represents the relative increase in noise power at the amplifier input when the switch is operated.

If the ENR is known, the ratio P_{NO_1}/P_{NO_2} can be measured and the noise figure calculated.

Note that it is not necessary to measure the absolute power levels at the network output, merely their ratio.

Therefore $\qquad F = \dfrac{ENR}{(Y - 1)}$

The switched noise source method can be used manually using Y Factor, however it normally forms the basis of an Automatic Noise Figure Indicator (ANFI).

A simplified schematic of this instrument is shown in figure 3.7

FIGURE 7   AUTOMATIC NOISE FIGURE INDICATOR

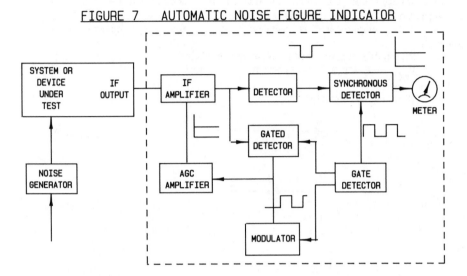

The first requirement for this type of system is a method of periodically turning the noise generator on and off. This is accomplished by a modulator driven by a gate generator, usually a free running square-wave generator (multivibrator) operating at a low audio frequency. The output of the receiver consists of square-wave modulated noise. As in the manual measurement, the two levels of this square wave represent the DUT noise and the DUT plus generator noise. This signal is amplified in the IF amplifier and detected in a square-law detector. A sample of this modulated noise signal is also detected in a second detector which is gated to provide an output only during the time the noise generator is turned off. When the noise generator is turned on, the detector is gated off. Its output is then converted to a dc signal proportional to this noise-off condition. This dc signal is then amplifier and applied to the IF amplifier in such a manner as to keep the noise off level at the output of the IF amplifier constant.

The square-law detector provides an output voltage proportional to the input power. This output will consist of two voltage levels (V_1 and V_2), proportional to the two levels of noise power at the input to the IF amplifier.

Noise-on voltage: $V_2 = a \, P_{NO_2}$

Noise-off voltage: $V_1 = a \, P_{NO_1}$

where a is the detector sensitivity constant.

The peak-to-peak amplitude of the square wave is measured by the synchronous detector and the resultant voltage is indicated by the meter.

Therefore, the meter voltage

$$V_m = c \, (V_2 - V_1)$$

where c = a gain constant

$$V_m = ac \, (P_{NO_2} - P_{NC_1})$$

Rearranging terms

$$V_m = ac \, P_{NO_1} \, \frac{P_{NO_2} - 1}{P_{NO_1}}$$

From the previous discussions

$$\frac{P_{NO\,2}}{P_{NO\,1}} = Y$$

Therefore, $V_m = ac \, P_{NO_1} \, (Y - 1)$

since $F = \dfrac{ENR}{Y-1}$

$$V_m = ac \, P_{NO_1} \, \frac{ENR}{F}$$

This equation indicates that if ENR is known and P_{NO1} is constant (a condition satisfied by the gated detector and AGC amplifier), the meter voltage is inversely proportional to the noise figure of the device under test. This is a highly desirable result since lower noise figures read closer to full scale and thus greater resolution is achieved.

Note also that because a square-law detector is used, the meter scale, when calibrated in dB, becomes logarithmic, and thus produces a natural expansion on the upper half of the scale (where low noise figures are indicated).

3.2. Gain and Linearity

If the most important characteristic of a receiver is its Noise Figure then the second most important characteristic must be its linearity since this is a measurement of the fidelity of the receiver.

The term fidelity refers to the ability to faithfully recover the transmitted data without distortion.

It can be seen from figure 3.8 that a typical microwave receiver consists of several amplification and frequency conversion stages.

FIGURE 8   MICROWAVE RECEIVER-SUPERHET

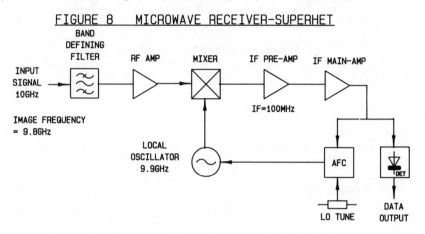

The following sections refer to methods of characterising these various stages.

3.3.   Amplifier Gain Measurement   Ref. (6)

Expressed simply, the power gain of an amplifier is the ratio of the output power to the input power.

"Scalar" gain is often measured using the equipment shown in figure 3.9.

FIGURE 9   SWEPT FREQUENCY SCALAR GAIN MEASUREMENT SET UP

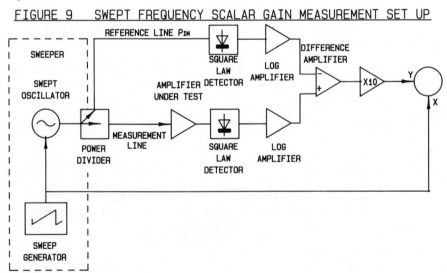

A sweeping oscillator unit is used to generate a swept frequency signal which is divided with two equal amplitude paths.

The measurement line signal is fed via the amplifier under test to a square law detector.

The reference line signal is fed directly to another square law detector.

The detector output voltages (V ∝ Pin) are fed via logarithmic amplifiers to a difference amplifier and the difference signal is amplified x 10 and fed to the vertical deflection system of an oscilloscope.

The horizontal deflection system of the oscilloscope is fed with a "sawtooth" waveform generated by the sweeper unit such that the X axis of the display will represent frequency.

Thus the system displays power gain versus frequency directly.

Unfortunately life is never simple and amplifier gain is not strictly a scalar quantity since the amplifier may introduce some phase shift and the gain becomes a vector quantity.

It is nowadays possible to make rapid measurements of vector gain by using a swept frequency network analyser and recording S parameters.

3.4. Amplifier Bandwidth Ref. (6)

The test equipment described for scalar gain measurement is generally known as a "scalar network analyser".

The system displays gain versus frequency and it is a simple matter to determine the amplifier bandwidth from this display. Amplifier bandwidth is usually specified to the -1dB or -3dB points, these points being where the midband gain has dropped by 1 and 3dBs respectively.

If the input to the amplifier is equal to P_{in} then the reference signal is also P_{in}.

The amplifier output power is P_{out}.

Then Reference power = P_{in}

Measurement line Power = P_{out}

The square law detector output will be

$V_1 = a\ P_{in}$ (reference)

$V_2 = a\ P_{out}$ (measurement)

where a = sensitivity constant.

The logarithmic amplifier outputs will be:-

$\log V_1 = \log a + \log P_{in}$

$\log V_2 = \log a + \log P_{out}$

The difference signal will be:-

$$(\log V_2 - \log V_1) = \log \frac{V_2}{V_1} = (\log P_{out} - \log P_{in})$$

$$= \log \frac{P_{out}}{P_{in}}$$

The Y deflection input voltage will be:-

$$V_Y = 10 \log \frac{P_{out}}{P_{in}} = \text{GAIN}$$

Specifications often include limits on the "gain ripple" (or flatness) within the amplifier pass band, this parameter may be measured readily using the same equipment.

3.5.   Amplifier dynamic range   Ref. (6)

The dynamic range of an amplifier is that range of output power over which the amplifier remains linear.

Fig 3.10   shows a typical amplifier transfer (gain) curve.

FIGURE 10   AMPLIFIER TRANSFER CURVE

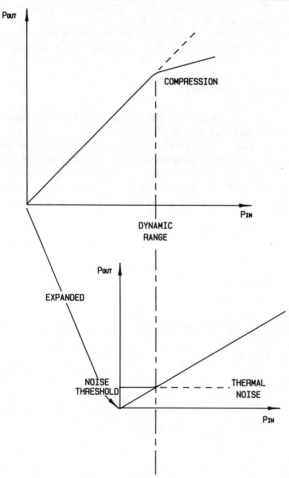

It can be seen that as the input power is increased, the output power increases linearly by the gain of the amplifier until a point is reached where the gain starts to drop. This is the upper end of the dynamic range and is usually referred to as the 1dB compression point.

This parameter is usually measured on a scalar network analyser by slowly increasing the input power then measuring the output power when the gain has dropped by 1dB.

If the input of the amplifier is terminated in a matched load, the only signal at the input of the amplifier will be the thermal noise generated by the termination.

i.e. Pn_1 = KTB watts

From the definition of noise figure

$$F_{(ratio)} = \frac{S_i/KTB}{GS_i/P_{NO}}$$

∴ P_{NO} = FGKTB watts

This means that inputs below a certain level will not affect the amplifier output signal since this will be dominated by the internally generated noise.

The dynamic range of the amplifier is therefore the ratio of the "1dB compression power level" to the "Thermal noise threshold power level".

i.e. Dynamic Range = $10 \log_{10} \dfrac{P_{out (1dB\ comp)}}{FGKTB}$

For example
A X band GaAs FET amplifier has the following characteristics.
Power Gain 20dB (100 times)
Bandwidth   1GHz
Noise figure   3dB
1dB compression point + 10dBm (10mw)

Dynamic Range = $10 \log_{10} \dfrac{10^{-2}}{2 \times 100 \times 1.38 \times 10^{-23} \times 293 \times 10^{9}}$

at 20°

= 71dB

3.6.   Mixer Measurements

3.6.1.   Conversion Loss   Ref. (7)

In a microwave receiver, the function of a mixer is to convert the received signal to a lower Intermediate Frequency.

This conversion process involves a loss of signal called the "conversion loss" and is defined as:-

$$L_c = 10 \log_{10} \frac{P(IF)}{P(signal)}$$

Conversion loss is normally measured using a Scalar network analyser as shown in figure 3.9 The test method is very similar to that for amplifier gain except that a frequency conversion process is involved. This implies that the square law detectors must be very broadband since the signal frequency may be at least ten times the IF frequency or a correction factor must be introduced to allow for different detector sensitivity at signal and IF frequency.

A typical value of conversion loss for a production X band balanced mixer is 5dB.

4.   DEMODULATION AND SIGNAL PROCESSING

A receiver system may have been designed with a good Noise Figure and adequate dynamic range. However the subsequent demodulation and signal processing stages may introduce distortion if they exhibit non-linearities.

A system or component may be defined to be linear if the output varies linearly with changes at the input, that is if the slope of the transfer characteristic is constant.

For example in a linear amplifier
$$V_o = A V_i$$
where A is a constant (the gain) and in a linear frequency discriminator
$$V_o = B \Delta f_{in}$$
where Δf_{in} = input frequency change
and B = gain factor

Any device with a non linear transfer function will introduce harmonic distortion.

A method in common use in amplifiers is the Two Frequency input test.

4.1.   Two Frequency Test for Amplifiers   Ref.(8)

The curve shown in fig. 4.11 is a special case of the non-linear transfer function

FIGURE 11   A PLOT OF OUTPUT POWER VERSUS INPUT POWER

$$\text{FOR } e_0 = 15e_1 - 2e_1^2$$

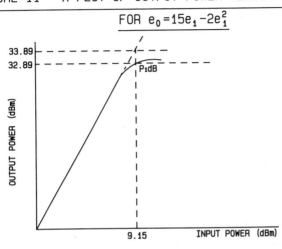

$$e_o = K_1 e_i + K_2 e_i^2 + K_3 e_i^3 + \qquad (4.1)$$

The linear gain may be defined as

$$G_o = 20 \log K_1 \qquad (4.2)$$

The 1dB gain compression point is defined as the signal level where

$$G_{1dB} = G_o - 1 \text{ dB} \qquad (4.3)$$

Consider a two-port with the following transfer characteristic (assuming $R = 50 \, \Omega$)

$$e_o = 15e_i - 2e_i^3$$

According to (4.2) and (4.3), the linear gain $G_o = 23.5$ dB and the 1 dB gain compression point is $G_{1dB} = 22.5$ dB. At this point the amplitude is limited to $A = (0.145k_1 / |k_3|)^{1/2} = 1.044$ V, and the output power is $P_{1dB} = 32.89$ dBm. A plot of P_{1dB} is shown in Fig. 4.11

Two-Frequency Input Test

Now consider an input signal $e_i = A(\cos\omega_i t + \cos\omega_2 t)$ that consists of two equal amplitude sinusoids at two different frequencies ω_i and ω_2

Applying e_i to (1) yields.

$$e_{o_3} = k_1A(\cos \omega_1 f + \cos \omega_2 t) + k_2A^2(\cos \omega_i t + \cos \omega_2 t)^2 + k_3A^3(\cos \omega_1 t + \cos \omega_2 t)^3$$

$$= k_2A^2 + k_2A^2\cos(\omega_1 - \omega_2)t + (k_2A + \tfrac{9}{4}k_3A^3)\cos \omega_1 t$$

$$+ (k_1A + \tfrac{9}{4}k_3A^3)\cos \omega_2 f + \tfrac{3}{4}k_3A^3\cos(2 \omega_1 + \omega_2)t$$

$$+ \tfrac{3}{4}k_3A^3\cos(2\omega_2 - \omega_2)t + k_2A^3\cos(\omega_{1+} \omega_2)t + \tfrac{1}{2}k_2A^2\cos2\omega_1 t$$

$$+ \tfrac{1}{2}k_2A^1\cos2\omega_2 t + \tfrac{3}{4}k_3A^3\cos(2\omega_1 + \omega_2)t + \tfrac{3}{4}k_3A^3\cos(2 \omega_2 + \omega_1)t$$

$$+ \tfrac{1}{4}k_3A^3\cos3\omega_1 t + \tfrac{1}{4}k_3A^3\cos3\omega_2 t \qquad (4.4)$$

From (4.4) it is seen that the output signal consists of components at dc. the fundamental frequencies ω_1 and ω_2, the second and third harmonics $2 \omega_1$, $2 \omega_2$ and $3\omega_1$,$3\omega_2$ and the second-order intermodulation products at $\omega_1 \mp \omega_2$ (the sum of the coefficients of ω_1 and ω_2 is 2). and the third-order intermodulation products at $2\omega_1 \mp \omega_2$ and $2\omega_2 \pm \omega_1$ (the sum of the coefficients of ω_1 and ω_2 is 3). In systems where the operating frequency band is less than an octave, all the spurious signals at $\omega_1 - \omega_2$ $2\omega_1$, $2\omega_2$, $2\omega_1 + \omega_2$, $2\omega_2 + \omega_1$, $3\omega_1$ and $3\omega_2$ fall outside the passband and can be filtered out by appropriate filters. But all the spurious signals at the frequencies $2\omega_1 - \omega_2$ and $2\omega_2 - \omega_1$ will fall within the passband and can distort the desired signal at the fundamental frequency ω_2 or ω_1. The input and output spectrum of e_i and e_o are shown in Fig. 4.12 a - b.

FIGURE 12
(a) INPUT SPECTRUM
(b) OUTPUT SPECTRUM OF A TWO-PORT WITH THIRD ORDER DISTORTION

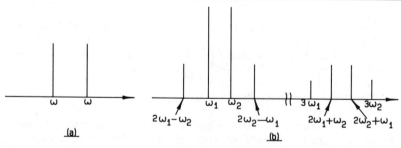

A useful measure of the third-order intermodulation distortion is the "intercept point", defined as the output power level P_1 at which the output power $P_{(2\omega_1-\omega_2)}$ at the frequency $2(\omega_1-\omega_2)$ would intercept the output power P_o at ω_1 (when the two-port is linear) if low-level results were extrapolated into the higher-power region as shown in Fig. 4.12. We remark that at low level, the output power P_o is directly proportional to the amplitude of the input signal while the output power $P_{(2\omega_1-\omega_2)}$ is directly proportional to the cube of the input amplitude. Thus the plot of each on a log-log scale (or dBm/dBm scale) will be a straight line with a slope corresponding to the order of the response, that is the response at ω_1 will have a slope of 1 and the response at $2\omega_1-\omega_2$ will have a slope of 3. Their intersection is the intercept point. We note that the actual amplitude of the output signal at ω_1 is $k_1 A+\frac{9}{4}k_3 A^3$ where $k_3<0$ for compressive two-ports. Hence at lower power levels ($k_1 A \gg \frac{9}{4}k_3 A^3$), the response of the output power $P(\omega_1)$ at ω_1 almost coincides with the response of the output power P_o at ω_1 when the two-port is assumed to be linear. At higher power levels the response of $P_{(\omega_1)}$ will be compressed and will deviate from the response of P_o as shown in Fig. 13. From (4.4) we have

FIGURE 13   DEFINITION OF THE INTERCEPT POINT

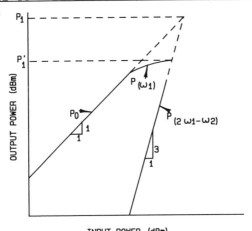

P₁ axis label: P_1, P_1'

OUTPUT POWER (dBm)

$P_{(\omega_1)}$

P_0

$P_{(2\omega_1-\omega_2)}$

INPUT POWER (dBm)

$$P_o = 10\log\left\{\left(\frac{k_1 A}{\sqrt{2}}\right)^2 \cdot \frac{10^3}{R}\right\} \quad dB_m \qquad (4.5)$$

$$P_{(\omega_1)} = 10\log \left\{ \left(\frac{k_1 A + \frac{9}{4}k_3 A^3}{\sqrt{2}} \right)^2 \frac{10^3}{R} \right\} \quad \text{dBm} \qquad (4.6)$$

$$P_{(2\omega_1 - \omega_2)} = 10\log \left\{ \left(\frac{\frac{3}{4}k_3 A^3}{\sqrt{2}} \right)^2 \frac{10^3}{R} \right\} \quad \text{dBm} \qquad (4.7)$$

Since P_1 by definition, $P_o = P_{(2\omega_1 - \omega_2)}$ by comparing 4.5 and 4.7 we obtain the theoretical amplitude A at P_1 as

$$A^2 \text{ (at } P_1) = \frac{4}{3} \frac{k_1}{|k_3|} \qquad (4.8)$$

and therefore

$$P_1 = 10\log \left(\frac{2}{3} \frac{k_1^3}{|k_3|} \right) \frac{10^3}{R} \quad \text{dBm} \qquad (4.9)$$

If R = 50 Ω

$$P_1 = 10\log \frac{k_1^3}{|k_3|} + 11.25 \quad \text{dBm} \qquad (4.10)$$

And from Fig. 4.13 we note that the response of $P_{(\omega_1)}$ intersects the response of $P_{(2\omega_1 - \omega_2)}$ at the P_1'. It is useful to relate P_1' to P_1. At P_1' we have $P_{(\omega_1)} = P_{(2\omega_1 - \omega_2)}$. By comparing 4.6 and 4.7, the amplitude A at P_1' is given as

$$A^2 \text{ (at } P_1') = \frac{2}{3} \frac{k_1}{|k_3|} \qquad (4.11)$$

and hence

$$P_1' = 10\log \left(\frac{1}{12} \frac{k_1^3}{|k_3|} \right) \frac{10^3}{R} \quad \text{dB m} \qquad (4.12)$$

$$= P_1 - 9 \text{ dBm}$$

4.2. Harmonic distortion, single frequency test

This method is often used in more complex networks involving frequency discriminators or AM detectors.

An input to the system under test is provided with pure sinusoidal modulation at a low frequency (of frequency or amplitude dependant on the system function). The demodulated output is fed to a selective voltmeter or fast fourier transform spectrum analyser. See Figure 4.14

FIGURE 14   THE MEASUREMENT OF HARMONIC DISTORTION

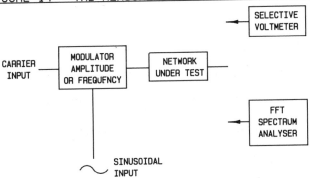

The harmonic level, with respect to the fundamental, may then be measured directly and the percentage harmonic distortion may be calculated.

i.e. If the second harmonic is 40dB down on the fundamental, the second harmonic distortion is 1%.

4.3.   Phase distortion - Group delay

The previous sections have described amplitude distortion in various aspects.

In some systems, information is conveyed in the phase of the carrier and in such systems it is important that this phase information is not distorted by networks having non-linear phase/frequency characteristics.

Group delay is defined as the rate of change of phase shift with frequency for any network.

Group delay D = $\dfrac{d \phi \ (rads)}{d \omega t \ rads \ sec^{-1}}$

Group delay is expressed in units of time and may be measured directly using a network analyser. An alternative test method is shown in fig. 4.15

FIGURE 15   SCHEMATIC DIAGRAM OF GROUP DELAY MEASUREMENT SYSTEM

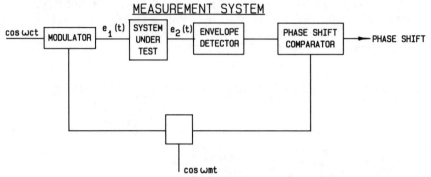

An amplitude modulated carrier is transmitted through the system under test. The phase shift of the modulated envelope is measured. In fig.4.15 the amplitude modulated input signal may be expressed as

$$e_1 = E_i (1+M\cos \omega_m t)\cos \omega_c t$$

$$+ E_i \cos\omega_c t + \frac{ME_i}{2} \cos (\omega_c + \omega_m)t +$$

$$\frac{ME_i}{2} \cos (\omega_c - \omega_m) t$$

where M is the modulation index.
ω_m is the modulating frequency
ω_c is the carrier frequency
E_i is the amplitude sufficient to ensure that e_i is a small signal
The output of the system under test with small signal gain K may be expressed as:-

$$e_o = KE_i\cos\omega_c(t-t_d)+ \frac{KME_i}{2} \cos(\omega_c+\omega_m)(t-t_d)$$

$$+ \frac{KME_i}{2} \cos(\omega_c -\omega_m)(t-t_d)$$

$$= KE_i \left[1+M \cos \omega_m(t-t_d)\right] \cos\omega_c(t-t_d)$$

where t_d is the group delay of the system under test, assumed to be constant from $\omega_c - \omega_m$ to $\omega_c + \omega_m$. The modulation frequency ω_m determines the bandwidth of the system within which fluctuations in group delay can be measured. The smaller ω_m, the more accurate t_d can be measured. However, a smaller ω_m results in a smaller phase shift and decreases the resolution. In practice ω_m can be chosen between 1 and 10MHz for small group delay measurement and 20kHz for larger group delay. Since the phase shift is measured only at ω_m, the carrier frequency ω_c can be swept over the entire bandwidth of the system, and thus the group delay over the system bandwidth can be measured. A relationship between group delay t_d and the phase shift of the modulation signal that is the envelope of the signal can be expressed as

$$t_d = \frac{\phi_c}{\omega_m}$$

REFERENCES

1. Connor F.R. 'Modulation' - Edward Arnold London 1975

2. Robins W.P. 'Phase Noise in Signal Sources'- IEE Tel. 9 Peter Peregrinus, London 1982.

3. Ondria J.G. 'A Microwave System for Measurements of AM and FM Noise Spectra' IEEE Trans MTT 16 No. 9 Sept. 1968.

4. Friis A.T. 'Noise Figure of Radio Receivers' - Proc IRE, 32, July 1944.

5. Sucher and Fox Handbook of Microwave Measurements Polytechnic Press of the Polytechnic Institute of Brooklyn 1963.

6. Laverchetta T.S. 'Handbook of Microwave Testing' - Artech House 1981.

7. 'Noise Figure and Conversion Loss of the Schottky Barrier Mixer Diode'
    IEEE Trans MTT,15, 1967

8. HA T.T. 'Solid State Microwave Amplifier Design', John Wiley & Sons.

Chapter 16

Digital system measurements

P. Cochrane

1   PROLOGUE

Whilst the microwave domain is currently dominated by analogue systems, the past five years has seen a rapid influx of digital applications spanning computers, signal processing, radar and communication. Systems occupying the spectrum up to 300 MHz have become commonplace and widely distributed. Above 300 MHz systems range from commercially available (generally < 600 Mbit/s) to development models (up to 2 Gbit/s) and leading edge research (from about 2-10 Gbit/s). Although the present sum total of digital systems (across all disciplines) only represent a 10-15% penetration into the "analogue world", they have now encroached well into the microwave spectrum and are set to become the dominant mode by the end of the millennium. It therefore seems both timely and wholly appropriate to now introduce the topic of digital system measurements as a chapter of this book.

As might be expected, today's digital system designer has made much use of the body of knowledge, experience and techniques accrued by his analogue counterparts over the past 50 or so years. Perhaps we should not be surprised by this fact as most human problem solving is very heavily based (especially in engineering and science) upon past similar or related experience. Moreover, digital systems do, in any case, depend to some significant degree on analogue technology in their realisation. Happily, therefore, we find that most of the chapters in this book have some bearing on "digital system design" and afford us some economy of effort.

In view of its introductory nature we commence this chapter with a brief comparison of the key differences between digital and analogue systems. We then go on to consider specific measurement techniques broadly pertinent to device, component, channel and system. However, because of the now very diverse nature of digital systems, it has proved necessary to define key parametric measurements and then to cite a few specific examples to illustrate particular realisations and applications.

2   INTRODUCTION

There are many who would cite economics as the key driving force responsible for the general move away from analogue towards an almost wholly digital world. Whilst in the broadest sense this is true, the "prime mover" has been micro-electronics. Since about 1960 the integrated circuit has produced a doubling of circuit density each year, resulting in a 2^{25} device/unit area today compared with a single transistor [1] in 1959! It is perhaps worth reflecting that without this incredible size reduction any significant digital operation at microwave rates would certainly have been precluded on the grounds of sheer physical size and power consumption of discrete circuits. As might be expected, the effect of progressive miniaturisation has seen gain-bandwidth products and switching speeds steadily progress with silicon bipolar technology now freely available for rates up to 500 Mbit/s [2]. More specialised items, capable of up to 2.5 Gbit/s, resulting from military and "leading edge" research programmes are also being realised in silicon [3,4]. However, Gallium-Arsenide (GaAs) integrated circuits now look poised to provide the next technological leap to 3-5 Gbit/s and perhaps beyond [5,6].

Before we contemplate our very specific topic of microwave measurements, let us note the key differences between analogue and digital systems listed in Table 1. We should also note the following essential differences:-

· In analogue systems the objective is to preserve the fidelity of a signal/process across a continuum of time. This is usually achieved by maximising some time average signal to noise ratio.

· In digital systems the objective is to preserve the fidelity of a signal/process at discrete time instants. This is usually achieved by maximising some peak signal to noise ratio.

Further important differences are also evident in the concise statement of analogue and digital system characteristics embodied in the 1948 works of Shannon [7]. Here digital system capacity is related to channel bandwidth W, average received signal power S, the average received noise power N, and system constant K, as follows:-

$$C = W \log_2 (1 + KS/N) \text{ bit/s} \qquad \ldots \ldots (1)$$

TABLE 1   Comparison of analogue and digital systems

Parameter	Analogue	Digital
Complexity — Circuit	Low	High
Complexity — System	Low	Very high (10-100x analogue)
Suitability for Integration	Low	High
Operating Mode	Linear	Non-Linear
Number of Signal States	Infinite	Finite
Bandwidth Occupancy	Narrow	Wide
Bandwidth Efficiency	High	Low
Signal Format	Essentially simple - sinusoidal - directly related to source	Essentially complex - sequence of pulses - indirectly related to source
Signal Definition	Relatively uncertain	Tightly constrained
Signal Degradation	Strongly dependent upon distance/number of operations	Almost independent of distance/number of operations
Phase Distortion	Generally insensitive	Very sensitive
Amplitude Distortion	◄———————— Sensitive ————————►	
External Interference	Highly sensitive	Relatively impervious
Failure Mode	Generally graceful - smooth transition from working to non-working	Usually catastrophic - sudden transition from working to non-working
Utility	Limited	Wide
Production	Difficult	Simple

For digital systems the upper bound to the capacity is achieved when the system constant K = 1 and for transmission of information at a rate R < C an arbitrarily small bit-error probability may be achieved by adopting a sufficiently complex coding scheme.

For analogue communication a similar relationship is also derived:-

$$R < W \log_2 S/N \qquad \ldots\ldots (2)$$

The difference in the interpretation of (1) and (2) however, is that in the analogue mode the bandwidth of the original signal may not be reduced without incurring distortion. This is in contrast to the digital case where such reduction may be compensated for by an increase in S/N without any distortion penalty [8].

It therefore follows that information capacity may be visualised in terms of a space composed of the variables: time T, S/N and the bandwidth W which are interchangeable for a given fidelity [8].

In the analogue domain the most common trade-off between bandwidth and S/N is realised using Frequency Modulation (FM), whilst in the digital case Pulse Code Modulation (PCM) is used. The relative performance advantages of these modulation schemes [9] are shown for comparison in Fig 1.

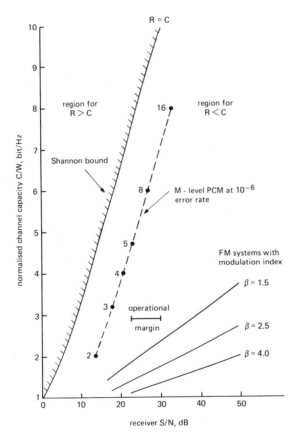

Fig 1   Shannon capacity bound

Whilst in principle the task of bit transfer in computer, communication and radar systems is straightforward, in practice precise design demands sophisticated models [10]. This in turn dictates sophisticated measurement techniques - some of which we will now consider.

3   SIGNAL FORMAT

In order to dispel at the earliest possible stage any confusion prompted by the quoting of MHz and Mbit/s, let us just note that; THERE IS, IN GENERAL, NO OBVIOUS RELATIONSHIP BETWEEN THE TWO! The widespread and indiscriminate use of multi-level (M-ary) signalling, coding, modulation and spectral filtering renders such a notion generally impracticable [9]. Only with a full knowledge of a specific system may we make any intelligent estimate of the spectral occupancy for a given bit rate. This fact is illustrated in the "scattergram" of Fig 2 for a range of digital systems. At the simplest level this situation may be demonstrated by considering the time waveform and frequency spectra of a computer and telecommunication transmission path - Fig 3. In the former case data is transported over a very short distance (within a machine) with a high S/N ratio with little necessity to constrain the bandwidth. Conversely, the latter case has to contend with very long distances, and as a result, a relatively low S/N ratio. There is thus a need to constrain the bandwidth to the minimum necessary [11].

Whilst the most common digital format is binary, it is often necessary to utilise multi-level signalling generated via some complex coding algorithm [10]. Equally, it is also often necessary to employ a variety of modulation techniques to facilitate bit transport between two or more distant points. Fortunately, most system design considerations, including measurement data and techniques, map with relative ease between the various levels of complexity. Throughout this chapter, therefore, we confine our descriptions to the binary case unless otherwise specified.

4   BASIC DIGITAL MEASUREMENTS

Let us now pose the obvious question; just what do we need to measure to engineer digital systems at microwave rates? A summary answer to this question is given in Table 2 via a matrix of device, media and system against measurement technique for both the analogue and digital case. It should be recognised that no matrix of this kind can ever be complete, and of necessity must rely upon implication to cover those techniques/devices/systems not explicitly included. Nevertheless, the matrix does encompass the majority of cases of interest and those that play a key role in our present study.

Close inspection of Table 2 reveals a high degree of commonality between analogue and digital device, media and system measurements. As those items (marked X) have received extensive attention in other chapters they can be dismissed in the present chapter as requiring only a passing mention. Of the remaining items, those pertinent to digital systems (marked *) fall into three broad categories:-

- Pulse Shape
  $\begin{cases} \text{Impulse Response} \\ \text{Pulse Response} \\ \text{Rise Time} \\ \text{Fall Time} \\ \text{Stability} \end{cases}$

- Pulse Position
  $\begin{cases} \text{Propagation Delay} \\ \text{Jitter} \\ \text{Long Term Drift} \end{cases}$

- Pulse Fidelity
  $\begin{cases} \text{Bit Errors} \\ \text{S/N Ratio} \end{cases}$

Each of these categories is now addressed along with specific application examples.

TABLE 2 Measurement requirements for analogue and digital systems

/ = Analogue
* = Digital
X = Analogue + Digital
- = Not Applicable

	Gain/Frequency	Phase/Frequency	Return Loss	CW-Continuous	CW-Burst	TDR-Step	Isolation/Frequency	Impulse Response	Pulse Response	Rise Time	Fall Time	Propagation Delay	Amplitude	Linearity	Power	Noise Power	Noise Figure	S/N Power Ratio	Bit Errors	Jitter	Subjective Assessment	Trial Data
DEVICE/FUNCTIONAL BLOCK																						
Transistor/Diodes	X	X	-	-	-	X	X	*	*	*	*	X	X	/	X	X	X	*	-	-	-	-
Amplifiers	X	X	X	-	-	X	X	*	*	*	*	X	X	/	X	X	X	*	-	*	-	-
Mixers/Switches	X	X	X	-	*	X	X	*	*	*	*	X	X	/	X	X	X	*	-	*	-	-
Digital Gates/Toggles	-	-	-	-	-	*	-	-	-	*	*	*	*	-	-	-	-	*	-	*	-	*
Integrated Digital Circuits	-	-	-	-	-	*	-	-	-	*	*	*	*	-	-	-	-	*	*	*	-	*
CHANNEL/MEDIA																						
Pair	X	X	X	-	*	X	X	*	*	*	*	X	X	-	X	X	X	X	-	-	X	*
Coax	X	X	X	X	*	X	X	*	*	*	*	X	X	-	X	X	X	X	-	-	X	*
Waveguide	X	X	X	X	-	X	X	-	-	-	-	X	X	-	X	X	X	X	-	-	X	*
Radio	X	X	-	-	-	-	X	-	-	-	-	X	X	-	X	X	X	X	-	-	X	*
SYSTEM																						
Communication	/	/	X	-	-	-	X	-	-	-	-	X	X	/	X	X	-	X	*	*	X	X
Computer	-	-	-	-	-	-	X	-	-	-	-	-	X	-	X	-	-	-	*	*	*	X
Radar	/	/	-	-	-	-	X	/	/	-	-	/	X	/	X	X	-	X	-	*	X	X

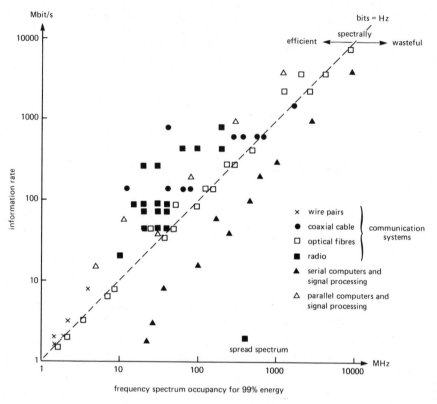

Fig 2   Bandwidth-bit rate distribution for modern digita
          systems

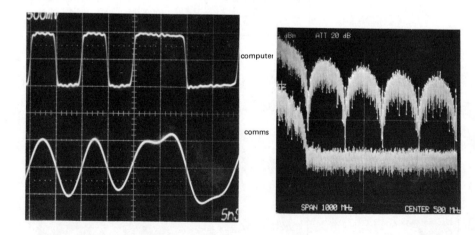

Fig 3   Example digital system waveforms and spectra

5   PULSE SHAPE

Whilst a pulse may be the product of a test on a
passive or active circuit, or be just the output of some
generator, the stimulus may take the form of; a step,
impulse, pulse or CW burst [12]. Mathematically these
stimuli are conveniently related as follows:-

Step $- u(t) <=> \dfrac{1}{jw}$ ...... (2)

Impulse $- u'(t) <=> 1$ ...... (3)

Pulse $- u(t+T/2)-u(t-T/2) <=> T \; Sinc(fT)$ ...... (4)

CW Burst $- cos 2\pi f_c t \,[u(t+T/2)-u(t-T/2)] <=>$

$$T/2 \,[Sinc(f+f_c)\,T+Sinc(f-f_c)\,T] \qquad ...... (5)$$

In reality these forms become distorted by the
constraint of a finite bandwidth, resulting in a smoothing
of the temporal shape. This process takes no heed of the
original stimulus as it merely convolves each by the same
time function [11]. It is sufficient therefore to
contemplate some final pulse and how it should be described
- how it was produced is, in this sense, of little
consequence!

5.1 Pulse Shape Definition

There are numerous national and international standards
related to the definition of a pulse [13]. Whilst at the
system end of the scale these employ simple go, no-go,
amplitude masks [14] of the form shown in Fig 4, the demand
for precision at the fundamental level leads to definitions
of the form given in Fig 5. This last form is of prime
interest as they define such parameters as:-

.   Width - 50% amplitude duration

.   Rise time - 10-90% or 20-80% amplitude

.   Fall time - 10-90% or 20-80% amplitude

.   Relative amplitude - average top line and base
     line separation

.   Absolute amplitude

.   Pre/Post-cursor under/overshoot - amplitude and
     duration

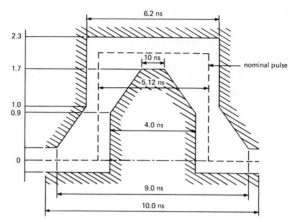

Fig 4   A CCITT [14] pulse mask

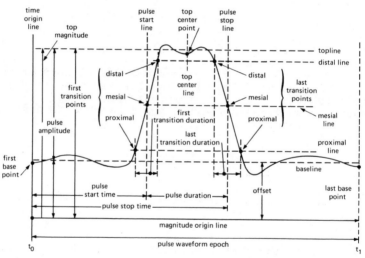

Fig 5   Defining features [13,15-17] of a pulse

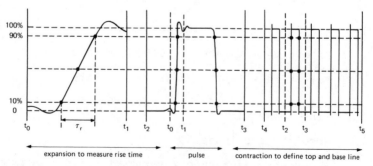

Fig 6   Rise time measurement - "by eye"

All of these measures, plus the many more defined in the standards cited [15-17], are relative to the notionally ideal rectangular pulse shape and can, in principle, be made available directly from a CRO display. In practice the presence of noise and other distorting phenomena introduce practical difficulties.

We should also note at this point that some applications demand references that are based upon the physical generation of pulses by pre-defined generators. These generally take the form of an electronically generated step or impulse driving a tightly specified uniform lossy transmission line [18]. However, as we are principally concerned with measurements, rather than specification and reference standards, these two topics are only mentioned here for background and future reference.

5.2 Rise/Fall Time Measurement

These transition measures of a pulse assume some significance via their direct indication of circuit speed (and thus bandwidth). In the case of logic elements it is essential, for circuits to operate well, that pulses (at least) observe the engineering "one third rule of thumb": 1/3 to rise, 1/3 flat, 1/3 to fall. This ensures a 1/3 period stability zone during which logical decisions '0 or 1' can be made; a necessary feature as this is usually controlled by a clock which is at worst a sinusoid, and perhaps at best of the 1/3 regime! To put this in a measurements perspective; a 1 Gbit/s system has a 1/3 period time of just 333 ps!

Practical pulses are neither rectangular nor trapezoidal, but distorted in the manner of Fig 5. Rise and fall time measurement thus become difficult as bit rate and pulse distortions increase. At the lower rates (< 100-300 Mbit/s) it is common to employ a 10-90% measure, whilst at higher rates (> 100-300 Mbit/s) additional pulse distortion (which can totally preclude 10-90%) encourages the use of 20-80%.

The key to this measurement is deciding where the 'Top Line' and 'Base Line' actually lie [15]. An engineering "by eye" approach uses waveform expansion and contraction as indicated in Fig 6 to make this decision. A more sophisticated solution is available on many microprocessor controlled instruments which perform a similar process by taking the probabilistic approach indicated in Fig 7. Typically averages of 100 or more can be employed to overcome the influence of additive noise and display/digitisation deficiencies. Modern equipments employing this method can return uncertainties < 0.5% when >> 100 averages are taken [13].

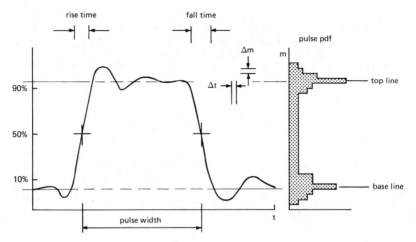

Fig 7   Rise time measurement - "probabilistic routine"

5.3   Pulse Width Measurement

The most commonly used width measures are the rms and 50% amplitude. Engineering approximations, such as assuming a Gaussian distribution and 'by eye' decisions on top and base-line, give some quick measure of these values. For any degree of accuracy however, the histogram approach of the previous section has to be employed. Derivations based upon the rise and fall time algorithm depicted in Fig 7 are generally favoured, yielding a similar accuracy [19,20].

5.4   Absolute Shape

The total definition of a pulse may be achieved through a comparison with purely mathematical definitions or via some electronic reference standard [18]. Forming a correlation or variance difference measure then defines the degree of fit. Problems associated with noise, absolute value and normalisation are readily overcome by post digitisation processing as indicated below:-

$$\text{Correlation} - \int_T g(t)\,[a + b\tilde{g}(t+\tau)]dt \qquad \ldots\ldots \ (6)$$

Reference   Measure Pulse
Pulse   DC shift   = a
             Scaling   = b
             Distortion = ~

$$\text{Variance} \ - \frac{1}{T}\int_T [g(t) - a - b\tilde{g}(t+\tau)]^2\,dt \qquad \ldots\ldots \ (7)$$

The factors 'a' and 'b' have to be adjusted to maximise the degree of fit as $g(t)$ and $\tilde{g}(t)$ are fixed entities.

Notice that the form of integral in both cases is closely allied to analogue detectors described in other chapters. The multiplicative and non-linear action is identical, but the low pass filter of the analogue case has been replaced by an integrator. We have also gone from a discrete spectral line (carrier wave) to a broad spectral continuum for a single event, and a "comb" spectrum for periodic events!

5.5 Dynamic Pulse Shape Variation

In many applications pulse shape can be expected to be constant over all repetitions. However, in active circuits time dependent non-linear effects can produce a significant variation from pulse to pulse. This is particularly true of systems driven with random bit patterns which introduce effects such as run-length related charge storage and DC wander which in turn lead to; bias drift, slice level variation, gain and bandwidth fluctuation [4,21]. Generally speaking this situation implies the use of a "pseudo-random pattern" to facilitate the selection of one or more specific pulse positions [22]. An aggregate or probabilistic approach may also be taken with min, mean, max and distribution of rise, fall times and pulse shape generated via suitable software routines.

A subjective picture of all these variations, plus those related to pulse position and jitter, is also possible using the "eye diagram" approach described later in Section 8.

5.6 An Engineering Example - Impulse Testing Coax

The structural imperfections present on coaxial cables introduce transmission impairments by virtue of the forward scattering of the signal. Traditionally such phenomena have been assessed by TDR, Pulse Echo, CW Burst Return Loss and Excess Attenuation [23]. Whilst each of these techniques is adequate for short lengths of cable, they suffer significant limitations on long lengths due to the increased dispersion and attenuation, especially at the higher frequencies [24]. At bit rates above 140 Mbit/s the coaxial cable scattering problem rapidly becomes the prime rate limitation to both speed and reach. For example, at 140 Mbit/s repeater spacings of 2.2 km are used, whilst at 560 Mbit/s this is reduced to 1.1 km, with a substantial increase in the impairment budget allocated to the echoes introduced by the scattering mechanism [25].

An equipment developed to provide a field evaluation of installed cables is shown schematically in Fig 8. Because the cable ends are not colocated, this equipment has to perform an impulse response test by "looping back" the test signal via a regenerative repeater [26]. A 50 V peak, 1 ns wide, sin^2 impulse is used to excite the cable which

presents the following characteristics for 560 Mbit/s
operation:-

- Total section length < 1.1 km

- Total equalised section loss < 80 dB

- Thermal noise level ~ - 90 dBm

- Received signal-forward-echo-ratio ~ 60 dB

With an input stimulus of 50 V peak the receiver is
presented with the following conditions:-

- Amplitude of received impulse S_0 > 5 mV

- Amplitude of thermal noise N_0 ~ 8 μV

- Amplitude of echo components E_0 ~ 5 μV

The echoes generated are clearly swamped by thermal
noise. To recover the echo information a combination of
correlation detection and signal averaging is employed.
Assuming a 40 dB Echo/Noise ratio is required, the number of
samples can be calculated from:-

$$\frac{E}{N} = \frac{E_0}{N_0}\sqrt{n} \qquad \ldots\ldots (8)$$

where the number of averages $n \simeq 2.6 \times 10^4$, by calculation,
but in practice have to be ~ 10^5 to overcome hardware
limitations. An impulse repetition rate of < 100 kHz is
used to avoid impulse and echo overlaps in the cable which
offers a nominal 3.5 μs path delay. The correlator employs
a programmable delay generator which can be stepped in 1 ns
and 100 ps intervals to give fine grain detail.

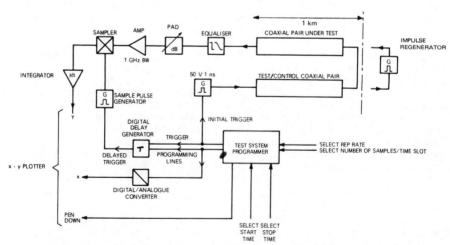

Fig 8   Through pulse test equipment

A typical measurement result produced on an installed cable is shown in Fig 9 - illustrating the presence of periodic and random echo components. System assessment in the field simply used a "go/no-go" impulse response mask, but in the laboratory used a point by point assimilation into a mathematical model.

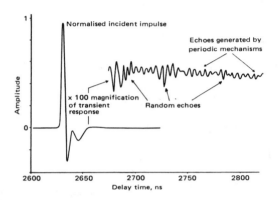

Fig 9　　An example transient response for an installed coaxial cable

It should be noted that the periodic echo transient exists for several hundred bit periods after its excitement by an impulse. So, what at first sight appears to be a negligible phenomenon (with a S/E ratio ~-60 dB) actually poses a fundamental limitation due to its aggregation over hundreds of bits. Interestingly, the frequency domain manifestation of the cable irregularities is a ripple on the attenuation/frequency curve of < 0.1 dB. Not surprising therefore when measuring a total path loss ~80 dB such a ripple is difficult to detect - though not impossible [27]. Overall, however, the described equipment offers significant advantages with a demonstrated ability to detect an impulse through a 180 dB path loss whilst returning a > 20 dB S/N ratio. Moreover, the time domain results produced are directly pertinent to digital systems calculation - in short "a digital test for a digital system" [28-31].

6　　PULSE POSITION

Uncertainties in pulse position come about through a fundamental inability to define the statics and dynamics of systems. As with pulse shape we are, in practice, confronted with stable, unstable, periodic and aperiodic pulse position variations [32-34]. In all cases some reference epoch has to be taken as a datum for measurement. This may in fact be the original stimulus (pulse or clock) or an aggregate at the point of measurement. Conveniently we find that a relative pulse position figure is often sufficient.

6.1 Delay

The measurement of delay falls into two distinct camps; relative and absolute. In turn each of these present two broad classes of problem; short and long distance/time. At the "simple end" of the measurement complexity scale a CRO may be used to measure delay by taking some amplitude feature that can be traced between two distinct epochs. Unfortunately, in many practical situations pulse shape changes due to passage through a system render such a simple approach inaccurate and unusable. Correlation offers a solution to this problem as indicated in Fig 10. The stimulus g(t) may be; a pulse; a burst of pulses; a continuous random or periodic sequence; or a burst/gating of a continuous random or periodic sequence. In the absolute delay case the correlation function is given by:-

$$\Upsilon_{rg}(\tau) = \int_T r(t)\,g(t-\tau)\,dt \qquad\qquad \dots\dots (9)$$

which is maximised when τ = the system delay.

For relative delay, which involves a comparison between two points in the received output, the correlation function is of the form:-

$$\Upsilon_{rr}(\tau) = \int_T r(t)\,r(t-\tau)\,dt \qquad\qquad \dots\dots (10)$$

again this is maximised when τ = the delay between correlated events.

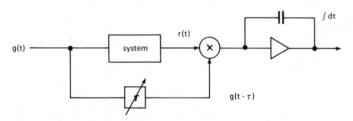

(a) Absolute delay (cross-correlation)

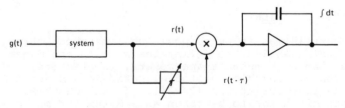

(b) Relative delay (auto-correlation)

Fig 10   Delay measurement by correlation

Extension of this approach to comparisons between two or more separate systems is clearly straightforward. The resolution varies directly as a function of the pulse width and shape, but in the best possible (narrowest functions) can return better than 1 ps accuracy [13].

Once more we should note the form of the integrals and their relation to the analogue realisation of similar measurement techniques; where the functions would be of the form $\cos \omega_0 t$ and $\cos(\omega_0 t + \emptyset)$, with a low pass filter instead of an integrator.

6.2 Jitter

This phenomenon is induced by both deterministic and random effects associated with pulses, bit stream patterns, passive and active circuits. Specifically, sources of jitter include; random noise, interference, cross-talk, non-linearities, circuit stability, finite thresholds and bias disparities [32].

Measurement of jitter poses some significant problems as the rate of change can be on a bit-by-bit, or pattern-by-pattern basis, implying a very rapid variation. Alternatively, relatively slow variations may be present due to random/cyclic drift of circuits. Unfortunately, the tendency is for several (if not all!) the jitter sources to be present simultaneously [35]. This makes measurement difficult - and very often source separation impossible.

There are two classes of measurement technique in general use; those concerned with direct operations on pulses and those employing some derived feature. In the "derived feature camp" it is necessary to produce a clock directly from the data stream via a suitable non-linear operation and filter [11]. The jitter variation and frequency spectrum may then be assessed directly using a coherent detector of the form shown in Fig 11(a) [36]. In some applications the data source and reference clock may be geographically distant to the receive terminal [37-39]. Two options are then exercised; the transmission of the reference clock via a second system or the generation of a local reference clock using a very high Q filter compared to that in the measure path - see Fig 11(b). This "comparative" approach is one that finds extensive application in practice and is echoed later in the pulse fidelity section.

In the "direct operations camp", correlators provide an instrumentationally simple solution for microwave applications, but require a knowledge of the pulse shapes in order than $\tau(t)$ (9) may be derived from $\Upsilon[\tau(t)]$. This is not always possible in practice, but even if it is the results are critically dependent upon shape and often extracted using a derivative route - $\Upsilon'[\tau(t)]$.

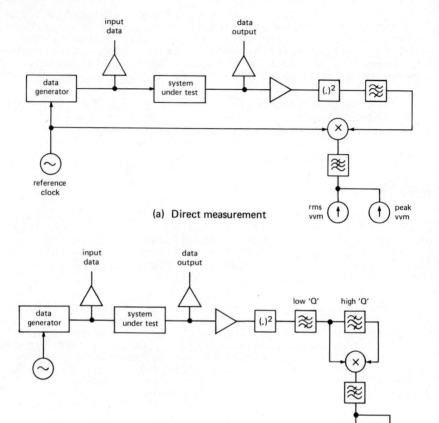

(a) Direct measurement

(b) Comparative measurement

Fig 11   Jitter measurement using a recovered clock

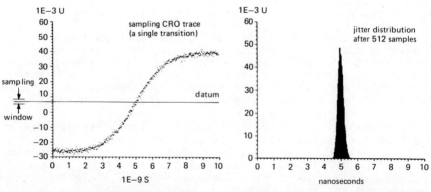

Fig 12   DPO display datum   Fig 13   DPO generated
            for jitter measurement   jitter pdf

6.3 An Engineering Example - Direct Jitter Measurement

A versatile and accurate jitter measuring technique overcoming many of the previously described limitations is available via the sampling CRO, digitisation and processing embodied in the Digital Processing Oscilloscope (DPO) [20,40]. By setting a single transition across the display and digitising relative to some convenient datum (50% amplitude), as indicated in Figs 12 and 13, a probabilistic histogram may be produced. This is achieved by merely filling storage locations each time a transition passes through. Uniquely this technique may be applied to clocks (carriers), deterministic, pseudo-random and random bit streams [41]. It also offers the facility of being able to separate out the random jitter component by triggering from a pattern clock or synchronising pulse [42].

Because a DPO digitiser typically operates on a full screen matrix of 512 x 512 locations, a considerable economy of processing time can be achieved by "windowing" the sampling area to the region of a transition. In addition the use of a weighted sum of samples about the reference datum within the window also increases the acquisition rate. The influence of the window width and number of scans employed is illustrated in Figs 14 and 15 respectively.

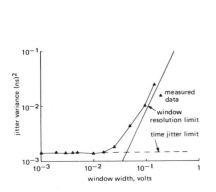

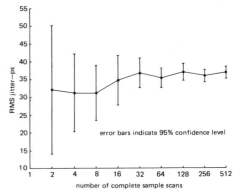

Fig 14   RMS jitter as a   Fig 15   RMS jitter as a function
         function of window   of acquisition rate

The resolution attainable at 2 Gbit/s with a commercial DPO is of the order 4 ps with a 95% confidence range of 1.6 ps. Experiments performed by the NBS in Colorado with experimental equipment achieved an order of magnitude improvement over these figures [20].

7   PULSE FIDELITY

The performance arbiter of any system is its fidelity; for analogue systems this is expressed as a S/N ratio, for digital systems it is the "Bit Error Ratio" (BER) [43]. In the final analysis the result of all digital system specification, design, construction and testing comes down to a single performance parameter - a number - the BER. As this is defined as; "the ratio of bits in error to total number of bits over a given period", its derivation, is in principle, via a simple counting process. In practice this generally turns out not to be the case as it is subject to a number of difficulties and constraints, which we now consider.

7.1   Error Generation

In an ideal world, digital systems would not make decision errors. A signal presenting itself at a logic element would sit clearly in the '1' or '0' region as depicted in Fig 16(a). At each clock transition a "faithful reproduction" would result with all '1' and '0' states correctly accounted for. Such a situation envisages an infinite bandwidth, zero noise, interference and distortion regime that is physically unrealisable (Heisenberg) [44]. In reality the input data and clock will, to some degree, be subject to all these influences, and furthermore, the logic elements will also be subject to a number of uncertainties due to the imperfections inherent in the electronic circuits used. Typically these will include; decision deadbands and hysteresis, bias and timing offsets, finite set up and hold times [11].

For all the above reasons all digital systems make decision errors - the problem is fundamentally unavoidable. However, in practice we generally have sufficient latitude to reduce the incidence of errors to a negligible level as depicted in Fig 17. Here the mean arrival time between errors for a given BER and bit rate is annotated with typical system "End of Life" performance targets. It is interesting to consider, for a moment, the various operating points cited:-

> Communications Systems: Whilst the human ear and eye have difficulty in perceiving error events with a BER below about 10^{-5} and 10^{-4} respectively, data users demand a higher level of performance. The human interface is inherently tolerant to errors, but the computer is not. An economic bargain has thus to be struck between engineering "forward error correction" and "message repeat" schemes for data communication against the simplicity allowable for the human interface. At present full international connections (of < 27,500 km) are specified to give a BER < 10^{-6}. This implies an average system BER of about

4×10^{-11} /km at the end of life with the worst possible combination of interconnections!

Computer and Control Systems: In this sphere the standards and objectives adopted vary widely. On the one hand, those systems controlling nuclear arsenals are designed to return exceptionally low error occurrence (and understandably so!) with "Mean Time Between Disaster" (MTBD) of millions of years. On the other hand, low cost home computers need not be so rugged and typically return a MTBD > 10 years. This time span is a good deal longer than the MTBF of the associated peripherals - and almost certainly the device itself! The span of system "End of Life" BER in this arena is $\sim 10^{-15}$ - 10^{-30} or better.

Remote Sensing and Surveillance: The wide contrast between the "target information rate" and "system data rate" generally allows very poor BER values to be used. A combination of complex coding, with "frame by frame" interpolation, checking and averaging permits BER $\sim 10^{-1}$ or more to be tolerated.

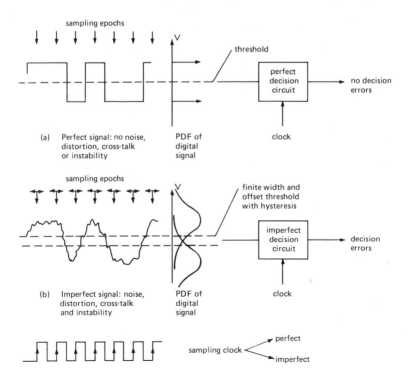

Fig 16   The decision process - perfect and imperfect

The purpose in explaining these different system
requirements is to put in perspective the very essence of
the BER measurement problem. For any degree of measurement
confidence we require at least 10, and preferably > 100,
error events [45] within a reasonable measurement period of
say < 1 day. Inspection of Fig 17 shows that such criteria
constrain BER measurement to the range 10^{-10}-10^{-14} for rates
spanning 1 Mbit/s-10 Gbit/s. Clearly, increasing the
measurement time by an order of magnitude only wins us an
extension to 10^{-15}-10^{-16} at best, whilst incurring a
significant measurement time penalty. It is thus common
practice to perform BER measurements in the range to > 10^{-14}
and then, of necessity, extrapolate beyond.

At the other extreme of the BER range there are
instrumentation and circuit difficulties that prevent
accurate measurement much above 10^{-2}-10^{-1}. Hence there is a
safe window for BER measurement with the majority of today's
systems (< 1 Gbit/s) constrained from 10^{-11} to 10^{-2}. This
allows a reasonable measurement time with > 100 events per
point recorded.

Throughout our discussion we have considered the
"average situation" with a calculated mean assuming a
uniform probability distribution. However, in the real
world error events are generally "bursty" - that is, they
arrive in clusters separated by quiet times [11,28]. As the
mathematics of this mechanism is extremely complicated, and
moreover, has yet to be resolved [32], we will continue to
take an average view in this chapter. In fact this problem
does not preclude any meaningful and valid work, but is more
closely linked to precise telecommunication network/system
design falling outside our present study.

7.2   The Basis of Bit Error Ratio (BER) Measurement

The basic principle involved in BER measurement is
depicted in Fig 18. A bit pattern generator is used to
activate the test and reference paths, their output signals
meet at the "Exclusive Or" gate (EX-OR) and are compared.
Any bit difference in 'A' and 'B' causes an output from the
EX-OR, recognising the occurrence of an error. By counting
the number of errors in a given number of clock periods the
BER may be calculated. Such a straightforward scheme finds
application at a subsystem and chip level [46] where the
path delay is insignificant. Moreover, the use of a
reference subsystem or chip (instead of a passive delay) to
allow a multi-functional comparison with a multiplicity of
inputs and outputs may also be conveniently realised in this
way [47]. This is a commonly adopted technique employed at
the manufacturing QA level when environmental and life tests
are performed.

At a system level it is generally true that the total
path delay, coupled with the likely non-colocation of the
ends, precludes the use of a simple passive delay as

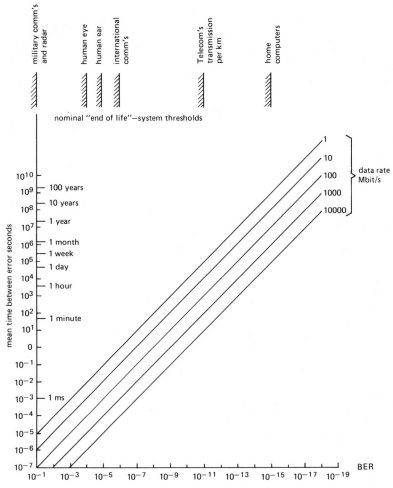

Fig 17   Mean time between decision errors

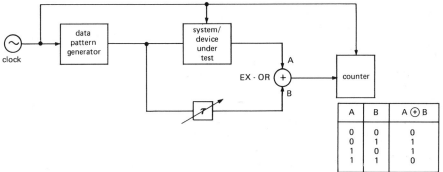

Fig 18   The basis of BER measurement

indicated in Fig 18.   The method has been used, but is prone
to differential delay, interference and crosstalk between
the two paths.   For these reasons the most commonly adopted
scheme employs autonomous pattern generators and detectors
as shown in Fig 19.

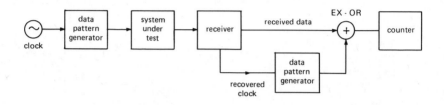

Fig 19   BER measurement with autonomous ends

The bit pattern generator usually takes the form of a
shift register configured to produce a deterministic or
(more commonly) Pseudo-Random Binary Sequence (PRBS) [22].
At the receiver the same sequence must also be generated by
the error detector, but in synchronism with that received at
the system output.   Again an EX-OR comparison on a
bit-by-bit basis, followed by a counter will produce the BER
[48].   The "trick is" to get the received and locally
generated sequences into synchronism.   There are principally
two schemes, which are based upon frame alignment and self
aligning algorithms.   In the frame alignment case some word
in the sequence is assigned to be the reference.   For a PRBS
this might well be the "all 1's" word or 000 ...... 01.   At
the error detector a recognition circuit inspects the
incoming sequence slipping the clock a bit at a time until
alignment is achieved.   To negate the influence of decision
errors, which will occasionally produce a loss of lock
state, the protocol adopted usually requires a high
percentage of alignment words to be wrong before a new
search is initiated.   Such a regime will tolerate a BER $\sim$
10^{-1} before the regular loss of alignment makes measurement
impossible.   For example, in one commercial equipment a
$2^{10} - 1$ PRBS is used with frame alignment on the "all 1's"
condition.   For the incorrect reception of this state, 1 or
more of 10 bits has to be in error.   However, loss of frame
alignment is based upon the BER count and deemed to have
occurred when more than 30×10^3 bits are detected to be in
error in any 80×10^3 bit block.   This arises at a mean
BER:-

$$p = \frac{3}{8} = 0.625 \qquad\qquad\qquad ...... (11)$$

The probability that an alignment word is wrong, for $2^{10}-1$ PRBS, is thus given by:-

$$P_{aw} = 1 - q^{10} = 0.99 \qquad \qquad \dots\dots (12)$$

$$\text{where } q = 1 - \overset{\downarrow}{p} = 0.375 \qquad \dots\dots (13)$$

Hence, on average, synchronism is not lost until 99% of alignment words are in error.

A simple self alignment scheme due to Westcott [49,50] very neatly avoids the pattern synchronisation problem by utilising identical pseudo-random sequence generators for both the transmit and receive circuits. In the receive case however, the feedback loop of the shift register is left open as indicated in Fig 20. This configuration results in the generation of an 'expected sequence' identical to, and in step with, the original message. Whence, under error free conditions no 'bit-by-bit' differences occur and thus no errors are detected. However, should a transmission error occur, then we might logically expect it to be detected by such an arrangement [51]. In fact, a single bit error produces a total error count of three (when two feedback taps are used) and a divider/counter decoder circuit is usually employed to give the true error count.

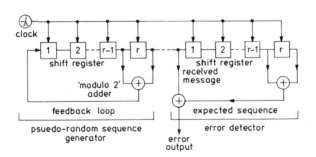

(a) PRBS generator and self synchronising error detector

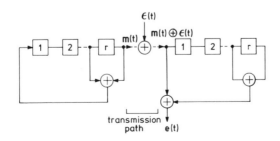

(b) Mathematical model and notation

Fig 20   The Westcott BER measurement scheme

In the case of a high density of bit errors being present within the received message, it is possible for a group or block of closely spaced errors to yield an incorrect error count. The successful application of the technique can thus be restricted in such a case. The manner in which this arises may be analysed with reference to Fig 20 using the following notation:-

PRBS time waveform - $m(t)$ $\qquad = m_o$ ...... (14)

Time shifted PRBS - $m(t-rT)$ $\qquad = m_r$ ...... (15)

Time shifted PRBS - $m[t-(r-1)T] = m_{r-1}$ ...... (16)

The output from the PRBS data source may thus be described as:-

$$m_o \qquad = \qquad m_{r-1} \oplus m_r \qquad \dots\dots (17)$$

The fundamental   Time displaced
message polynomial   message polynomials

Assuming some error polynomial $\epsilon(t)$ introduced by the system under test, the error detector receives the following input:-

$$m_o' = m_o \oplus \epsilon_o \qquad \dots\dots (18)$$

The output of the 'Exclusive OR' 'bit-by-bit' comparator is thus:-

$$e = m_o' \oplus \left\{ m_{r-1}' \oplus m_r' \right\} \qquad \dots\dots (19)$$

$$= \left\{ m_o \oplus \epsilon_o \right\} \oplus \left\{ m_{r-1} \oplus \epsilon_{r-1} \oplus m_r \oplus \epsilon_r \right\} \qquad \dots\dots (20)$$

$$= \left\{ m_o \oplus m_{r-1} \oplus m_r \right\} \oplus \left\{ \epsilon_o \oplus \epsilon_{r-1} \oplus \epsilon_r \right\} \qquad \dots\dots (21)$$

but $m_o = m_{r-1} + m_r$ (see equation 17)

$$\therefore e = \left\{ 0 \right\} \oplus \left\{ \epsilon_o \oplus \epsilon_{r-1} + \epsilon_r \right\} \qquad \dots\dots (22)$$

Whence $e = \epsilon_o \oplus \epsilon_{r-1} \oplus \epsilon_r$ $\qquad \dots\dots (23)$

The operation of the detector circuit is now apparent. If our error polynomial contains components that are spaced such that they do not overlap at the 0, r-1, or r positions, then the true error count is multiplied by a factor of 3 (in this particular example). However, if the error polynomial does overlap at any of these points, then a mutilated output sequence would be expected. A condition for the successful application of the technique is therefore a digital system error polynomial that does not contain closely spaced or 'burst' errors [5]. However, systems are usually specified and designed to operate (the military excluded) at very low BER $\sim 10^{-6}$–10^{-10} and beyond. Thus for a good deal of experimental and systems work, the usefulness of the technique is not devalued by such a limitation as indicated by the computed accuracy estimates of Table 3.

True BER p	Measured BER p_o'	Approximate Measurement Inaccuracy X%
10^{-1}	1.87×10^{-2}	18.7
10^{-2}	10^{-4}	2.0
10^{-3}	10^{-6}	0.2
10^{-4}	2×10^{-8}	0.02
10^{-5}	10^{-10}	0.002
10^{-6}	10^{-12}	0.0002

TABLE 3: Measurement accuracy with error probability

This problem is generally insignificant and the scheme is commonly used for BER up to 10^{-2}. For higher values of BER it is possible to introduce an additional shift register to decode error bursts [52].

Equipments using the described techniques are available [53,54] at rates spanning 0.1-2 Gbit/s, with range extenders to 4 Gbit/s and bit patterns of 2^7-1, 2^{10}-1, 2^{15}-1 and 2^{23}-1. Beyond this rate research systems have been reported up to 10 Gbit/s [55], but these tend to be based upon delay line techniques (to replace the D types) and discrete logic and/or analogue elements.

7.3 Pseudo-Error Ratio Measurement

An interesting alternative to the above techniques allows an assessment of system performance on a self comparative basis - without the need for a known or reference sequence [56]. This approach relies upon two digital receivers configured as per Fig 21. One of the receivers is considered "perfect-or reference", whilst the second has its performance degraded in a known manner [57]. The output from the EX-OR gate therefore generates "pseudo-errors" related to the degree of degradation introduced between the two receivers [58].

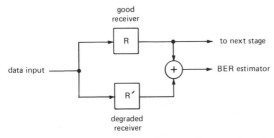

Fig 21 · Pseudo-error detection principle

The practical implementation of this scheme could use the degradation of:-

• Pulse shape   - by filter adjustment

• Timing jitter - by static or dynamic off-set

• Decision point - by static/dynamic off-set

Of these options decision point static or dynamic off-set appear to be the most popular. The practical realisation is relatively straightforward and easy to control (Fig 22) whilst the recorded output gives a reasonable agreement with theory [59,60].

Obviously the "Achilles heel" of this technique is the need for a reference receiver of known characteristics. Providing this, relies upon the far more complex techniques previously described.

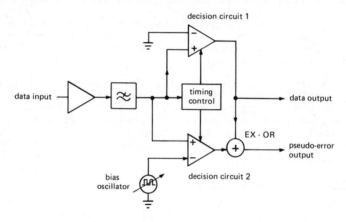

Fig 22   A practical pseudo-error scheme

7.4 An Engineering Example - Precision BER Measurement

In the telecommunications network regenerative repeaters are required at regular intervals to compensate for the transmission media losses. The associated decision errors accumulate additively and as a result system designers have to qualify repeaters at BERs below 10^{-10}. The need for a high level of precision in the related measurements is then dictated by the necessity to extrapolate beyond the minimum working value, defined by the CCITT [43], to some value giving sufficient engineering margin for temperature variation, component tolerances, ageing and transmission path imperfection. For example, a regenerator in a 565 Mbit/s coaxial system may be designed to give a BER of 10^{-11} with a 7 dB S/N margin for satisfactory operation over a 15 year life. Depending upon the line code used [11] and the specific system design this 7 dB margin ensures a start of life error rate of $< 10^{-55}$.

Clearly this is beyond the range of any practical
measurement technique - hence the need for extrapolation and
great precision.

Electrical noise is generally the prime source of
random decision errors in digital systems. In many cases
this is generated by thermal effects and exhibits Gaussian
amplitude statistics. For our present purpose we shall
constrain our treatment to this one error generating
mechanism in the context of a binary transmission system.
Thus for a repeater section we may deduce the following
error probability [11,31] expression:-

$$P_e(v) = \frac{1}{\sqrt{2\pi}\,\sigma} \int_v^\infty \exp\,[-\tfrac{1}{2}(x/\sigma)^2]dx \qquad \ldots\ldots (24)$$

where after integration $(v/\sigma)^2 = \dfrac{\text{the peak signal power}}{\text{average noise power}} = \dfrac{S}{N}$

For the case of small tail areas (probability of error
or BER) this complementary error function may be linearised
[11] by applying a double logarithmic transformation to give
a convenient approximation for graphical presentation:-

$$\frac{S}{N}_{dB} = 10\,\log_{10}\left\{\log_e P_e^{-1}\right\} + 3\,dB \qquad \ldots\ldots (25)$$

Examples of this curve are given in Fig 23 where it can
be seen that the error probability scale rapidly compresses
as the S/N ratio increases. This characteristic has rather
perverse implications as it points to a need for greater
accuracy in a region where it is most difficult to generate
a large number of errors and thereby accrue a good
statistical average.

It is clear from (24) that the error probability in our
scheme so far, is only a function of the S/N ratio. To
measure the error probability at a given S/N value thus only
involves an adjustment of the signal or noise level. A
straightforward method of doing this is to employ a variable
attenuator at the input to a regenerator/section to vary the
signal level independently of the noise. However, this
approach does suffer from a number of practical limitations
associated with the repeater electronics. To be specific,
it assumes that the repeater performance (other than error
rate) is not influenced by the signal amplitude. But in
practice additional timing jitter can result due to the
reduced S/N ratio, and finite aperture decision/slice
thresholds, can lead to a magnified error count. A further
difficulty also arises in the case of systems using metallic
cable or optical fibre where pulse dispersion is related to
the path loss. This is usually compensated for by shaped or
adaptive agc circuits that change the overall channel
response according to the received pulse amplitude. Any
controlled experiment is thus difficult as both signal and
noise variations result.

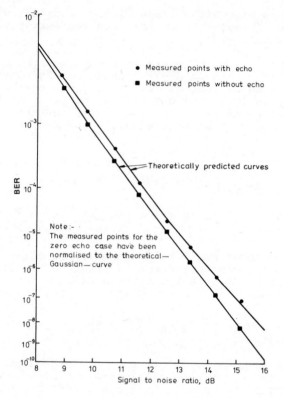

Fig 23   Predicted and measured BER for a cable section
with periodic echoes

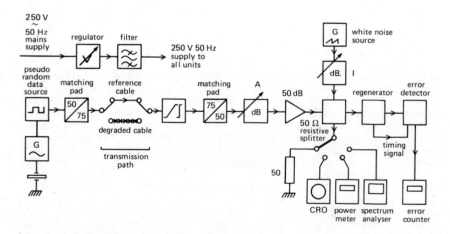

Fig 24   Precision BER measurement scheme

In contrast to the above, adjustment of the noise level presents few problems, provided a Gaussian source of known characteristics is available. A simple method of introducing a controlled noise source is shown in Fig 24. This is preferred to alternative methods such as signal attenuation and reamplification as it avoids problems associated with amplifier non-linearity and overload. It also reduces the number of experimental variables present and allows traceable calibration of all components.

Injecting Gaussian noise into the transmission path shown reduces the effective S/N ratio as follows:-

$$S/N = \frac{S}{N_o + N_i} \qquad\qquad \dots\dots (26)$$

where $S = v^2$ the peak signal power at a decision epoch

N_o = the ambient system noise power

N_i = the injected noise power.

$$\frac{S}{N}\bigg|dB = \frac{S}{N_i}\bigg|dB - 10\ \log_{10}\left[1 + \frac{N_o}{N_i}\right]dB \qquad \dots\dots (27)$$

Provided we make $N_i \gg N_o$ then this reduces to:-

$$\frac{S}{N}\bigg|dB \doteq \frac{S}{N_i}\bigg|dB \qquad\qquad \dots\dots (28)$$

If this condition does not apply, then it is necessary to adjust the results using a correction factor:-

$$F_c = 10\ \log_{10}\left[1 + \frac{N_o}{N_i}\right]dB \qquad\qquad \dots\dots (29)$$

The particular equipment configuration given in Fig 24 overcomes the following key practical problems:-

• Temperature variations: It is important to avoid wide temperature changes when completing long runs of measurements, especially when low error counts were being recorded (ie a change of temperature from 290 to 291 K gives a change in the relative noise power of 0.015 dB).

• Power meter scale errors: If the S/N ratio is measured for each attenuator setting (I), this results in a discernible scatter due to the accuracy of the power meter scale and changes between ranges. An obvious solution is to adopt a comparative measure based upon a particular setting of the attenuator (I) and only one reference point on the power meter scale.

- Attenuator inaccuracy: The signal-to-noise ratio is measured for a particular value of attenuation (I) and thereafter used as a reference point with no other power readings necessary. Using a precision attenuator [61] ensures a minimisation of the scatter of recorded results. This procedure also affords the considerable advantage of referring all measurement points back to a common reference point with a defined uncertainty.

- Amplifier overload: Overload of the receive amplifier is only possible by the incoming signal. Correct adjustment of the attenuator (A) to give a fixed output level for all test situations prevents this condition.

Although the test system itself may be well engineered, some variance in the recorded results can be attributed to external influences such as current surges propagating down power supplies and radiation from other equipment and heavy plant. Such effects generally manifest themselves as quite distinct bursts of errors which cannot be treated satisfactorily by a statistical analysis [43]. This may be overcome as follows:-

- Measurement period: Important measurements should be performed at electrically quiet times.

- Number of errors recorded: The experimental uncertainty in defining error probability of digital systems has been investigated by Miles and Crow [45] on the basis of wholly stationary statistics. According to their results it is sufficient to record less than 100 errors (typically 39) for an uncertainty of 10%. In practice this is found to be a somewhat optimistic estimate [62].

- Averaging of results: Even if all the precautions and criteria described are observed, there may still be further sources of error present in the recorded results. To overcome these effects experiments should be repeated a number of times and block averages produced.

- Normalising results: The precise definition of the operating point of a practical digital transmission system, relative to the theoretical ideal, invariably involves a significant measurement error ~ 0.5 dB [63]. To overcome this a normalisation procedure can be used to place the reference results along the theoretically ideal Gaussian curve. This involves no more than a lateral shift (S/N-wise) of all the results and thus allows normalised comparisons to be made between different transmission conditions.

Over a wide range of system operating conditions the described equipment and measurement techniques return a 90% confidence S/N accuracy of ±0.025 dB. This figure is attainable over the error rate range of 10^{-3} to 10^{-9} for comparative measurements [64]. In terms of a system base operating condition this is equivalent to a S/N = 23 dB and resulting error rate $\sim 10^{-45}$.

8   SUBJECTIVE TECHNIQUES

In the preceding sections the techniques considered produce objective single parameter measures applicable to device, component, channel and system. However, it is very often the case that we wish to take a "global or aggregate view" - simultaneously assessing a number of interdependent parameters and events. Notably, this is possible using the "eye and constellation diagrams" to create a complete subjective picture.

8.1 The Eye Diagram

During the alignment of a digital transmission channel it is essential to gauge the level of performance prior to commencing detailed tests. The "eye diagram" - so called because of its resemblance to the human eye - is the most convenient measure.

Generating an "eye" is straightforward - as depicted in Fig 25. Given that the data may be random, pseudo-random or deterministic, a CRO will be unable (for random data) or able (for pseudo random and deterministic data) to provide a synchronised display of 1's and 0's. If the trigger is fed from the data clock, which is always available, a number of trajectories are overlaid (in the manner indicated in Fig 25) to produce an eye. For a perfect wave the eye will be clean and well defined, but if pulse shape variations, jitter, echoes and thermal noise are present the aggregate view will be of the form given in Fig 26. For multi-level (M-ary) signals the eye diagram becomes much more complex, but no less useful, as indicated in Fig 27.

Eye closure due to impairments can thus be estimated on the basis of the width and height boundaries inside the eye - decisions being made about the centre "cross-hairs" of Fig 25. First order estimates of system performance penalties can then be made using the ratio of max/min eye height and width [65].

At a device, component or sub-system level the eye diagram produced will generally be of a "square nature" as there are no significant bandwidth constraints imposed. In addition the thermal noise level will also be insignificant. The eye then serves the purpose of highlighting any echo and distortion effects due to poor line matching and electronic circuit bias.

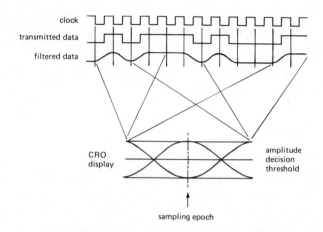

Fig 25   Eye diagram construction

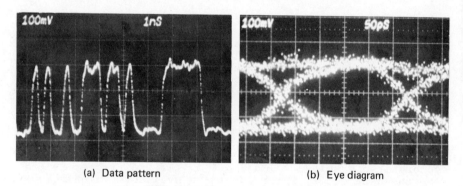

(a) Data pattern   (b) Eye diagram

Fig 26   A 4 Gbit/s data stream and eye

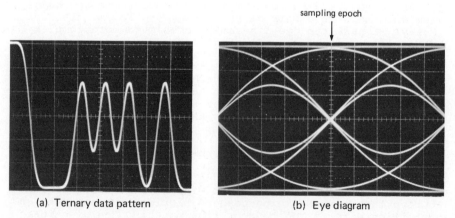

(a) Ternary data pattern   (b) Eye diagram

Fig 27   A "perfect" three level eye

8.2   Constellation Diagrams

The modulation techniques commonly used to transport digital information on radio carriers include, PAM, FSK, PSK and QAM, plus many variants and refinements thereof [9,66]. Each of these leads to a particular signal space occupation of varying complexity and decision threshold pattern [67] as shown in the examples of Fig 28. When displayed on a CRO such signals appear as point sources of light (constellations) in the precise locations indicated. But with the addition of transmission distortions (as in the eye diagram case) the point definition is reduced to a blur [68], with position and thus the decision process becoming uncertain (Fig 29).

With a "trained (human) eye" it is possible to detect additional phenomenon (in this aggregate picture), such as; poor carrier (PLL) tracking at the receiver (the whole constellation appears to slowly rock to and fro); poor agc tracking (the constellation pulsates); non-linear distortion [67] (the constellation becomes assymmetric). The constellation diagram thus has a dual role as both an alignment aid, as well as a performance estimator in the same manner as the "eye".

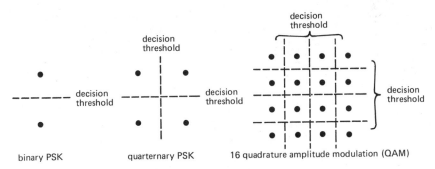

Fig 28   Example signal constellations

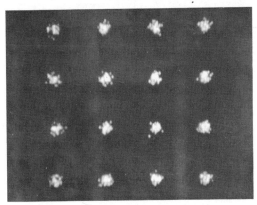

Fig 29   16 QAM constellation with additive noise

8.3 Subjective to Objective

Using the DPO approach to display, digitise and process both the eye and constellation diagrams it is possible to dynamically compute system performance estimates. This has been done on a number of systems to provide an in-service performance monitor, but is inadequate for laboratory and precise work.

9 A FINAL NOTE

This chapter has attempted to provide a comprehensive review and introduction to the topic of digital system measurements. All the techniques described are in use today, and most are based upon commercial equipment. Within the next three years telecommunications systems capable of 1.6-2.4 Gbit/s will be commercially available [69]. The following ten years may also see systems available up to and beyond 5 Gbit/s using GaAs technology [70,71]. Those measurement techniques using system related techniques such as; correlation, pulse testing and BER for example, will obviously be available. However, those reliant upon CRO techniques pose a serious problem. At present commercial sampling CRO's offer a sampling window ~20 ps, but the period of a 5 Gbit/s binary signal is 200 ps! "The sampler is only 10 x faster than the system". A fempto-second sampling oscilloscope is thus urgently required for these future system developments. Without such a device time domain system alignment will become extremely difficult, and precise objective measurements meaningless!

10 ACKNOWLEDGEMENTS

I am indebted to a number of colleagues at BTRL for providing assistance in preparing this chapter, in particular Stuart Walker for the 4 Gbit/s results, Dave Monro and Peter Adams for their patient proof reading and helpful comments. Acknowledgement is also made to the Director of British Telecom Research Laboratories for permission to publish this contribution.

11 REFERENCES

1   JONES T: 1980, Microelectronics and Society, Open University Press.

2   WELBOURN A D: 1982, Proc IEE, 129/5, 157-172.

3   MARKOC H and SOLOMON P M: 1984, IEEE Spectrum, Feb, 28-35.

4   HAWKER I: 1985, BTTJ, 3/1, 70-78.

5   MELLOR P J T: 1984, BTTJ, 2/1, 60-68.

6   YOSHIKAI N, KAWANISHI S, SUZUKI M and KONAKA S: 1985,
     IEE Elec Lett, 21/4, 149-150.

7   SHANNON C E: 1948, BSTJ, 27, 379-423 and 523-656.

8   COCHRANE P, WESTALL F A and CRAWFORD D I: 1984,
     IEE E&P, Jan, 37-42.

9   LAWRENCE V B: 1983, IEEE COM-SOC, Tutorials in Modern
     Communication Theory, Pitman.

10   SLEPIAN D: 1973, Key Papers in the Development of
     Information Theory, IEEE Press.

11   BYLANSKI P and INGRAM D G W: 1980, Digital
     Transmission Systems, Peter Peregrinus for IEE.

12   CCITT: 1984, Compendium of Cable Measurement Methods,
     ITU Geneva.

13   NAHMAN N S: 1983, IEEE Trans, IM-32/1, 117-124.

14   CCITT: 1984, Red Book Recommendations, 1-G703.

15   BRITISH STANDARD 5698:   1979, Pt 1 and 2.

16   IEEE STANDARD 181:   1977.

17   IEC STANDARD 469-1/2:   1974.

18   NAHMAN N S: 1970, IEEE Trans, IM-19/4, 382-390.

19   GANS W L: 1976, IEEE Trans, IM-25/4, 384-388.

20   GANS W L: 1980, Proc IEE, 127(H)/2, 99-106.

21   HAWKER I: 1983, Proc IEE, 130(G)/6.

22   GOLOMB S W: 1965, Shift Register Sequences,
     Holden-Day Inc.

23   CRANK G J: 1978, POEEJ, 71/3, 167-174.

24   CRANK G J and HATHAWAY H A: 1974, POEEJ, 66/4,
     246-252.

25   CATCHPOLE R J, CRANK G J and COCHRANE P: 1978, IEE
     Colloquium Digest 1978/43, 4/1.

26   COCHRANE P: 1978, POEEJ, 71/3, 175-180.

27   AHMAD S and STEPHENS W J B: 1977, IEE EUROMEAS 77 Conf
     Pub 152, 95-97.

28   CRANK G J and COCHRANE P: 1980, POEEJ, 73/3, 145-152.

29   COCHRANE P: 1979, Proc IEE, 126/1, 26-28.

30   BATES R J S and COCHRANE P: 1980, <u>Proc IEE</u>, <u>127/1</u>, 16-21.

31   CATTERMOLE K W and O'REILLY J J: 1984, Mathematical Topics in Telecommunications, Vol 2, Pentech Press, 319-337.

32   KEARSEY B N and McLINTOCK R W: 1984, <u>Jnl IERE</u>, <u>54/2</u>, 70-78.

33   HART G: 1984, <u>ibid</u>, <u>54/4</u>, 155-162.

34   COCHRANE P, KITCHEN J A and POWELL W: 1981, <u>IEEE</u>, <u>ICC-81</u>, (<u>CH 1684-5</u>), 38.2.1-4.

35   WALKER S D, CARPENTER R B P and COCHRANE P: 1983, IEE <u>Elec Lett</u>, <u>19/6</u>, 193-194.

36   BYLASTRA J A: 1977, <u>ATR-11/2</u>, 37-45.

37   HURST G C, HOLMES W H and ZAKAREVICIUS R A: 1977, IEE <u>Elec Lett</u>, <u>13/3</u>, 81-82.

38   BALDINI J J, HALL M W and BATES R J S: 1982, <u>IEEE</u>, <u>GLOBECOM 82</u>, (CH 1819-2), C7.3.1-7.

39   CANNAVO S and PASQUINI V: 1984, <u>Alta Freq</u>, <u>L III/4</u>, 250-257.

40   REED R C: 1977, <u>ibid 27</u>, 101-103.

41   COCHRANE P, BARLEY I W and O'REILLY J J: 1979, IEE <u>Elec Lett</u>, <u>15/24</u>, 774-775.

42   ROGERSON S P: 1985, <u>IEE</u>, <u>MTTS-85</u>, Conference paper to be published November 1985.

43   McLINTOCK R W and KEARSEY B N: 1984, <u>Jnl IERE</u>, <u>54/2</u>, 79-85.

44   HEISENBERG W: 1927, <u>Zeitschrift fur Physik</u>, <u>43</u>, 127.

45   MILES M J and CROW E L: 1977, <u>URSI</u>, <u>Conf Lannion</u>, Measurements in Telecommunications, 125-130.

46   LEE S C and BASS A S: 1982, <u>IEEE Jnl</u>, <u>SC-17/5</u>, 913-918.

47   HAWKER I, SHEPPARD M J, SCOTT D L and FLAVIN P G: 1985, <u>Proc IEE</u>, <u>132 G/2</u>, 60-63.

48   HUCKETT P and THOW G: 1983, HP Telecom Symposium, (South Queensferry, West Lothian, Scotland), 6/1-6/12.

49   WESTCOTT R J: 1969, UK Patent 1,280,390.

50   MALLETT C T and COCHRANE P: 1979, <u>IEE</u>, <u>ESSCIRC 79</u>, <u>Conf Pub 178</u>, 14-16.

51 NEWCOMBE E A and PASUPATHY S: 1982, Proc IEEE, 79/8, 805-828.

52 HAN S B: 1973, USA Patent 3,914,740.

53 JOLY R, LIECHTI C and NAMJOO M: 1983, IEEE, Jnl, SC-18/4, 402-408.

54 HEDE C: 1978, ibid, 66-70.

55 SCHWARTE R et al: 1978, 8th European Microwave Conference, 61-65.

56 GOODING D J: 1968, IEEE Trans, COM-16/3, 380-387.

57 BIC J C, DUPONTEL D and LAINEY G: 1977, URSI Measurements in Telecommunication Conf, Lannion, France, 101-106.

58 FEHER K and BANDARI A: 1976, IEEE, Canadian Comms & Power Conf, 121-124.

59 KEETLY J M and FEHER K: 1978, IEEE Trans, COM-26/8, 1275-1282.

60 BATES R J S: 1984, OFC Cannes, 395-399.

61 CRANK G J: 1971, BPO Research Report 209.

62 COCHRANE P: 1979, University of Essex PhD Thesis.

63 COCHRANE P: 1978, CNET Lannion Vacation School Notes, Universite de Rennes, France.

64 COCHRANE P: 1982, IEE Colloq Digest 1982/3, 7/1-7/8.

65 BELL LABS: 1971, "Blue Book" - Transmission Systems for Comm's.

66 FEHER K: 1981, Digital Communications - Microwave Applications, Prentice Hall.

67 SALEH A A M and SALZ J: 1983, BSTJ, 62/4, 1019-1033.

68 BIGLIERI E: 1984, Trans IEEE, COM-32/5, 616-626.

69 COCHRANE P: 1984, Proc IEE, 131 F/7, 669-683.

70 BOSCH B G: 1979, Proc IEEE, 67/3, 340-379.

71 LIVINGSTONE A W: 1982, Microelectronics Jnl, 13/1, 29-33.

Chapter 17

Measurements on active devices

R. D. Pollard

1. INTRODUCTION

A substantial proportion of modern microwave systems require the use of semiconductor devices and the proper design of these circuits necessitates a knowledge of their characteristics. The measurement problem which this presents is of a different nature to that normally encountered by the metrologist since the interface between the device under test and the measurement system is poorly defined. In addition, the devices concerned are almost invariably non-linear and consequently the usual small signal measurements are inadequate to fully describe the behaviour under realistic operating conditions.

This lecture is intended to provide background information to identify the problems of semiconductor device measurements and will look at only three distinct examples as a vehicle for highlighting the problems. The first section deals with the characterisation of two-terminal devices such as detector diodes, transferred electron devices, IMPATTs, etc. Many of the basic concepts of de-embedding are brought out by this means. Secondly, a detailed examination of a specific technique for the measurement of the small-signal S-parameters of a packaged microwave transistor. Finally, in order to look at some of the problems associated with large signal behaviour, a description of the principles of load pull measurements is provided.

A microwave solid-state device usually consists of a small semiconductor chip, often less than 1 mm square. Although many devices are used in chip form, the majority are mounted in packages for ease of handling and to prevent contamination. This fact compounds the problems of device characterisation since a decision has to be made as to exactly what comprises the "device" and which part of the (essential) mounting structure must be considered to be part of the "circuit" in which the device is embedded. For a device in chip form the greatest uncertainty is usually due to the bond wires with which it is connected to the circuit and these are usually modelled separately. Although the variability of wire bonding is no less for a packaged device, there is no useful way in which information about the chip itself can usually be extracted and, consequently, planes of reference at the terminals of the package are usually chosen.

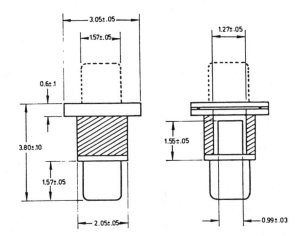

Figure 1: S4 (pill-prong) type diode package
(dimensions in mm).

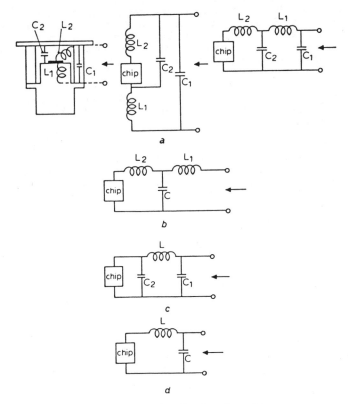

Figure 2: Package equivalent circuits.

2. CHARACTERISATION OF 2-TERMINAL DEVICES

Microwave diodes, if not used in the form of chips, are usually mounted in "pill"-type packages consisting of a ceramic cylinder separating gold-plated metal caps (Figure 1). The package is mounted in a circuit which may be co-axial, waveguide or microstrip; the details of the mount will have a considerable influence on the impedance presented by the active device. An example is the subdivision of the regions around a device mounted at the end of a coaxial line and their equivalent circuits (1).

2.1 Package equivalent circuits

The equivalent circuit for a package takes many forms, usually because it is the result of curve fitting of experimental data to a prototype circuit. In general, the greater the useful frequency range over which the circuit is required to operate, the larger the number of elements which will be found necessary in order to model the package. The model developed by Owens and Cawsey (2,3) can be simplified using a variety of impedance transformations to LCL, CLC or LC equivalents which can be used over narrow frequency bands and which do, at least, identify the principal resonant effects (Figure 2). It is usually dangerous to attempt accurate correlation between element values and package dimensions because of the nature of the modelling process.

A variety of techniques have been used to evaluate the parameters of the equivalent circuits. All employ measurements using a network analyser on packages with "known" contents. Some examples are:

(a) Construct a short circuit package (gold wire directly bonded to pedestal) and an open circuit package (bonding lead connected to insulating pad on pedestal). Measurements at low frequency and microwave resonant frequencies will allow identification of the principal reactive elements.

(b) Employ a diode with well characterised low frequency (1MHz) reverse bias capacitance voltage characteristic. It is necessary to assume that the diode capacitance is invariant with frequency.

(c) Variation of low-field conductance on application of a magnetic field (geometric magnetoresistance) is frequency independent and provides a reference inside the package which can be measured externally using d.c. techniques (4).

(d) Least squares curve fitting of prototype equivalent circuit data to measured data as a function of frequency.

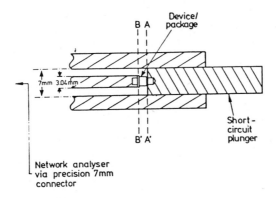

Figure 3: Diode measurement mount in co-axial line.

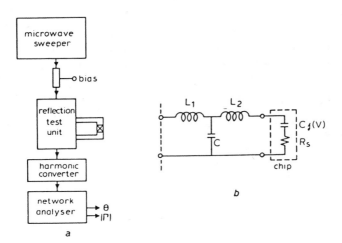

Figure 4: a. Network analyser small signal impedance
measurement.
b. Equivalent circuit of package and chip.

2.2 Small signal impedance of 2-terminal devices

The diode chip (bonded in a package) and set up in a co-axial mount is measured on a network analyser system (Figure 3). Bias must be supplied to the active device and this is usually done by means of a bias tee behind the calibrated port of the network analyser (Figure 4). In this example, it will be assumed that the device under test is an IMPATT and that the reverse bias depletion capacitance is invariant with frequency. Assuming a three parameter package model (L_1, C, L_2), the procedure is as follows:

(a)   A set of low frequency capacitance measurements are made on an open circuit package and on a packaged device biased well below breakdown. This gives C and $C_j(V)$.

(b)   Assuming that the forward biased diode is approximately a short circuit, measurement of the resonant frequency under these conditions yields L_2.

(c)   With some known value of voltage and hence $C_j(V)$, the measurement of the resonant frequency will give L_1. (It is probably better to use a computer program to adjust the equivalent circuit values to fit the measured impedance over a range of frequencies for different values of $C_j(V)$. The function to be minimised is

$$F = \alpha \sum_{j=1}^{n} (|\Gamma_m| - |\Gamma_c|)^2 + \beta \sum_{j=1}^{n} (\theta_m - \theta_c)^2$$

where α, β are weighting factors, $|\Gamma|$ is the input reflection coefficient magnitude, θ the input reflection coefficient phase and m, c designate the measured and calculated values respectively (5).)

(d)   The device can now be biased into the operating regime and the chip impedance calculated from the impedance measured at the terminals of the mount.

The method described does not allow for losses, which can be estimated from a measurement with the device biased at breakdown. The technique requires great care and places a great deal of reliance on very accurate measurements since the significant information is represented by a small change in the measured impedance at the test port.

2.3 Large signal impedance

Any active device has characteristics which are functions of bias level, signal frequency and r.f. signal amplitude. The small signal approximation has assumed that the device is linear and that, in consequence, the impedance is independent of r.f. drive. At any useful power level, however, the large signal characteristics must be considered. In general, unlike the small signal case, non-sinusoidal voltages may be present and a full analysis should take account of this (6). When analysing this general case, it must be assumed that the total current and voltage at the diode terminals are comprised of harmonic components related to each other at each harmonic frequency in a manner determined by the circuit in which the device is embedded and the signal amplitude. At each harmonic, a linear circuit is constructed, driven by the device, the properties of which are specified under the particular conditions by a describing function (7).

In practice, it is usual to require that an oscillator circuit operates with (nearly) sinusoidal signals. The large signal impedance of a device can be measured on a network analyser in the same manner as the small signal by including a suitable high power source and controlling and measuring the signal level. It is necessary to pad the signals sent to the harmonic converter and ensure that harmonic and spurious responses are small. It should also be noted that the load seen by the device at frequencies other than the measurement frequency will affect the measured impedance and that this must be borne in mind when designing the test mount. This approach is only valid is a few specific cases and load pull techniques (see below) are more commonly used (Figure 5).

The most difficult case, that of oscillators, fortunately is the one requiring as near as possible a pure sinusoid and is usually implemented by designing with the device embedded in a high-Q circuit. The principle of most large signal oscillator characterisation techniques is based on the analysis of Kurokawa (8) which uses a simple but general oscillator model. It may be shown that, under conditions for steady state oscillations

$$Y(\omega) + \overline{Y} = 0$$

where a parallel model is used. (A similar result holds for a series model.) Consequently, in order to determine the device conductance and susceptance at a particular value of bias, frequency and signal amplitude, it is necessary only to know the circuit admittance when the device is operating as a steady state oscillator under those conditions. The outline procedure is (9, 10, 4):

(a)   Bias the device and adjust the circuit to obtain steady state sinusoidal oscillations at the desired frequency.

(b)   Measure the output power (P) and the frequency (use a spectrum analyser to check that harmonic and spurious content is negligible).

(c)   Remove active device and measure the circuit impedance under the same conditions.

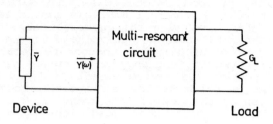

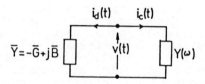

Figure 5: General schematic of an oscillator circuit.

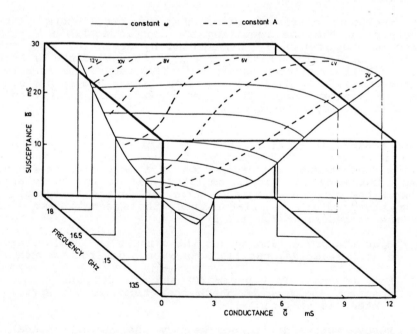

Figure 6: Measured device surface for GaAs transferred electron device.

(d) Deduce $\bar{G}$ and $\bar{B}$ from the circuit impedance and the signal voltage amplitude from

$$P = \tfrac{1}{2}A^2|\bar{G}|$$

The results of such a measurement are a complex mass of data representing device conductance and susceptance as functions of amplitude and frequency (see example in Figure 6). This information provides a useful design tool when used in conjunction with a computer graphics display and a circuit simulator.

3. MICROWAVE TRANSISTOR S-PARAMETER MEASUREMENTS

Although a wide range of schemes exist for the characterisation of microwave transistors, there are usually problems associated with calibration of the test jig and hence in defining the plane of reference. The majority of applications at frequencies below 18 GHz employ packaged transistors and many uses of chip transistors require the testing of a few (packaged) samples from each batch. Measurements are usually made with an automatic network analyser (ANA) and thus, although error correction is essential to good results, the problem becomes one of de-embedding.

The key to accurate transistor measurements lies in the design of a test fixture which can be calibrated at precisely known planes. The accuracy of the final results is directly related to the quality of the artefacts used for calibration and the repeatability associated with the use of a precision connector. It is the transition from the calibration medium to the region in which the transistor is located which ensures that a measurement based on calibration at the ANA system connectors will be highly inaccurate.

The following describes the design principles and use of a universal transistor test fixture which is based on 7 mm, 50 ohm co-axial lines (11). The fixture employs interchangeable inserts to accommodate a range of package styles. The outer halves of the fixture are co-axial and equipped with APC-7 connectors and can therefore be calibrated with the usual type of high quality standards. The discontinuities associated with the insert are then separately measured (Figure 7). It suffices for most purposes to model the discontinuity by means of a pi-network (Figure 8). Once the values of the elements of this network have been determined for a particular insert, the routine calibration and use of the fixture is considerably simplified.

The measurement procedure is as follows:

- Separate the fixture and the connect the halves to the ANA. It is now assumed that the open ends of the fixture are the effective analyser test ports.

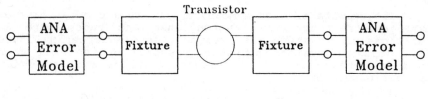

Port 1   **Port 2**

Figure 7: Schematic of network analyser measurement
of transistor in a test fixture.

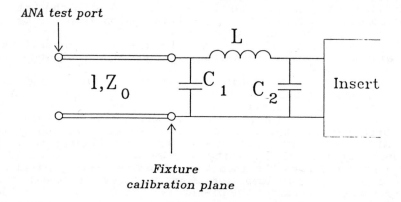

Figure 8: Equivalent circuit model of discontinuity
at device plane in transistor test fixture.

● Calibrate the analyser at the fixture reference planes using loads, shielded opens, shorts and a through connection. This enables the derivation of the parameters of the 12-term error model.

● Re-assemble the fixture with the insert in place and measure a check device to verify that the calibration procedure has been carried out correctly and that the correct insert model has been used.

A transistor may now be placed in the fixture, bias applied and S-parameter measurements made.

The de-embedding process is straightforward. Assuming that the S-parameters of the device and insert have been measured (after 12-term error correction):

● convert S_{meas} to T_{meas} (transfer scattering matrix)

● set up T matrices for the pi-networks using $-C_1$, $-L$, $-C_2$

● cascade $T_1 \bullet T_{meas} \bullet T_2$ to yield T_{dev}

● convert T_{dev} to S_{dev} yielding the S-parameters of the transistor alone.

Such a fixture, because it is low loss, is also suitable for device noise figure measurements (12).

The derivation of the insert parameters is somewhat delicate but need only be carried out a single time for each design of insert. The parameters are derived from measurements of

(a) A short circuit insert accurately defining the plane of the device to be tested;

(b) An open circuit insert to characterise the residual transistor package leads; and

(c) A "zero length" insert providing a through connection intended to characterise the reduced height aperture in which the transistor is located.

Simple models are used to provide initial estimates for the element values (13) which are then refined by curve fitting. The self-consistency of the results indicate that the model is a good approximation over the 2 to 18 GHz frequency range.

More recently, requirements for on-wafer r.f. measurements for monolithic circuit applications have resulted in the development of a reliable technique for direct probing of chips (14,15). The results appear repeatable and self-

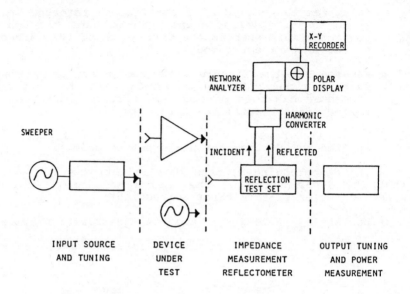

Figure 9: Basic arrangement for microwave load pull
            measurement.

consistent but there is still a considerable amount of work remaining to be done on the provision of accurate calibrations and "standards" in integrated circuit media.

4. MICROWAVE LOAD PULL

S-parameter characterisation, although in principle easy to carry out, is generally of little value with power transistors except in a few specific cases (16). Load pull characterisation is an approach which has resulted in many successful power amplifier designs. The measurement is essentially that of the load impedance under actual operating conditions and results in a series of constant output power contours plotted on a Smith Chart. A measurement is thus required at each power level and frequency of interest (17). A basic system is illustrated in Figure 9. The impedance measurement is done directly using a dual directional coupler and a network analyser. The system is calibrated by replacing the device under test with a short circuit, replacing the load with a source and adjusting the line stretcher until a short circuit is indicated on the network analyser display. (The directivity error of the couplers can be compensated using a sliding load if found to be necessary.) The accuracy of the measurement approaches that of a manual network analyser used at a single frequency. The principal sources of error are the mismatch uncertainty at the test port and the output power measurement inaccuracy. The latter is due mainly to loss in the tuner, directional couplers and the necessary bias tees.

Although the system can be operated with almost any type of tuner, it is important to note that the choice of tuner design limits the magnitude of the load reflection coefficient that can be presented to the device under test. Design considerations for tuners include loss, usefulness in one- or two-port applications and possibility of automation. The types of tuner available include stub tuners, slug tuners, moveable probe tuners and active tuners. The latter are necessary if tuning very near to or outside the unit circle is required. A basic active tuner consists of an amplifier, attenuator, phase shifter and either a directional coupler or circulator. An interesting form of solid state tuner which may readily be computer controlled has been described by Leake (18). An "equivalent load pull" may be set up without the need for an output tuner by driving both input and output ports of the device under test at the same frequency and adjusting the relative amplitude and phase (19).

5. REFERENCES

1.   Getsinger, W.J., 1966, IEEE Transactions, MTT-14, 58-69.

2.   Owens,   R.P.   and   Cawsey,   D.,   1970,
     IEEE Transactions, MTT-18, 790-798.

3.   Owens, R.P., 1971, Electronics Letters, 7, 580-582.

4.   McBretney,   J.   and   Howes,   M.J.,   1979,
     IEEE Transactions, MTT-27, 256-265.

5.   Fletcher, R. and Powell, M.J.D., 1963, Computer J., 6,
     163-168.

6.   Foulds,   K.W.H.   and   Sebastian,   J.L.,   1978,
     IEEE Transactions ED-25 646-655.

7.   Gustafsson, L., Hansson, G.H.B. and Lundstrom, K.I.,
     1972, IEEE Transactions, MTT-20, 402-409.

8.   Kurokawa, K., 1969, Bell Syst. Tech. J., 48, 1937-55.

9.   Pollard,   R.D.   and   Howes,   M.J.,   1977,   Solid-
     State and Electron Devices, 1, 146-150.

10. Jeremy,   M.L.   and   Howes,   M.J.,   1974,
     IEEE Transactions, ED-21, 488-499.

11. Pollard, R.D. and Lane, R.Q., 1983, IEEE International
     Microwave Symposium Digest Boston, Ma.), 488-500.

12. Lane, R.Q., 1978, Microwaves, (August 1978), 17, 53-57.

13. Lane, R.Q., Pollard, R.D., Maury, M.A., Jr. and Fitzpa-
     trick, J.K., 1982, Microwave J., (October 1982), 25,
     95-109.

14. Strid,   E.W.   and   Gleason,   K.R.,   1982,
     IEEE Transactions, MTT-30, No. 7, 969-975 and ED-29,
     No. 7, 1065-1071.

15. Strid, E.W. and Gleason, K.R., 1984, IEEE International
     Microwave Symposium Digest, (San Francisco, Ca.), pp
     93-97.

16. Chaffin, R.J. and Leighton, W.H., 1973, IEEE Interna-
     tional Microwave Symposium Digest, 155-157.

17. Cusack, J.M., Perlow, S.M. and Perlman, B.S., 1974,
     IEEE Transactions, MTT-22, 1146-1152.

18. Leake, B.W., 1982, IEEE International Microwave Sympo-
     sium Digest (Dallas, Tx.), 348-350.

19. Takayama, Y, 1978, NEC Research and Development, (April
     1978), 50, 23-29.

Chapter 18

Antenna measurements

R. W. Yell

1 INTRODUCTION

An antenna is a structure which controls the scattering of energy from one medium to another. It acts as an interface between a bounded medium, such as a waveguide, and free space. Ideally it should have the property of accepting energy from one well-ordered medium and creating the wanted radiation pattern with a minimum of loss. Antennas are designed to have particular properties of pattern and power gain according to the application and the types of design are seemingly without limit. The complexity of multi-element arrays is often such as to preclude an absolute determination of their performance by calculation. Even some relatively simple designs, such as exponential or pyramidal horns, can not as yet be rigorously determined. Thus it is necessary to establish performance, or at least confirm predictions, by measurement against stable, reliable and determinable standards.

In this chapter some of the concepts connected with antennas will be presented along with some discussion of anechoic chambers and measurement methods.

2 NEAR-FIELDS AND FAR-FIELDS

The energy radiating from an antenna passes through several distinct stages before the final pattern emerges. The near-field is the complex region in the vicinity of a radiating element and contains electric and magnetic components in addition to those which make up the radiation field. The radiation field comprises E- and H- components in space quadrature and in time phase. The ratio of E/H is the wave impedance. In addition to this radiation field there are five other components, four of which are electric and one magnetic, and these are associated with the induction and static fields respectively. These latter decrease in magnitude with distance as the inverse-squared and inverse-cubed respectively. Thus they become negligibly small at a few wavelengths distant from the antenna. These are the reactive near-field components which give the true near-field condition which is rarely encountered other than close to the antenna structure.

For a point source which radiates uniformly in all

directions, an isotropic radiator (Fig.1), the reactive near
-field components decay very rapidly - within a wavelength
or so - and only the radiating field remains decaying in
magnitude as the increase in distance from the source.
Albeit a potentially useful device the isotropic radiator is
unrealisable in practice.

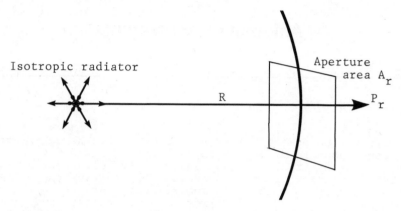

Isotropic radiator

Aperture
area A_r

R

P_r

Fig.1   An isotropic radiator

Most antennas have a structure which carries currents
which can be considered as an array of elementary sources
spread out over the structure. At the extremities there are
then, sources seperated by a distance D (Fig.2) radiating a
spherical wave into space. Also at the centre of the
aperture, assuming a simple geometry, is a similar source. A
point distant R from the antenna receives energy from the

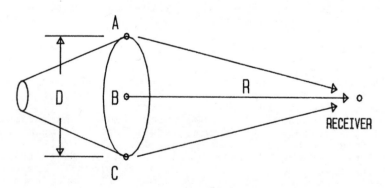

A

D

B

R

RECEIVER

C

Fig.2   The effective multiplicity of elementary
        sources on a radiating structure

sources at A, B and C. If R is large compared with D and the
wavelength then the path differences are negligibly small
and the various wavefronts arrive with a common phase. If R
is small then for a path difference of $\lambda/4$ between sources A
and B (A and C will be the same path length) then:

$$R^2 + D^2/4 = [R + \lambda/4]^2 \quad \dots\dots\dots\dots\dots\dots(1)$$

which becomes, for D very much greater than:

$$R = D^2/2\lambda \quad \dots\dots\dots\dots\dots\dots\dots\dots(2)$$

The distance at which a 90° phase shift obtains is known as the <u>Rayleigh Distance</u> and is a concept taken from Optics. Thus it can be seen that because the radiating elements have physical size the radiated field will exhibit standing waves which disturb the inverse distance field decay characteristic until far away from the antenna. This region of the field is the <u>radiated near-field</u>.

It is commonplace in microwave antenna practice to use $2D^2/\lambda$ (that is four times the Rayleigh distance) as a minimum working distance for antenna measurements. However at this distance the path phase differences are 22.5° and the relation of radiated power being proportional to $1/R^2$ is accurate to only 5%. For 1% uncertainty at least $10D^2/\lambda$, or 20 Rayleigh distances is the required separation.

The development of the field from a single mode in a waveguide is illustrated (Fig.3) as is the build-up of the radiation pattern(Fig.4).

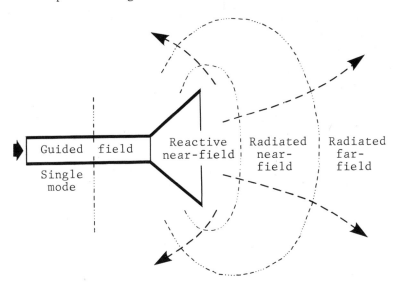

Fig.3   Field regions around an antenna

The radiating field is called the Fraunhofer region and that between the reactive near-field and the <u>Fraunhofer</u> region, where the phase effects of the radiated field are significant, the <u>Fresnel</u> region.

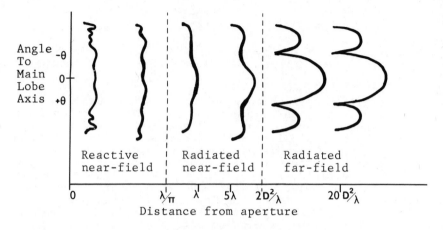

Fig.4   Radiated pattern formation

3   STANDARD ANTENNAS

3.1 Isotropic Radiator

The isotropic radiator (Fig.1) even though impractical is nevertheless used conceptually as a reference against which most other antenna standards are compared. For a total radiated power of P_t that through an aperture A, which is P_r, at a distance R is given by:

$$P_r = [A_r/4\pi R^2].P_t \quad \ldots\ldots\ldots\ldots\ldots(3)$$

If a radiator has directional properties represented by a power gain G compared with an isotropic radiator then:

$$P_r = [A_r/4\pi R^2].P_t.G \quad \ldots\ldots\ldots\ldots\ldots(4)$$

Thus it can be seen that the power gain of an aperture is related to the area by:

$$A_r = [\lambda^2/4\pi].G \quad \ldots\ldots\ldots\ldots\ldots(5)$$

3.2   Half-wave Dipole

One of the most fundamental and realiseable antennas is the half-wave dipole and this finds wide application at RF and microwave frequencies. The dipole (Fig.5) takes the form of a two colinear conductor device fed at the centre by a transmission line. At resonance the length of the conductors is approximately a half of the free space wavelength and the current distribution is approximately sinusoidal. The form of the polar diagram (antenna pattern) is shown in Fig.6. and the beam width to the half power points is 78°. The gain of a simple dipole can be shown to be 2.16dB. The dipole has developed into many forms and a comprehensive survey is given in (1).

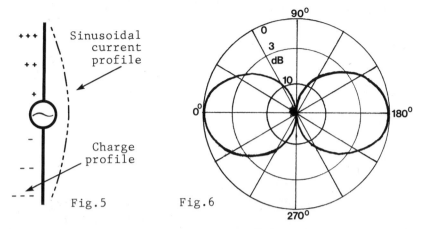

Fig.5   Fig.6

Fig.5 Half-wave dipole, centre fed

Fig.6 Radiation pattern of dipole antenna at 700MHz

3.3 Waveguide Horn

An open-ended waveguide radiates effectively and as the
aperture dimensions are comparable with the wavelength the
radiation pattern is very broad and the power gain about 6dB.
To increase the gain, and thus decrease the beamwidth, the
waveguide end can be 'flared'; if this is done to both walls
a pyramidal horn results. Horns can be made with gain values
ranging from about 8dB to 30dB depending on frequency and
acceptable size. They are an important antenna for coupling
power to free-space at frequencies above 1GHz.
     Waveguide horns have proved to be very suitable for use
in standards work as the wavefront in the aperture is similar
to a plane wave in free-space if one can disregard the sine
distribution of field across the aperture and some curvature
of the field caused by the horn taper. Over a small area at
the centre of the aperture the wavefront is essentially plane
subtending an an angle of about 14° about the phase-centre;
the distortion in the field magnitude is less than 0.5%. This
is adequate for most field measurement. Waveguide horns are
relatively cheap to produce, stable, low-loss and can handle
substantial powers; these properties are very desireable in
standards devices.
     Jull (2) has indicated a method of calculating the gain
of waveguide horns from the aperture and flare dimensions.
This is not an exact solution and present theoretical
analyses only allow accuracies of about ±0.2dB for gain
calculation without making any allowance for reflections at
the throat and aperture discontinuities which significantly
influence the gain/frequency characteristic. Thus measurement
is essential for precise characterisation of performance.

4 ANECHOIC CHAMBERS

The ideal environment for making measurements on an antenna is a large unobstructed volume which is free of any reflecting objects and electromagnetically interfering signals. The anechoic chamber developed as a practically realisable environment which, within limits, achieves many of the requirements. The 'free-space' condition is simulated by creating a room in which the reflections from the walls straying into the test area are small in amplitude compared to the direct test signal. The low reflectance is achieved by the use of an electromagnetic wave absorbent layer, often pyramidal in shape, covering all of the reflecting surfaces within the chamber.

The earliest chamber designs were <u>rectangular</u> (Fig.7a)

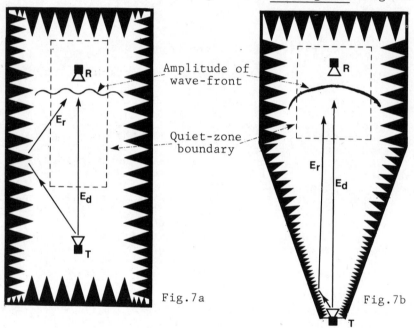

Fig.7a   Fig.7b

Fig.7   Anechoic chamber designs, a) rectangular and b) tapered, indicating the outlines of the quiet-zone and the wavefront amplitudes within

and reflectance levels in S-band were of the order of 20dB, a low value by modern standards, due largely to the poor performance of the absorber materials. Various chamber shapes were tried to overcome the limitations of the absorbers with varying degrees of sucess. In 1967 Emerson (3) devised the <u>tapered</u> chamber and this proved to be a breakthrough. Nowadays both types are widely used.

4.1   Rectangular anechoic chambers

The design of rectangular chambers for conventional far field measurements requires that the transmitting antenna T (Fig.6) is placed near one end wall and the receiving antenna, R, is placed at such a distance that the far-field criterion appropriate to the measurement is satisfied. The width of the chamber is ideally chosen such that the angle of incidence for the ray resulting in a reflection from the side wall is 70° or less. The receiving antenna is placed about half the width of the chamber from the back wall so coupling to those absorbers is minimal. These considerations indicate an optimum length to width (and height) ratio of about 3:1; a 2:1 aspect ratio will give lower side wall reflection. The best performance is achieved by placing the longest absorbers on the end walls and in the regions around the first reflection points on the side walls.

4.2   Tapered anechoic chambers

The tapered anechoic chamber consists of a pyramidal tapered absorber lined section joined to a cubical absorber lined section (Fig.7b) and may be thought as a large horn antenna terminating in a large waveguide in which a single mode plane wave is to be generated. In order not to generate higher order modes, which would disturb the quiet-zone, three regions must be considered; the position of the source antenna, the lossy side walls of the pyramidal section and the transition to the cubical section.

The absorber requirements for the end wall are much the same as for the rectangular chamber. The funnel region needs an absorber which is better suited to propagation parallel to the surface rather than normal incidence which is the usual requirement.

4.3   Suitability of anechoic chambers

The usefulness of an anechoic chamber is determined by the size and 'flatness' of the field within the quiet-zone which is the volume in which the reflections from the walls are below the direct radiated value of the field. The measure of the quiet-zone performance, the reflectivity, will be a function of position within the zone as well as frequency of operation and a value assigned to reflectivity represents the average value of the error which can be expected for a measurement made in the chamber. The volume of the quiet-zone depends on the chamber design and though ill-defined usually reflects that of theabsorber lining.

Due to asymmetry along the transmission path the tapered chamber cannot easily be applied to some antenna measurements such as bistatic measurements of radar cross-section or back scattering. Also the launching funnel region has a substantial influence on the characteristics of the launching antenna which thus largely precludes its use for standards work.

The assessment of the reflectivity level of chambers is discussed by Appel-hansen (4) and Kummer and Gillespie (5).

5   MEASUREMENTS ON ANTENNAS

There are a number of definitions relevant to antenna parameters some of which are as follows:

the power gain in a specified direction is 4π times the ratio of the power radiated per unit solid angle in that direction to the nett power accepted by the antenna from its generator. When the direction is not specified the gain is taken to be the maximum value.

the directive gain or directivity in a specified direction is 4π times the ratio of the power radiated per unit solid angle in that direction to the total power radiated by the antenna.

the radiation efficiency of an antenna is the ratio of the power radiated by the antenna to the nett power accepted at its input terminals. It may also be expressed as the ratio of maximum gain to directivity.

the radiation pattern is the graphical representation of the distribution of the radiated energy as a function of direction about the antenna. The graphical display is usually in spherical coordinates.

These parameters can be determined by sampling the field radiated from an antenna with a calibrated probe using either near-field or far-field methods (6). In essence the measurement technique is similar for either method. The test device is supported in an orientable frame and energised at the required range of frequencies. The radiated energy is sampled by a probe placed at an appropriate distance as is appropriate to the measurement method. The signal detected by the probe is analysed for amplitude, and phase if needed, and this data is stored either directly as a graphical plot or digitally if further computation or processing is needed. A schematic diagram of a test range is shown in Fig.8. The requirements of the anechoic environment and scanning ranges are discussed elsewhere in this chapter. A comprehensive orientation mechanism can rotate the test piece in azimuth and elevation or, as an alternative to the latter, roll about an axis perpendicular to the azimuth such that all points on a spherical shell surrounding the antenna can be sampled. The roll over azimuth configuration gives the smallest scanning sphere.
The receiver used for measuring the amplitude and phase of the signals detected by the probe will need to have a dynamic range in excess of 60dB, with good resolution and accuracy. The precision phase measuring capability is needed for near-field work.
Probably the most important parameter of standard antennas is the boresight gain and a number of methods have been devised for determining this parameter. Absolute gain measurements are those in which the gain is determined from a power ratio derived from power measurements or directly from a precision attenuator. Several methods will be

desribed including an aperture scanning method.
A schematic diagram of a typical test range is shown in Fig.8.

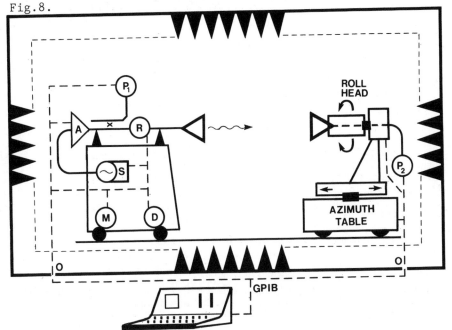

Fig.8 Configuration of an antenna test range within a screened anechoic chamber. P_1 and P_2 are power detectors, R is a calibrated rotary-vane attenuator, S is the signal source, A is a power amplifier, M is the carriage drive, D is the displacement monitor and O is the optical-fibre link through the screen to the computer.

5.1 Three antenna method

The basis of this method is a measurement of the transmission between two polarisation matched antennas, G_1 and G_2, which is expressible in terms of the antenna gains by the Friis formula

$$P_r = P_t . G_1 . G_2 (\lambda/4\pi R^2) \quad \ldots\ldots\ldots\ldots\ldots (6)$$

P_t and P_r are the transmitted and the received powers, λ is the wavelength and R the separation. As can be seen only the product of the gains can be obtained. However from the measured combinations of three such antennas one obtains three simultaneous equations which can be solved uniquely. In practice a directional coupler is inserted between the source and the transmitting antenna to monitor and thus keep the transmitted power constant; this simulates a matched source condition. The received power is noted for each pair of antennas also for the case in which the waveguides are

connected directly together. Small corrections are required
due to imperfect matching of the waveguide systems. These
require a knowledge of the reflection coefficients of the
antennas which may conveniently be made on a network
analyser.

5.2 Purcell's method

This method, otherwise known as the 'mirror method'
requires only one antenna operating in the monostatic mode,
the other being a virtual antenna produced by reflection in
a plane mirror. In this case equation (1) is used with G_1
equal to G_2. The ratio of the powers in the equation could
be measured with a slotted line placed behind the horn or by
means of a directional coupler. Some disadvantages are that
both losses and any significant departures from planarity of
the mirror will cause errors. Also the mirror must be large
enough so that diffraction at the edges are negligible.

5.3 Extrapolation method

Test ranges within anechoic chambers may not be of
sufficient length to get into the far field of standard
antennas with high gain. For example, to reduce proximity
corrections for standard gain horns to less than 0.1dB needs
a test range of about $16D^2/\lambda$ where D is the largest aperture
dimension and λ the wavelength. At such distances extraneous
reflections in an anechoic chamber can be significant and
obscure the potential accuracy of the measurement. The
extrapolation technique devised by Newell et al (7)
overcomes these problems and is the most accurate method
known for determining absolute gain. The extrapolation
theory accurately describes the complex signal transmitted
between two, arbitrary, antennas taking into account the
mismatch factors, the proximity effects and the multiple
reflections between the two antennas which occur at short
range. The transmission can be accurately represented by a
polynomial, eight terms usually being sufficient, and from
this, by extrapolation, the far-field (asymptotic) gain can
be determined. This latter is a task which is conveniently
done by a computer. A typical gain with range curve (Fig.9)

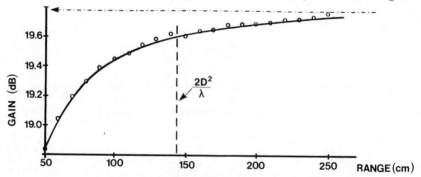

Fig.9 Typical gain/range curve for a waveguide horn

reveals the precision of the method, the circles indicating measured gain values.

At frequencies below 1GHz antennas with broad beamwidth will be difficult to measure by this technique because of oscillations in the transmission characteristic caused by excessive ground reflections. These oscillations generally have longer periods than the multipath reflections and can be dealt with by determining the phase as well as amplitude of the transmitted signal and by a method due to Newell et al (8) the spurious ground reflections can be accounted for.

5.4   Transmission RVA method

This method, a variant on the extrapolation technique, utilises a rotary-vane attenuator (RVA) placed between the source and the transmitting antenna   to maintain the power received constant. This removes the linearity requirements of the power meters and places the potential accuracy of the method in the calibration of the attenuator. For constant source and receiver powers and attenuator readings of L_o and L_r respectively for the direct connection of receiver and transmitter and at range R, then:

$$G = [20(\log 4\pi R/\lambda)-(L_o-L_r)]/2 \quad dB. \quad ....(7)$$

where G is the gain of one antenna and both antennas are assumed to be identical and operating in the far field. It is possible to work at reduced distances using extrapolation methods. The experimental set-up is shown in Fig.8.

5.5   Radar cross section method

In this method the unknown antenna is used as a target in a monostatic radar system and the received signals due to the absorption cross-section and scattering cross-section are separated so that the former can be determined on an absolute basis. Appel-Hansen (9) describes a method by which the received signal can be reradiated with variable phase by the use of an adjustable short circuit. A maximum received signal will occur at the radar receiver when the signals due to reradiation and scattering are in phase and a minimum when they are out of phase. The absolute values of the two cross-sections can be obtained if the radar is first calibrated with a target of known cross-section such as a conducting sphere. The two cross sections are finally resolved by placing a matched termination on the target antenna which will supress the reradiation of the received signal and leave only the scattered component. The gain of the antenna is calculated using:

$$G = (4\pi.\sigma_r)/\lambda \quad ....(8)$$

where σ_r is the backscatter cross-section of the reradiation.

5.6   Near-field scanning method

In recent years antenna metrology has concentrated on near-field techniques with the objective of improving aerial characterisation at VHF and UHF or of very large arrays (5). Near-field techniques offer an attractive alternative to conventional test ranges in that long range is not required. In this method, a probe antenna is used to sample the phase and magnitude, for two polarisations, of the radiated field in the near-field of the test antenna that is at a few wavelengths from its surface. The probe antenna scans over a well defined surface which can be a plane in front of the test antenna or a cylinder or sphere enclosing the antenna. These imaginary surfaces are divided up into a matrix of cartesian or polar coordinates as appropriate and the probe samples the field at this matrix of points. The objective of the near-field to far-field transformation technique is to express the initially unknown radiated field as a series of elementary waves. The excitation coefficients are determined and the required far-field properties are found by mathematically extrapolating the apparent measurement distance. The number of samples needed to define the near-field adequately can be considerable such that the measurement and computational time is substantial. It is therefore desireable to have a highly efficient measurement scheme and consequently to select the measurement geometry to suit the antenna under test. Planar scanning is well suited to horn antennas and similar devices which have fairly narrow beamwidth.

The principal features of the technique are:

the theory is rigorous, general and applicable to any antenna

gain, polarisation and complete vectorial patterns can be obtained for any distance, near-field or far-field from the antenna

measurements may be made indoors as close as desired to the antenna in a simple anechoic environment

reasonable uncertainty of measurement is attainable

the near-field data can be used directly for diagnostic tests such as determining the faulty elements in arrays (which would only appear as an indeterminate aberation in a far-field pattern).

There are some disadvantages to the near-field technique. One is that the method relies on the calibration of the probe antenna and for this it is necessary to revert to one of the other methods. A further disadvantage is that an automated system, based on a precision scanning mechanism, is required from which vast amounts of data must be gathered and processed. The greatest demands are made on the scanning system which must be precise in its position to about two hundredths of a wavelength for positioning accuracy to be negligible as a contribution to the uncertainty.

REFERENCES

1. 'Transmission and Propagation', The Services Textbook of Radio, 1958, 5, HMSO.

2. Jull, E.V., 1973, 'Errors in the Predicted Gain of Pyramidal Horns', Trans. IEEE, AP-21,

3. Emerson, W.H.,and Sefton, H.V., 1965, 'An Improved Design for Indoor Ranges', Proc. IEEE, 53, 1079-1081.

4. Appel-Hansen, J., 1973, 'Reflectivity Level of Radio Anechoic Chambers', Trans. IEEE, AP-21, 490-498.

5. Kummer, W.H. and Gillespie, E.C., 1978, 'Antenna Measurements', Proc. IEEE, 66, 483-507.

6. 'IEEE Standard Test Procedures for Antennas', The Institute of Electrical and Electronics Engineers, Inc., ANSI IEEE Std. 149-1979,1979.

7. Newell, A.C., Baird, R.C. and Wacker, P.F., 1973, 'Accurate Measurement of Antenna Gain and Polarisation at Reduced Distances by an Extrapolation Technique', Trans. IEEE, AP-21, 418-431.

8. Repjar, A.G., Newell, A.C. and Baird, R.C., 1983, 'Antenna Gain Measurements by an Extended Version of the NBS Extrapolation Method', Trans. IEEE, IM-32, 88-91.

9. Appel-Hansen, J., 1979, 'Accurate determination of gain and radiation patterns by radar cross-section measurements', Trans.IEEE, AP-27, 640-646.

10. Johnson, R.C., Ecker, H.A. and Hollis, J.S., 1973, 'Determination of far-field antenna patterns from near-field measurements', Proc. IEEE, 61, 1668-1694.

Chapter 19

Power flux density measurements

A. E. Fantom

1. INTRODUCTION

The need to measure power flux density derives to a
large extent from health and safety considerations. This
type of measurement has become relatively common with the
more widespread use of high-power rf and microwave
equipment, both in the home and at work. The potential
applications of radio-frequency and microwave heating were
first recognised following the development of high-power
radar systems during the Second World War. Practical use
was made in industrial processes during the 1950s and in the
commercial and domestic areas during the 1960s and 1970s
respectively. The subsequent rapid growth of applications
provoked scientific and public concern regarding possible
dangers and tolerable exposure levels. The main hazard is
believed to be the thermal effect, and the hazardous nature
arises partly from the fact that the body's detection
systems do not respond to these frequencies. Furthermore,
the sensors which enable the heating effect to be detected
are located in the surface layers of the skin, whereas rf or
microwave radiation causes internal heating. Particularly
sensitive organs are the eyes, since these do not have a
blood supply to carry the heat away. Prolonged exposure can
lead to the formation of cataracts. Non-thermal
interactions with biological systems have also been reported
(1,2), but there is some controversy over the question of
whether these constitute a significant hazard to human
health.

2. EXPOSURE LIMITS

Despite the world-wide use of rf and microwave energy,
no definitive world-wide exposure standards have been agreed
and only a minority of nations have made their own
protection standards. These nations comprise mainly those
of North America and some from Eastern and Western Europe.
The most notable feature of early microwave standards was
the very large discrepancy between the exposure limits of
East and West. In some instances, mainly for longer
exposure durations, this discrepancy was as great as three
orders of magnitude, although in later years the Eastern and

Western limits have tended to come somewhat closer to each other. The gap is still substantial, however. It arises partly from differences in philosophy in assigning safety margins and partly from some experimental data on non-thermal effects which scientists in the West have been unable to reproduce.

For many years the ambient level of microwave radiation which was regarded as safe in the Western world was 10 mW/sq cm or 100 W/sq m in SI units. Originating in the late 1940s, this level took no account of frequency-dependent effects such as body resonances, which lie in the range 30 MHz to 900 MHz. Considerable efforts were devoted in the 1970s to computer simulation of the body tissues and from these the rate of energy deposition and the frequency sensitivity were determined. As a result of this work a number of recommendations relating to exposure to rf and microwave radiations appeared (3,4,5) and they had much in common. The most significant of these was the recommendation which came from the Americal National Standards Institute (ANSI) (see Fig.1) and which was the result of the deliberations of a special committee of ANSI chaired by Professor Arthur Guy of the University of Utah (3). This new standard was based on two important factors.

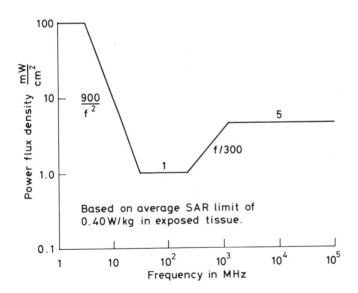

Fig.1   The ANSI safety standard

The first was that a Specific Absorption Rate (SAR) of 0.4 W/kg can be absorbed and dissipated without difficulty by the circulatory system of the human adult body. The second was the possible resonance effects within the body and the

way in which these would need to be compensated by reducing
the allowable exposure level at those frequencies. Among
the resonances considered were: whole body resonance
(typically 70 MHz), head cavity resonance (typically
375 MHz), and chest cavity resonance (typically 900 MHz).
At the lower end, where the human body is much smaller than
the wavelength, the permitted level is increased, although
it is curtailed below about 3 MHz as an additional safety
factor. Electric shock, as opposed to generalised sensation
of warmth, is unlikely at frequencies above 100 to 200 kHz.
The new exposure standard was ratified by the American
Conference of Government Industrial Hygienists (6). In the
United Kingdom a number of guidelines exist on allowable
power flux density exposure limits (7,8,9) and a draft
document containing revised proposals for the UK standard
has been issued by the National Radiological Protection
Board (NRPB) (10). In most respects the new recommendations
have much in common with the ANSI standard, but with small
differences of detail. The Commission of the European
Communities is preparing a Directive which will lay down
basic standards for the health protection of workers and the
general public against the dangers of microwave radiation
(11). At the lower frequencies the electric and magnetic
fields, rather than the power flux density, are the
important quantities, but levels are still expressed in
power flux density by assuming that plane wave conditions
apply. Thus:

$$\text{PFD in mW/sq cm} = \frac{E^2}{3770} \text{ or } 37.7\,H^2 \quad \ldots(1)$$

where E is the electric field in volt/metre and H is the
magnetic field in ampere/metre. Some authorities have
recommended different limits for E and H at the lower
frequencies, since heating is primarily associated with the
electric field. If the source of power is pulsed or
modulated, then the mean power flux density is used when
considering safety levels, since the thermal time constant
is normally much longer than the pulse repetition period.

3. POWER FLUX DENSITY MONITORS

A great many PFD monitors are available commercially
and the details of these have appeared in the scientific
literature (refs 12-18). Virtually none are true power flux
density measuring devices. Most respond to the electric or
the magnetic field component, although some behave as
directional antennas. The detectors are normally thin film
thermopiles or Schottky barrier diodes. Among the simplest
types is the short dipole or the small loop. Narrow band
monitors make use of resonant dipoles. Moderate bandwidths
can be obtained by feeding a short dipole into a high
impedance load, or by shunting the dipole with a capacitor,
but for really large bandwidths using diodes resistive
loading is used to reduce resonances. Most radiation hazard

monitors aim at constant sensitivity against frequency, but it is possible to tailor the frequency response to a particular safety standard by making the sensitivity proportional to the reciprocal of the permitted level. The scale of the meter is then calibrated to read percentage of the permitted level rather than power flux density. Omnidirectional probes consist of three mutually orthogonal elements. Each element provides a dc output signal proportional to the square of the electric (or magnetic) field strength incident on the element. The sum of the dc signals from the three orthogonal probe elements provides a measure of the total energy or equivalent power density, independent of direction of polarisation of the rf signals. A peak hold facility is useful for recording the maximum reading, obviating the need for the operator to observe the reading continuously. In most accurate instruments the meter is separate from the measuring head. This avoids the problems caused by disturbance to the fields by the meter unit, but such instruments are less easily carried, which is a drawback in some situations, such as work on aerial masts.

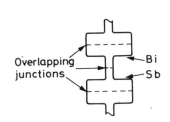

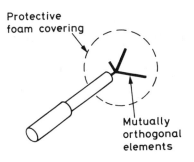

Fig.2
Distributed thermopile

Fig.3
Omnidirectional monitor

The detectors of one common type of E-field probe (Figs. 2 and 3) are thin film thermocouples, the dc signal being proportional to the power dissipated in the thermocouple elements. The broadband characteristics are obtained by distributing the thermocouples along the length of the element at spacings of less than a quarter wavelength at the highest frequency to be measured. The probe elements are composed of thin films of overlapping antimony and bismuth deposited upon a thin plastic substrate. The geometry creates alternate cold and hot junctions (Fig. 2). The hot junctions are formed at the centre of the narrow strips having relatively high resistance, thereby allowing for the dissipation of power and the resultant increase in temperature. The wider sections have a low resistance and thus function as cold junctions. The probe may be viewed as a group of series-connected small resistive dipoles or as a

very low Q resonant circuit. Omnidirectional properties are
obtained by using three probe elements mutually at right
angles and inclined at an angle of 54.7 deg to the handle.
As shown in Fig.3, the ends (rather than the centre) of the
three probe arrays intersect at a point. This orientation
makes the apparent centre of the array shift approximately
1.5 cm as the angle of incidence of the energy varies from
zero to 45 deg with the axis of the array. The three probe
arrays are contained within a 10 cm diameter sphere of
foamed polystyrene. Because of the light coupling into the
field, very little perturbation due to scattering is caused.
The leads that carry the dc outputs from the probe elements
to the metering instrumentation are high resistance films,
in order that these should disturb the fields as little as
possible.

Several detailed examinations of monitors have been
carried out (19,20). In general they support the use of the
more sophisticated devices for serious professional use.
Even these, however, can reveal unexpected and potentially
serious problems, such as erratic behaviour in the field
(21). The PFD monitor characteristics which may need to be
investigated are: frequency range, dynamic range,
linearity, isotropy and polarisation effects, response to
pulsed or modulated power, ability to withstand overloads
(whether switched on or switched off), and fail-safe
mechanism giving warning when defective.

4. USE OF POWER FLUX DENSITY MONITORS

If the approximate frequency of the source of radiation
is known, a narrow band probe may be used for testing, but
otherwise a broadband probe is required. In carrying out a
hazard survey two basic measurement situations arise:
unintentional radiation and intentional radiation. Examples
of the first of these are microwave ovens, dielectric
welding machines for plastics, equipment for the
vulcanisation of rubber, and medical diathermy apparatus.
Leakage measurements for some types of equipment have
prescribed testing methods. For example, microwave ovens
are covered by BS5175 on the testing of appliances using
microwave energy for heating foodstuffs. This standard
requires that the effective centre of the detection element
shall be 50 mm distant from the potential emission centre or
measurement point. Since probes are normally calibrated
under plane wave conditions, it is possible at the lower
frequencies for an E and an H probe to give different
readings under near field conditions, where the ratio of E
to H is not equal to the impedance of free space. It is
advisable to take measurements with separate E- and H- field
probe devices and to assume the worst-case PFD equivalent.

Intentional sources of radiation are mainly radio and
radar transmitters. When surveying hazards due to these
sources, a rough calculation of the expected power flux
density over the whole of the region in question is first

made. This necessitates considering three regions: the near field, the intermediate field, and the far field. If such calculations indicate that there may be a potential hazard, then a more detailed computation is carried out and the power flux density is measured at a number of positions. If the measured power flux density agrees with the computed values, then a combination of measurement and computation is accepted as the basis for determining the boundary of the potentially hazardous region. If the values do not agree, then the reason for the disagreement must be found and rectified. Safety measurements have to take into account not only ground reflections but also buildings, masts, cranes, gantries and other real-life situations which may affect power density levels.

5. CALIBRATION OF POWER FLUX DENSITY MONITORS

Calibration of power flux density monitors normally consists of setting up a standard field in which the monitor is placed. Techniques vary widely with frequency (22,23), the choice for a particular application depending on the size of the required working area in relation to the wavelength. For a broad band probe (e.g. 200 kHz to 26 GHz) several different set-ups are usually necessary. The quantity which is determined is the probe correction factor, defined as:

$$\text{correction factor} = \frac{\text{actual PFD}}{\text{indicated PFD}} \qquad \dots (2)$$

Uncertainties are usually in the range 7 to 26 percent, excluding any errors due to interpolation between calibration frequencies.

5.1 Calibration Below 500 MHz

At frequencies below 500 MHz the most common method of calibrating power flux density probes is by means of a Crawford type TEM (transverse electromagnetic) transmission cell (24-27) (Fig.4). This type of cell, which has largely replaced the older parallel-plate line (28), is a totally enclosed device consisting of a transmission line with a

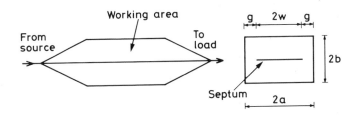

Fig.4 TEM cell for use below 500 MHz

rectangular outer conductor and an inner conductor in the form of a flat plate or septum. The input and output ends of the line are tapered so as to form a transition to ordinary coaxial cable. The fields lie entirely in the transverse plane and the field impedance (E/H) is that of free space (377 ohm). One can calculate exactly the characteristic impedance using a technique which involves conformal transformation, but the expression contains Jacobian elliptic functions and a simpler approximate expression is normally adequate for practical purposes (29):

$$Z_o = \frac{Z}{4\,[\,(w/b) \;+\; (2/\pi)\ln(1 \;+\; \coth(\pi g/2b)\,)\,]} \qquad \ldots(3)$$

where Z is the impedance of free space, and dimensions b, w, g are defined in Fig.4. Assuming the cell is terminated in its characteristic impedance, the electric field in the working area is obtained approximately by dividing the voltage V between the plates by the plate separation b. The voltage V is obtained from the power P transmitted through the cell (in watts) and the characteristic impedance Z_o. The power flux density may then be expressed as

$$PFD \;=\; V^2/(b^2 Z) \;=\; PZ_o/(b^2 Z) \qquad \ldots(4)$$

The working area is normally limited to a third of the plate separation b, and the non-uniformity over this area is typically a few percent. One of the problems in the manufacture of TEM cells is the prevention of standing waves inside the cell caused by reflections from the output taper. In some designs provision is made for adjusting the longitudinal position of the septum of the tapers in order to minimise the standing waves. An error is also caused by the proximity of the plates to the probe, which sees multiple reflections of itself in the plates. The magnitude of this error may be estimated by calibrating a probe in two different sizes of cell and comparing the results. Uncertainties of the TEM cell are usually in the range 0.5 dB to 1 dB. Of this the contribution due to non-uniformity is typically 0.1 dB, the proximity effect 0.1-0.2 dB, standing waves up to 0.5 dB, and power measurement 0.1 dB. The upper frequency limit of the TEM cell is determined by the sharp resonances due to the propagation of higher modes.

5.2 Calibration From 500 MHz To 1.5 GHz

For operation above 500 MHz a cell composed of a rectangular waveguide may be used (Fig.5). The impedance differs from that of free space and is dependent on frequency. Attempts have been made to make the TEM cell usable in the overmoded region by placing at suitable positions in the cell absorber material of the type used in

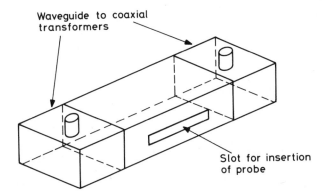

Fig.5 Rectangular waveguide cell

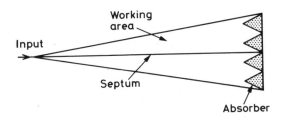

Fig.6 Tapered absorber-loaded TEM cell

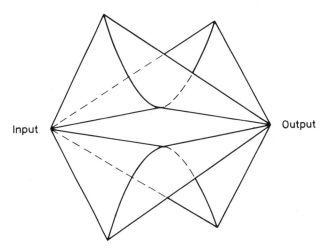

Fig.7 Broadband cell

anechoic chambers, but the fields are then difficult to
calculate and such absorber-loaded cells are used only as
comparison chambers (30). An alternative form of
absorber-loaded cell is a long tapered line, of similar
cross section to the Crawford cell, terminated in an
absorber (Fig.6).

Some work has been carried out on the behaviour of
parallel plate lines in the overmoded region by analysing
them in terms of leaky modes. The resonances are damped by
radiation but deep nulls are found in practice at certain
frequencies. A much improved performance is obtained by
using specially shaped conductors, and a new type of
electromagnetic transmission cell (31,32) based on this
principle has been developed at NPL (Fig.7). The spacing
between the plates is tapered in the transverse direction,
leading to a reduction in the Q-factors of the resonances
owing to the fact that the energy in the resonant fields can
escape from the cell more easily. The two conductors are
brought into close proximity at the two ends, without
touching, and coaxial cables (not shown) are attached to
these points. The working area is at the centre of the
cell. The cell has an extremely wide bandwidth, typically
200 MHz to 6 GHz for a spacing at the centre of 14 cm. The
fields are calculated by representing the structure by four
intersecting cones. The discontinuity due to the
intersection of the cones causes standing waves in the
working area, but these can be cancelled to a good
approximation by mismatching the load.

5.3 Calibration Above 1.5 GHz

At microwave frequencies above 1.5 GHz the conventional
technique for calibrating power flux density monitors is to
position the probe in front of a calibrated horn antenna in
an anechoic chamber. This is known as the standard antenna
method. For a horn antenna with a gain of G and reflection
coefficient Γ the power flux density at a given distance r
is given by

$$PFD = \frac{P_i G (1 - |\Gamma|^2)}{4 \pi r^2} \qquad \ldots (5)$$

where P_i is the input power to the horn. Good quality
exponential horns with a gain of about 16 dB are used
because of their good match (VSWR 1.02), relatively smooth
variation of gain with frequency, and mechanical and
electrical stability. The measurements are carried out in
the far field, that is at a distance greater than $2d^2/\lambda$,
where d is the aperture width of the antenna. The on-axis
gain of the horn antenna is determined previously by either
the three-antenna method or the identical-antenna method. A
schematic layout of a practical measurement system is shown
in Fig.8. The anechoic chamber is 5 metres long, 2.5 metres
wide and 3 metres high and is lined on five sides with

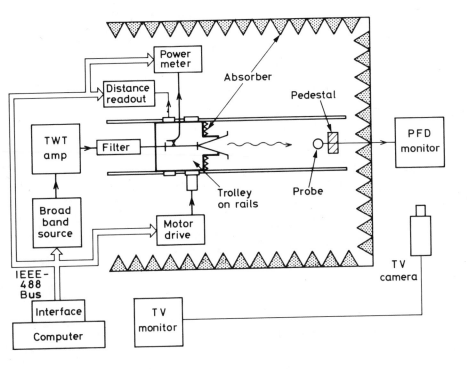

Fig.8 Probe calibration by the antenna method

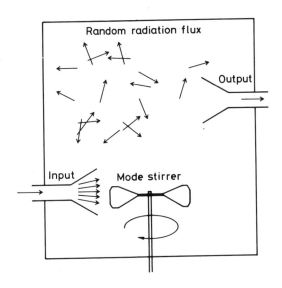

Fig.9 Stirred mode cavity

electromagnetic absorber (33). One end is left open to
allow convenient access to the signal generator and
measuring equipment, which is carried on a precision trolley
system. The trolley can be driven down the chamber on rails
which are straight and level to within 0.25 mm. The chamber
has a quiet zone of about 0.5 metre cube with a reflectivity
level of -40 dB down to 1.5 GHz. The reflectivity level
improves with frequency to about -60 dB at 10 GHz. The
measurement uncertainties are made up as follows: horn gain
0.15 dB, distance measurement 0.02 dB, power measurement
0.1 dB, random uncertainty 0.01 dB.

5.4 Alternative Calibration Methods

    A device which has found application in many fields is
the stirred-mode cavity (34,35) (Fig.9). This may be
regarded as the microwave equivalent of the integrating
sphere used at optical frequencies, but at the longer
microwave wavelengths it is necessary to include a mode
stirrer - for example, a rotating reflector - in order to
randomise the fields inside the cavity. The time-averaged
field is then isotropic and uniform. Stirred mode cavities
have been used experimentally for calibrating power flux
density monitors. The cavity contains two horns which are
used as the input and output ports respectively. The power
flux density can be calculated from a knowledge of the
output power and the wavelength using the formula:

$$PFD = \frac{8\pi \text{ (output power)}}{\lambda^2} \qquad \qquad \ldots (6)$$

It is not necessary to know the directional properties of
the output horn because of the isotropic nature of the
fields, but the horn must be matched to the load, which will
normally be a power meter. This method enables lower source
powers to be used than would otherwise be possible. For
example, typically a power of 1 watt can produce a power
flux density of around 1 mW/sq cm. For the more
conventional techniques, at least 10 watt would be needed.
Provisional results have shown agreement of approximately
15 percent with the antenna method in the frequency range
10 GHz to 40 GHz.

    An alternative to the rectangular waveguide cell shown
in Fig.5 is a dielectric-loaded rectangular waveguide. This
consists of a rectangular waveguide whose narrow walls are
lined with a low-loss plastic. If the thickness and
permittivity of the dielectric are correctly chosen, the
dielectric-loaded cell can be arranged to produce a plane
wave in the air-filled region with a wave impedance equal to
that of free space. However, the necessary conditions can
only be fulfilled at a single frequency. Another guided
wave structure which has been demonstrated is a dielectric
tube, which acts in a somewhat similar way to an optical
fibre at higher frequencies.

REFERENCES

1. BRANSKI, S. and CZERSKI, P., 1977, 'Biological effects of microwaves', Dowden, Hutchinson and Ross, USA.

2. GRANT, E.H., 1981, 'Biological effects of microwaves and radio waves', Proc. IEE, 128A, 9, 602-603.

3. GUY, A., 1981, 'Safety level with respect to human exposure to radio frequency electromagnetic fields', American National Standards Institute (ANSI) Standard C95.

4. CORNELIUS, W.A. and VIGLIONE, G., 1979, 'Recommended permissible levels for exposure to microwave and radiofrequency radiation (10 MHz to 300 GHz)', Australian Research Laboratory Report ARL/TR009.

5. STUCHLY, M.A., 'Health aspects of radio frequency and microwave radiation exposure (Part 2)', Department of National Health and Welfare, Ottawa.

6. American Conference of Government Industrial Hygienists (ACGHI), Reported in Microwave News, September 1981, 4-5.

7. 'Safety precautions relating to intense radio-frequency radiation', HMSO, 1961.

8. 'Exposure to microwave and radiofrequency radiation', Medical Research Council Report MRC70/1314.

9. 'The protection and medical surveillance of personnel exposed to radio-frequency (RF) radiation', Report D/DDC Med(RAF)/13/32B/MA4a (RAF), April 1980.

10. HARLAN, F. and DENNIS, J.A., 1983, 'Microwaves and radio frequency exposure standards', NRPB Radiological Protection Bulletin, 52, 9-15.

11. Official Journal of the European Communities, C249, 231, September 1980, 7-9.

12. ASLAN, E., 1971, 'Electromagnetic leakage survey meter', J. Microwave Power, 6, 2.

13. ASLAN, E., 1972, 'Broad-band isotropic electromagnetic radiation monitor', IEEE Trans. IM-21, 4.

14. HOPFER, S. and ADLER, Z., 1980, 'An ultra broad-band (200 kHz-26 GHz) high sensitivity probe', IEEE Trans. IM-29, 4, 445-451.

15. ASLAN, E., 1972, 'Broad band electromagnetic radiation monitor', IEEE Trans. IM-21, 4, 421-424.

16. ASLAN, E., 1975, 'A low frequency H-field radiation monitor', URSI Annual Meeting, Boulder Co USA.

17. ASLAN, E.E., 1970, 'Electromagnetic radiation survey meter, IEEE Trans. IM-19, 4, 368-372.

18. ASLAN, E., 'A low frequency H field radiation monitor', J. Microwave Power, 11, 2, 155-156.

19. NESMITH, C.W., 1982, 'Performance evaluation of rf electric and magnetic field measuring instruments', Bureau of Radiological Health (USA) Report FDA 82-8158.

20. HERMAN, W.A., 1979, 'Inexpensive microwave survey instruments: an evaluation', BRH Report FDA 80-8102.

21. LANGLET, I., 1980, 'Investigation of some rf radiation hazard meters', IERE Conference on Electromagnetic Compatibility at University of Southampton.

22. DONALDSON, E.E., FREE, W.R., ROBERTSON, D.W. AND WOODY, J.A., 1978, 'Field measurements made in an enclosure', Proc. IEEE, 66, 4, 483-507.

23. MA, M.T., KANDA, M., CRAWFORD, M.L., and LARSEN, E.B., 1985, 'A review of electromagnetic compatibility/interference measurement methodologies, Proc. IEEE, 73, 3, 388-411.

24. CRAWFORD, M.L., 1974, 'Generation of standard EM fields using TEM transmission cells', IEEE Trans. EMC-16, 4, 189-195.

25. HILL, D.A., 1983, 'Bandwidth limitations of TEM cells due to resonances', J. Microwave Power, 18, 2, 181-195.

26. CRAWFORD, M.L., WORKMAN, J.L., and THOMAS, C.L., 1978, 'Expanding the bandwidth of TEM cells for EMC measurements', IEEE Trans. EMC-20, 3, 368-375.

27. BABIJ, T.R., AND TRZANSKI, H., 1976, 'Accuracy limitations of the Crawford cell', CPEM digest p. 52.

28. ROSEBERRY, B.E., AND SCHULZ, R.B., 1965, 'A parallel strip line for testing RF susceptibility', IEEE Trans. EMC-7, 142-150.

29. TIPPET, J.C. AND CHANG, D.C., 1976, 'Radiation characteristics of electrically small devices in a TEM transmission cell', IEEE Trans. EMC-18, 4, 134-140.

30. CRAWFORD, M.L., WORKMAN, J.L., AND THOMAS, C.L., 1977, 'Generation of EM susceptibility test fields using large absorber loaded TEM cell', IEEE Trans. IM-26, 3, 225-230.

31. Patent application no 8222382, Aug. 1982.

32. FANTOM, A.E., 1982, 'A new calculable RF and microwave field standard with extremely broad bandwidth: theory and preliminary results', NPL Report DES 73.

33. EMERSON, W.H., 1973, 'Electromagnetic wave absorbers and anechoic chambers through the years', IEEE Trans. AP-21, 4, 484-491.

34. CLARKE, R.N., 1982, 'Electromagnetic stirred mode cavities (SMCs)', NPL Report DES 75.

35. CORONA, P., LATMIRAL, G., and PAOLINI, E., 1980, 'Performance and analysis of a reverberating enclosure with variable geometry', IEEE Trans. EMC-22, 1, 2-5.

Chapter 20

Automation in microwave measurement

S. Arnold

1   INTRODUCTION

The use of automation in microwave measurement is fast
becoming the rule rather than the exception. A result of
this radical change is that microwave measuring instruments,
techniques and procedures are being modified more rapidly
than ever before. In consequence, automation may currently
be viewed as the single greatest influence on microwave
metrology.

The mechanism of such change is the comparatively cheap,
readily available and technically advanced integrated
circuits that provide processing, control and data storage.
The drive for such change is the many benefits offered by
automation.

Automation has been demonstrated to provide two
principal advantages over manual methods in microwave
measurement. From a technical point of view, it is possible
to implement important techniques that have only become
viable through the automation of measurement processes.
From a financial point of view, the life-cycle costings of
measurement activities have been significantly reduced.

Either one of these benefits would have justified a
strong shift towards automation in microwave measurement;
together they have guaranteed a profound change in the
implementation, the capability and even the application of
microwave measurements. From the microwave metrologist in a
laboratory to the consumer in the microwave-product market,
many changes directly attributable to the introduction of
automation in microwave measurement may be identified.

With the rapid shift towards automation in order to
derive benefits, associated problems have been identified.
Many of these are the teething problems of any rapid
evolution in technology and, once recognised, may be easily
avoided. Others are endemic in an automated measurement
environment. Thus, an appreciation of both the advantages
and potential pitfalls associated with automation is
essential to the successful future development and use of

automated techniques and measuring instruments in microwave engineering.

2   ACCURACY

A major technical impact of automation derives from the use of techniques for enhancing the accuracy of measurement - techniques that would otherwise be impracticable.

For the most part, the reduction in uncertainty of measurement comes from methods of calibration that permit mathematical correction to be applied to observed data. This improvement in accuracy, therefore, depends on the ease with which data may be reliably acquired, operated on and stored. This in turn depends directly on the availability of powerful real-time processing and control.

While the benefits of such systematic correction are general to electrical metrology, they have proved of especial benefit at microwave frequencies. This is because residual errors of measurement are significant in most methods of microwave measurement. Their principal origins are two-fold.

Firstly, characteristics used to good effect in instruments at lower frequencies become progressively less satisfactory as frequencies reach the microwave region. The increasing effect of parasitic components and of the electrical size of devices results in a degradation in their useful performance. Examples of this are seen in the performance of mixers and detectors.

Secondly, the difficulty in realising designs for active and passive networks used in instruments increases with frequency. This is particularly evident where wide-bandwidth performance is sought. The difficulties encountered in the analysis and synthesis of new designs and in their manufacture lead to characteristics that fall well short of the ideal. The result is a compromise in overall performance or a restriction of the useful bandwidth. Multiple-port devices typify the general problem.

The principle of correction for systematic errors depends on calibration of the measuring instrument against a reference standard(s), i.e.

Systematic error (unknown function of instrument) =
     Observation -
       True value of reference standard(s) (having a known
   uncertainty) +
         Random error (function of the specific instrument and
         the independent variables)

The systematic errors due to failings in the instrumentation are thus determined, and then applied to subsequent observations made using the same instrument under similar conditions, i.e.

True value of unknown network + Random error = Observation - Systematic errors

This procedure always leads to a relative measurement with respect to the reference standard(s). These standards can take the form of internal references, transfer standards or calculable references.

Internal references generally offer a limited number of calibration points (often only one), chosen for accuracy and convenience. Examples are to be found in instruments such as power meters, frequency counters and spectrum analysers. Transfer standards, however, can provide a range of reference values, though the cost of stable standards and their calibration still places limitations on the number of reference points. Common examples are reflection standards, frequency standards and reference antennas, such as a standard-gain horn. Where they can be realised, calculable reference standards provide a continuum of calibration points, with reference values computed to suit the chosen independent variables of measurement. The many forms of impedance standard used to calibrate a twin-channel superheterodyne network analyser are a good example of calculable standards.

The choice of reference standard depends on the accuracy conferred by each type, on their convenience in use, on the correction procedure selected and on cost. The periodicity with which the calibration process must be performed depends on the stability of the instrument at its calibrated measurement interface. Changes of independent variable, such as frequency, power and temperature, or of configuration, such as interconnection paths or connector type, may invalidate an existing calibration.

At its simplest, correction consists of determining, storing and applying to an observation constants that are functions of the independent variables of measurement, in a manner established by a theoretical analysis of the failings of a measuring instrument or system. Thus, in the general case, a measurement y is related to an observation x in the form:

$$y = f_0(z) + f_1(z)x + f_2(z)x^2 + \ldots\ldots\ldots$$

where

z is an independent variable
f_0 is a function representing offsets, e.g. internal frequency reference as a function of temperature
f_1 is a function representing scaling, e.g. power-sensor efficiency as a function or frequency

f_2 is a function representing a first-order non-
linear component, e.g.mixer conversion efficiency
as a function of power.

Of greater complexity still are methods requiring large
quantities of correction data in order to solve complex
field or network equations. Correction parameters may be
vectors that are functions of one or more independent
variables. In such cases, powerful processing, integrally a
part of an instrument's architecture, enables the result of
operations on observed data to be available as a quasi-
real-time output. Probe compensation for near-field antenna
measurements (Paris et al.(1)) and the many correction
procedures, using several standards, applied to network-
analyser measurements (Warner (2)) testify to the value of
lengthy and exacting correction techniques.

Automatic correction to compensate for instrument
deficiencies may be compared with manual correction methods
in which tuners, isolators, attenuators and phase shifters
compensate for microwave errors, and gains and offsets are
adjusted in IFs. With automation, the residual systematic
errors are determined explicitly and eliminated
mathematically. The large number of observations
practicable with an automated method contrast with the
limited number normally made manually. Additionally, the
skill required to calibrate out errors automatically is
usually significantly lower than for manual methods that
achieve equivalent accuracy.

Improvements in accuracy may also be derived through
real-time signal conditioning in both the microwave and IF
stages of an instrument. Monitoring of signal levels and
automatic control of gain or bandwidth confine signal
processing to linear or well characterised regions of
instrument performance, e.g. spectrum analysers. Warnings
or blocking of the measurement process outside acceptable
regions of operation guard against erroneous use.

Automatic monitoring of signal level, coupled with a
knowledge of the noise characteristic of the measuring
instrument, may be used to control the number of
observations made. Statistical processing of the data will
then improve the accuracy where signal-to-noise ratio
becomes significant. Simple averaging of a sequence of
repeated observations is a common and effective method of
reducing uncertainty.

3   REPEATABILITY

Automation has proved to be a powerful force for
introducing discipline into the activity of microwave
measurement. It requires that measurement procedures be
devised and stated, that the associated software processes
then be coded, and that this be integrated in the target
hardware and finally proved. This process does not tolerate

ill-defined specification, ambiguous interpretation or vague implementation. The measurement procedure is thus rigidly preserved in the hardware configuration of the measuring instrument or system and in the software controlling it.

Correctly structured, these procedures control and guide both the user and the measurement equipment along tried and tested paths. This control of the actions of the user and the dissociation of the user from the activity of making observations improves the repeatability of a measurement procedure. The observations and their presentation are independent of individual skills and attitudes, and the 'green-fingered' approach to microwave measurement is largely eliminated. In consequence, measurement repeatability is improved, both between personnel and for the same operator over a period of time. An important and difficult-to-quantify component of the measurement uncertainty is thus minimised.

Microwave measurement often comprises tedious repeated execution of a defined procedure. This repetition is itself a significant source of error in manual measurement. Repetitive processes are ideally suited to automation, and the repeatability of both trivial and complex measurement procedures is improved through the delegation of routine activity to process control.

4   DATA HANDLING

One of the great strengths of automated microwave measurement is its intrinsic capability of data handling. Following its acquisition in a digital form, observed data may be processed with ease. After most manual microwave measurements, data has to be ordered, organised and presented in a required form and manner. For the most part this is a purely mechanistic process of collection, calculation, analysis and presentation, with no decision-making required. The routine of these operations can be effortlessly performed with the aid of the computation and storage capabilities that either complement, or are found within, automated instrumentation.

Circuit properties can be readily computed from observed parameters to provide parameters that are difficult to measure directly. Stability criteria, spectral characteristics and specialised microwave sub-system performance are typical of parameters that require the acquisition and reduction of large quantities of observed data in order to characterise a network both conveniently and meaningfully.

The transformation of data broadens the scope of application of many microwave instruments. For example, appropriate sets of observations using the same tunable automatic receiver can be instantly analysed to provide data on spectral content, signal power, modulation

characteristics and amplitude and phase noise
characteristics. Similarly, frequency-domain to time-domain
transformation enables a network analyser to behave as a
time-domain reflectometer with convenience and accuracy.

The mass storage and fast computation of data means that
device or component selection is an effortless and error-
free process following automatic characterisation. The
grouping of matched sets of devices such as mixers, or the
selection of related characteristics such as antenna feeds
possessing related electrical lengths, ceases to be a
prohibitively time-consuming procedure. In general,
automation in microwave measurement advantageously affects
the viability of manufacturing designs that are strongly
dependent on fabrication-sensitive parameters.

Automated characterisation is an important element of
process monitoring and yield improvement in microwave device
and network manufacture. The ease of data manipulation
simplifies the evaluation of parameter trends within and
between batches of microwave components. Measured
parameters can be compared with specified values, and
compliance with complex criteria can be automatically
verified and presented in commercially acceptable formats.

The data-acquisition and handling requirements in
antenna measurement are generally so complex that automation
is essential. It is particularly important in near-field
measurement, where the control to effect accurate
positioning of the probe antenna, the acquisition of near-
field data, the transformation of data to a far-field
pattern and the inclusion of calibration data in these
calculations to correct for probe-antenna effects all
combine to present a vast data-handling and computation
problem. Similar problems exist in polarisation, boresight
and radome measurements.

5   SPEED

The improvements in the time required to perform a
measurement are an obvious advantage of automatic methods
over manual methods. A typical reduction in the time taken
to complete a measurement procedure is an order of
magnitude, though particular examples demonstrate much
greater improvements.

The most immediate value of speed is financial.
Productivity and work in progress both benefit from rapid
measurement. Faster, and therefore more, measurement can
lead to enhanced quality control. Nevertheless, clear
technical advantages are gained by an increase in speed.

Time-dependent characteristics, otherwise lost to the
non-automated observer, can be acquired through automation.
A rapid sequence of observations can capture parameters that
change in a time period too short for manual recording of

results. Power-up situations or frequency shift of pulsed
oscillators represent this type of measurement. Conversely,
the measurement of slowly changing parameters is most
efficiently monitored through continual observations with a
defined periodicity. Dielectric properties, e.g. moisture
content in process monitoring, atmospheric microwave
propagation characteristics and the frequency stability and
spectral purity of transmitters are examples of monitoring
situations. Most difficult is the isolated event;
continuous observations are made, but data is retained or
analysed only in the event of a significant change in the
observed parameter. Many microwave-based warning or fault-
correction systems rely on this automatic measurement
technique.

The speed of signal acquisition and processing is
currently such that, even where highly complex mathematical
operations are required to correct observations or process
results, the time to presentation of a required parameter is
negligibly short. Hence, when an engineer closes the
measurement, data-processing and information-display loop,
it is possible to adjust, in real time, network parameters
that may be either difficult or impossible to observe when
manual instrumentation is used. For example, critical
filter alignment, such as minimum deviation from a linear
group delay over a defined frequency band, can be performed
with the aid of a simple graphical representation of the
required performance criterion.

Successive measurement, continuously updating a
graphical display of transformed data, instantly provides a
researcher with comprehensive information on the effect of
network changes. This can provide an insight into network
performance or system operation that is difficult to achieve
with non-automated equipment.

6   ESTABLISHING MEASUREMENT UNCERTAINTY

Acquiring an adequately large sample of observations and
performing the statistical analysis of this data is an
impracticable method of establishing the measurement
capability of most manual methods. Thus, manual-measurement
uncertainties are normally established from a theoretical
analysis of instrumentation performance. The supporting
traceability is established through a limited number of
measurements against traceable standards.

Automation assists both in establishing traceability and
evaluating uncertainty by statistical methods and in
presenting these in simple and readily understood terms.
The objectivity and repeatability of each of many automated
observations and the compatibility of automation with
information processing offer a ready means of empirically
defining the measurement process. In this manner, all
factors influencing the measurement, including human ones,

are represented in a determination of instrument or
measurement-system capability.

Sets of observations on traceable microwave standards
are performed at regular intervals, using the application
software and the hardware configuration that are used in
routine measurement. These data sets are stored and
subsequently analysed to derive the mean and standard
deviation of each measured parameter as a function of the
independent variables of measurement.

The repeated and on-going execution of this procedure is
a powerful method of characterising and monitoring the
performance of a microwave measurement system. The
statistical reduction of constant-size samples of
observations, acquired with a regular period and under the
normal conditions of equipment use, will expose performance
variations with time. Each analysis may be performed on
independent samples. For a more frequent and a weighted
analysis of performance, data can be reduced after the
observation of each new set of data, using the most recent
data sets that constitute the selected sample size.

Shifts with time in the mean are indicative of
variations in the systematic errors of measurement (Hinton
(3)). The relationship between the direction and rate of
change and the independent variables of measurement is
frequently indicative of the cause of the detected
variation. Similarly, increases in the standard deviation
expose the existence of mechanisms leading to an excessive
lack of repeatability. Mechanical wear in connectors,
cables and switches or electrical-component degradation can
be identified from characteristic changes. Impending
catastrophic fault situations can sometimes be predicted
from unusual or rapid changes in mean and standard
deviation.

7   SOFTWARE

Software is fundamental to many aspects of automatic
microwave instrument design and use. Functions and
activities traditionally vested in hardware and in manual
control are performed through operations defined in real-
time software.

Principally, the software defines a schedule of
activities that constitute a complete measurement procedure
associated with a specific instrument configuration in order
to achieve a required measurement function. It replaces
much of the documentary information specifying and
instructing user actions.

Using an equipment's available communication channels,
the software can create an environment that interfaces an
operator with the complexities of the instrumentation and
its activity during measurement tasks. Interactive

communication, status information and real-time intervention
are important elements in this interface.

Through the hardware architecture of processors and
controllable instrument functions, the software establishes
the instrument conditions necessary to making an
observation. The setting of gains, bandwidth, signal paths
and independent variables of measurements are a part of this
initialisation process.

The software supervises the sequence and timing of the
actions that constitute the making of an observation. This
real-time activity controls such factors as critical timing,
number of samples and transfer and storage of data, and
monitors detectable error conditions.

The operations on this observed data, the reduction of
measurement information and the analysis of results are all
data-processing activities defined in software and executed
as an integral part of the measurement schedule. In this
phase, correction algorithms and data transforms are
executed.

Software-controlled real-time data display, hard-copy
data output formatted to user needs or long-term data
storage completes the measurement process.

Support activities such as instrument self-check
software, fault-dianostic routines and configuration data
complement the range of application software.

In its turn, application software or firmware is built
on a base of assemblers, compilers, interpreters, language
translators, debugging aids, editors and operating systems,
which constitute the program development and support
environments.

For the most part, the specification, design, coding,
testing, documentation and configuration should conform to
the conventions, methodologies and quality-assurance
practice applied to any software embedded in real-time
processor-based equipment. The activities constituting the
software life cycle and the principles of good programming
are entirely appropriate to modern microwave instrument
design and application (MOD(4), Kerola(5), Myers (6)).

In general, the cost of program development for
automated measurement methods continues to be
underestimated. Costs to complete the software for
automated instruments can equal the hardware cost. However,
present trends in standardisation of instrument interfaces
and control codes reduce cost, and cheaper memory encourages
the use of high-level languages, which, though less
memory-efficient , are simpler and quicker to write.

Programs should be structured to make best use of existing proved modules of software. In instruments, blocks of firmware code can be accessed by user-generated code, which then defines sequences of high-level functions linked to perform a specialised or unique application. At system level, good structuring is the key to the re-usability of software modules, such as application-library modules or standard user-developed routines.

For infrequent or small-volume measurement purposes, the costs of programming must be weighed against the advantages of automation. In microwave research and development, where either the measurement method or the network being evaluated is subject to frequent changes, the inflexibility of automation is most apparent. Periods of continual reprogramming result from this, and a cost/benefit balance must be struck.

For user-developed measurement assemblies, the user must be prepared for high initial programming costs before any benefit is derived. In the early stages, when program development is at a peak, little or no useful measurement capability makes the return on capital employed look unattractive. The activity of debugging the software (and the hardware) is necessarily performed on the measurement system, and occupies long periods of equipment time that might otherwise be available for routine measurement work. The need for this activity must therefore be taken into account in establishing the utilisation level of the measurement equipment. It is equally important that the user should appreciate that continual improvements and new techniques often cause the programming activity to be an ever-present load on automated measurement equipment.

When released, the software becomes an integral part of a microwave instrument. It contributes to its intrinsic performance, and defines its method of application in a measurement procedure. Thus the range, uncertainty and traceability of measurement apply to a specific configuration of both hardware and executable software, the latter being dependent on the configuration of the software-development environment. Approvals and authorisation associated with microwave measurement are consequently as dependent on software configuration control as on hardware.

8   ERROR CONDITIONS

Establishing the presence of an error condition in the measurement equipment or process can be difficult. Fortunately many error conditions are catastrophic and easily recognised. Automated microwave measurement equipment, however, can produce erroneous data as efficiently as good data, which, with lower-technical-skill supervision, may pass undetected. This problem is exacerbated by the increased dissociation of the operator from the measurement process. The lack of intermediate data presentation can lead to unobserved operator errors, such as quality of RF connections. Application software can contain checking routines for valid data or timing bounds: increasingly, instrument designers are embedding software in equipment to monitor correct operation.

Removal of tedious repetition by the use of automation leads to a sense of a lack of involvement and influence: this may simply replace monotony by detachment as the prime source of operator error. The development of software that communicates freely and sensibly with the user can lead to a level of involvement that maintains a lively interest in the process. Presented in a suitable form, usually graphical, this communication can be a valuable indication of the presence of error situations.

A particularly difficult type of error to detect can stem from the correction techniques applied to observations. An error in the correction process will repeatedly miscorrect all subsequent observations and introduce a residual error. Such errors often do not readily betray their existence, since the distortions introduced are generally neither gross nor unreasonable.

Sources of this type of error can lie in the accuracy or stability of the reference standards used for correction. Machining and plating tolerances, damage and mechanical instability or inaccurately defined standards will all introduce characteristic forms of systematic error.

Errors may be undetected during the testing of software. They are only betrayed under particular circumstances; independent checks against a limited range of application conditions can appear satisfactory. Timing is a common source of such errors, and is particularly associated with the control and synchronisation of measurement peripherals. Faster processing and parallel processing aggravate the problem. Software containing complex logic or mathematics may hide paths containing errors that are only activated under specific combinations of measurement conditions.

These and other sources of error in automated measurement are compounded by the unjustified trust often placed in the integrity and accuracy of the computer-printed 'output, often to an excessive number of figures. Careful checking and monitoring of the measurement process against control networks or stimuli help to minimise measurement errors. Such control standards need have no traceability other than that conferred by the automatic instrumentation. Used before, and possibly after, each sequence of observations, control standards provide a real-time evaluation of total system accuracy and, equally important, documentary evidence of this. Remeasurement of

reference standards used in a correction process is of
limited value, since it can only provide assurance of
repeatability in the measurement.

9   INTERFACING

Only rarely do the physical and electrical
characteristics of a microwave instrument exactly complement
an unknown network being measured. Interfacing and
interconnection of a network or reference standards to
instrumentation is thus an essential element in most
automated measurement equipment. The desire to minimise
manual operations encourages the introduction of signal
paths containing electronic or electro-mechanical switches.

The resulting connector pairs, flexible transmission
structures and switches are generally the source of
increased uncertainty or systematic error. At worst, they
can be unreliable and introduce gross non-repeatability,
which then compromises the whole measurement process.
Interfaces must thus balance the conflicting requirements of
instrument reconfiguration and ease of use with the need for
a stable and minimal effect on an observed parameter.

Correction techniques implicitly rely on repeatability
and stability at the plane of interconnection to the
standards. The greater the change and disturbance
throughout a correction and measurement procedure, the
greater the chance of consequential increased uncertainty in
observation. A theoretically effective correction technique
can be totally negated by the practical consequences of
interconnections. The interconnection repeatability of the
transmission structure of a network being measured will thus
influence the acceptability of a correction method and the
form and number of associated standards.

The need for good engineering design and good
measurement practice to be associated with the mechanical
and electrical interfacing of the networks being measured to
the instrumentation can hardly be over-emphasised.

10 MEASUREMENT INTERVALS

Automated RF measurement techniques rely on sequences of
observations at discrete values of independent variables
such as frequency. In some circumstances care must be
exercised in choosing the interval or distribution of the
independent variable(s).

A primary consideration is that the interval for the
independent variable must ensure that significant changes in
the measurement parameter are not missed between points of
observation. This is especially true if interpolations are
to be performed on the observed data.

Various operations performed on uncorrected or corrected data, such as Fourier transforms, can require specifically related sets of independent variables in order to facilitate the presentation of a required parameter.

The rapid acquisition and manipulation of large amounts of data, while undoubtedly an outstanding advantage of automation, can lead to the measurement and computation of a network's performance as a function of a large number of values of independent variables simply because it is possible. The result, if not confusion and possible error, can be to waste much time during the measurement and analysis of essential data.

The compromise that must be struck to avoid the danger of an insufficient or an excessive number of observations needs careful consideration when an automated microwave measurement procedure is being developed.

11 EFFECTIVE MEASUREMENT PROCEDURES

The strengths of manual measurement methods can prove to be weaknesses when used with automatic instruments, and advantages such as increased repeatability, lower uncertainty and better error detection may be lost.

Interchanging the dependent and independent variables, as used in a manual measurement, may be of benefit. Search techniques, interpolation and other forms of mathematical data manipulation can strongly influence the approach adopted in order to determine the desired parameter.

Automating an existing manual measurement technique without reviewing alternative approaches can result in inefficient and sometimes ineffective automatic measurement techniques.

12 MEASUREMENT CAPACITY

Automation encourages reliance on intensively used integrated assemblies of measurement and control equipment, which may be complex, expensive and few in number.

Automatic microwave measurement equipment is generally more expensive than the manual equipment it can replace. If assembled as a system, it is viewed as a single integrated structure dedicated to an area of application. The cost of establishing or increasing such resources requires that large quantised steps of investment be made compared with manual instrumentation. The general tendency, whether with a single or multiple system resource, is for users to overload automatic measurement equipment in order to avoid the next quantum of investment.

Since measurements are performed faster than by equivalent manual methods, a large volume of measurement work can flow through a small specialised resource. This speed of measurement, together with the high cost of further equipment, channels routine measurement and any associated measurement development through a single equipment.

Peak load and equipment failure rate, both difficult parameters to predict, are important factors in evaluating the total resource capacity needed. The larger and more complex the measuring system, the shorter the time between failures and the longer the time required to diagnose a failure and return a measurement system to operational status. Unnecessary equipment complication should be strongly avoided, since it leads to rapidly escalating costs.

It is thus a combination of the high capital cost and consequent overloading and the technical complexity of the equipment that leads to the introduction of potential bottlenecks where none may have existed with the more flexible and less concentrated nature of a manual measurement resource. In general, careful planning in the loading both of the routine measurement and the development of new measurement schedules and a comprehensive maintenance policy will minimise this potential danger area.

13   OBSOLESCENCE

The potential flexibility of software-based measurement instruments and the continuing advances in processing hardware have combined to create a competitive high-technology software-based instrument market. In such a market, where large changes can occur rapidly, obsolescence is a particular problem. The user may consequently find a microwave measurement instrument or system out-of-date only a short way into its anticipated life.

Currently, many changes in automated microwave instrumentation are due to the nature and method of implementation of the automation and to changes in its objectives. Automation offers improvements in ease of use, reliability and serviceability, and instrument changes seek to maximise these properties. Changes in the intrinsic measurement capability of the instruments are less subject to radical change. Thus, an automated measurement equipment can be outdated because of the techniques associated with automation, while its range and uncertainty of measurement remain unsurpassed.

Although automation offers long-term security through greatly reduced dependence on individual staff skills in the measurement activity, when engineering skill is transposed into equipment 'intelligence' that security depends on maintenance, support and ungradability for software as much as for hardware. The upgrading of assemblers, compilers, operating systems, libraries or instrument firmware is an essential part of service activities, in order that faults can be corrected or enhancements made to the software.

The basis of upgradability is standardisation and modularity. This is true in discrete instruments, but especially necessary in measuring systems. A microwave system comprising many items of measurement hardware and associated, expensively developed, software can be seriously compromised by a single obsolete or unsupported part. Automated instrumentation should therefore be based on a hardware and software architecture that allows progressive replacement of obsolete modules by compatible items. Increasing rationalisation and standardisation in instrument communication protocol and buses, in programming language and in system software makes such a concept possible.

14 SKILLS FOR AUTOMATION

The growth of automation in microwave measurements is characterised by the removal of manual operations and intervention from almost every aspect of the total measurement activity. There has been a shift to defining in software the procedural information hitherto contained in documentation, and to controlling by software the acquisition and reduction of data and the presentation of information traditionally performed by technicians.

Increasingly, the knowledge and technical skills of metrologists, instrument designers and measurement technicians are embodied in firmware and software, which is either embedded in, or associated with, target measurement hardware. Consequently, microwave-measurement engineers must now be familiar with, if not competent in, the disciplines of software design, coding and testing. Thus scarcity of skilled technicians to perform non-automated microwave measurements may be replaced by an equal scarcity of the staff who develop automated measurement techniques and the associated software.

The ideal qualities in staff operating automatic measurement equipment, whether in the laboratory or on the production floor, are often different from those required for manual measurement. The combination of machine minder and skilled technician, recognising, and possibly understanding, faults that may be in the equipment, the software or the network being measured, is a blend not commonly found.

The relationship between man and machine, a combination of flexible and analytical thinking with restrictive, highly-efficient, predetermined logic, can be a deciding factor in the effectiveness with which automated microwave instrumentation can be used.

15 CONCLUSION

Improvements in microwave network performance have stemmed from advances in network analysis and design that rely on numerical methods based on modelling and optimisation. Rigorous and accurate characterisation of devices, components, transmission structures and radiating structures is an essential pre-requisite of such computer-aided design.

New fabrication technologies that offer performance and cost advantages provide little opportunity for subsequent physical network adjustments to accommodate inaccurate measured design data or the parameter distributions in manufacture. Accurate process-monitoring-based automation is vital to these fabrication processes. Where adjustment in the fabrication process is possible, e.g. laser trimming, the speed, high accuracy and rapid data reduction are essential for monitoring and control.

Once realised, these advanced and complex microwave components and networks must be subject to rigorous and accurate measurement in order to demonstrate their conformance to demanding performance criteria.

Computer-aided design and manufacture of microwave networks is, in consequence, fundamentally dependent on measurements that are increasingly reliant on automation. An appreciation of the technical and economic strengths of this automation in microwave measurement and an understanding of its potential weaknesses are important factors in meeting the demands and challenges of microwave engineering.

16 REFERENCES

1. Paris, D.T., Marshall, W., and Joy, E.B., 1978, 'Basic Theory of Probe-Compensated Near-Field Measurements', IEEE AP-26,3

2. Warner, F.L., 1977, 'Microwave Attenuation Measurement', Peter Peregrinus

3. Hinton, L.J.T.

4. MOD, 1977, 'Requirements for the Documentation of Software in Military Operational Real-Time Computer Systems', JSP 188

5. Kerola, P., and Freeman, P., 1981, 'A Comparison of Life-Cycle Models, Proc. IEE 5th ICSE

6. Myers, G.J., 1980, 'The Art of Software Testing', Wiley

Chapter 21
Uncertainty and confidence in measurements
L. J. T. Hinton

1. INTRODUCTION

The importance of measurement in the development of technology has long been recognised. What is unfortunately less appreciated is a need to have a proper understanding of the measurement process and of the meaning of reported results. A measurement of any kind is indeterminable in some way and the exact difference between an indicated value of an instrument and the true value of the quantity being measured cannot be known and is not in fact knowable. Nevertheless the statement of a measurement result is not complete unless the residual indeterminacy, after all possible corrections have been made for errors, is expressed in a quantitative manner as an uncertainty that is obtained from a knowledge of the measurement process.

The way in which uncertainty in measurements should be assessed and reported has given rise to much debate however both in this country and abroad. A lack of uniformity in uncertainty statements is a constant source of trouble both for the expert metrologist who needs to compare his results with those of other workers in his field and for the technician in industry who needs to understand the meaning of the certificate of calibration for the instrument he is using. For the most accurate measurements at the national standards level there are advantages in reporting a list of all component uncertainties with the relevant details, for progress is often made by detecting previously unknown errors. However even at this level of measurement there is also a need to combine the components of uncertainty using a stated procedure to provide a resultant total uncertainty so that the overall progress in realising a unit of measurement can be readily appreciated. On the other hand at much more modest levels of accuracy, the technician in industry may be using his calibrated instrument to see if a product of manufacture is within specification. He needs not only to be clear about the corrections to be made to his instrument's indications, but also he must understand what is meant by the uncertainty statements on the certificate of calibration in terms of confidence.

Since the early years in the operation of the British Calibration Service (BCS), it was recognised that there was a need for established procedures for the treatment of

uncertainties which could not only be used by calibration
laboratories in the various fields in which BCS was or
might be operating in the future, but which could also be
suitable for the national standards laboratories at the
apex of such a service. The early work in BCS was undertaken by Dietrich who developed a statistical approach for
the combination of uncertainties based on the assumption of
probability distributions for all components of
uncertainty (1). An important early contribution to the
expression of uncertainty in measurements also came from the
work of a Ministry of Defence Working Party under the
chairmanship of Harris (2). Based on this ground work,
Harris and Hinton prepared in 1977 a practical code of
practice for use in the electrical fields of measurements (3). In this chapter the basis for the BCS
publication is described and some examples of its application are given.

2. CONCEPTS AND TERMINOLOGY

Measurement as a science has its own concepts and terms
and it is important that these be clearly understood. Some
reference has already been made to this chapter's main
theme 'uncertainty'. This term (4) quantifies the
indeterminacy in the measurement process by a statement of
the range of values within which the true value of the
quantity being measured is estimated to lie. It is usually
expressed in bilateral terms, ie ±, either in units of the
measured quantity or as a proportion of the measured or
some specified value, for example as percentage, 1 in 10^x,
parts per million (ppm) etc. Because these limits cannot
be known with certainty they need to be qualified with a
statement of confidence. It may perhaps be thought reasonable
to assume that the absence of qualification indicates that
the person assigning the uncertainty limits has the utmost
confidence (corresponding to an estimated 100 per cent
probability) that they embrace the true value, but this may
not be what is intended, or, if it is, it may not be
properly justified. Furthermore for some measurement
processes involving relatively large contributions to
uncertainty it may be more appropriate to work at a lower
level of confidence, such as 95 per cent probability that
the true value lies within the uncertainty limits (that is a
1 in 20 chance that it lies outside these limits). It is
important to realise that without information being provided
about the meaning of uncertainty limits or how they are
derived it is not possible to assess properly the
significance of the difference between measurement results.
The term 'error' (in a measurement) is sometimes
confused with the term 'uncertainty'. Error is the
difference between the result of a measurement and the true
value of the quantity measured. However, as the true value
is not knowable it is approximated in practice by the value
of the quantity measured that is established by traceability
to national standards. That part of the error that has a
known magnitude and sign is thus an amount for which a

correction can be made;   such an error is given a
positive sign if the measured value is greater than the
true value (4).

When a measurement is repeated a number of times
under substantially the same conditions, then, provided
the instrument has an adequate ability to resolve small
differences it is common experience that the results will
not all be the same due to the presence of small
independent random variables, such as power source
instabilities or detector noise. If a large number of such
measurements are made and the frequency of occurrence of
indicated values recorded, then it is found that the
envelope for the probability density distribution will
approximate to the well-known bell-shaped Gaussian or
'normal' curve. In Figure 1 a sample of all possible
results is shown as four values under the distribution
curve. It will be noted that the arithmetic mean value
for the sample of four results does not coincide with that
for the whole population of possible values (the central
value of the distribution curve). Each time a set of
measurements is made it can be expected that a different
mean value will be obtained, and the difference between
a sample mean and the population mean is an expression
of the 'random component of uncertainty' or 'random
uncertainty'. Probability is involved and theory can
provide limits about a sample mean value that will embrace
the population mean with a stated level of confidence. This
is discussed further in Section 3, but it should be noted
straight away that no corrections can be made for random
errors.

In practice, it is found that a relatively small
number of contributions to random uncertainty which are of
similar range can combine to produce a distribution of
values which is a close approximation to the normal curve
of Figure 1, even if their individual density
distributions are not normal. In practice the individual
distributions are often truncated so that the resultant
distribution curve will then be truncated at limits
separated by the sum of the individual contribution ranges.

To relate the mean indicated value of a quantity and
the associated uncertainty limits obtained from a
measurement sample to the value of the national standard of
that quantity involves the concept of traceability. A
calibration certified for an instrument issued, by one of
the approved laboratories of BCS will provide the user with
a formal and authoritative record of the measurements
undertaken and the results obtained. From the
information on such a certificate corrections can be
applied to the instrument indications and thus relate the
user's measurements to national standards. The
corrections, however, will be subject to the uncertainties
reported on the calibration certificate. These
uncertainties are the maximum values of residual errors
and remain unchanged when the user repeats his
measurements with the instrument and are thus a 'systematic
component of uncertainty' or 'systematic uncertainty'.

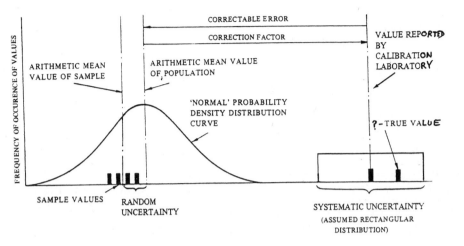

Fig. 1   Illustration of measurement uncertainties

In Fig. 1, it is assumed for simplicity that there are
no systematic errors in the instrument indications other
than those correctable by calibration. This is discussed
further in Section 4.

Whereas the probability density distribution for random
uncertainty approximates to the normal form and is measure-
able, the distribution for systematic uncertainty has to
be deduced from either a prior knowledge of the
contributions or conservative assumptions have to be
made. Knowledge of the nature of these contributions
will vary considerably. Often all that is known is an
estimated range of uncertainty for each contribution;   and
that knowledge may vary from an authoritative statement
on a certificate of calibration to the instrument user's
own estimation of realistically safe limits for some
measurement influence factor in his laboratory, such as
the ambient temperature. For each such estimated
uncertainty it is probable that there will be a greater
frequency of occurrence of values at the centre of the
range than at the uncertainty limits, and the distribution
may in fact approximate to a normal form. Thus it is safe
to assume that the form of the distribution is rectangular,
that is, an occurrence of values equally probable anywhere
between (±) limits. One may of course have some knowledge
of the probability distribution for systematic uncertainty
and then of course that knowledge should be used. For
example, there may be a theoretical basis for the
distribution of a contribution, or the results of prior
experimental work may be available.

In the following sections the methods for calculating
random and systematic uncertainty will be described,
followed by combining systematic uncertainty and random
uncertainty to give a single value of total uncertainty with

a statement of confidence in the result. But before
leaving the general topic of concepts in measurement and
calibration it is important to appreciate that the
classification of components of uncertainty into either
'random uncertainty' or 'systematic uncertainty' (dependent
on whether or not the contribution is determined by the
number of measurements made) may only be correct at the
point and time of application of the measurement process.
For example, referring to Fig. 1 the uncertainty from
the calibration laboratory is classified as 'systematic'
as it effects all the user's measurements. But at the time
the instrument was calibrated it could easily be that part
of the total uncertainty reported on the calibration
certificate was dependent on the number of measurements
made by the calibration laboratory, that is 'random
uncertainty'. Thus the classification is not an
absolute one but reflects traditional practice. Because
this may be misleading the Bureau International des Poids
et Mesures recommends (5) that the classification be
avoided but the terms 'random uncertainty' and
'systematic uncertainty' are well established. Nevertheless
with the present terminology it is important to remember
that some uncertainties can be random for one person but
systematic for another person.

3.   RANDOM UNCERTAINTY

For many electrical measurements and particularly at
dc and low frequency, random uncertainty can be negligible
compared with other contributions to uncertainty but this
needs to be confirmed by some preliminary measurements
before it is assumed. When the results of repeated
measurements are not the same the first step should be
to calculate the arithmetic mean or average of the values
obtained. The spread in values or range, reflects the merit
of the measurement process, but a more useful statistic
is the standard deviation of the sample. This is the root
mean square difference from the mean value of the sample
values. If there are n results for the measured variable
x_i where $i = 1,2...$ n and the sample mean is $\bar{x}$ then

$$S^2 = \frac{1}{n} \sum_1^n (x_i - \bar{x})^2 \quad\dots\dots\dots\dots (1)$$

where S^2 is the variance and S is the standard deviation of
the sample. Each time a set of n measurements is taken a
slightly different value for S will be obtained.
The best estimate of the standard deviation for a
whole population of results, σ (est), based on a single
sample of n measurements is given by

$$\sigma(est) = \left[\frac{1}{(n-1)} \sum_1^n (x_i - \bar{x})^2 \right]^{\frac{1}{2}} = S \left[\frac{n}{n-1} \right]^{\frac{1}{2}} \quad\dots\dots (2)$$

It will be noted that the difference between σ (est) and S is only significant when the number of measurements is small. When the number of measurements is very large (for example in excess of 100) then σ(est) approaches closely in value to the actual standard deviation of the whole population, σ. It can be shown that for limits of ±Kσ about the mean value of a normal distribution, the percentage of values that will fall within these limits are as given in Table 1.

TABLE 1.   Normal distribution confidence limits

Range limits	(±Kσ)	Percentage of population within limits	
	0.675 σ		50.0
1 σ		68.3	
	1.96 σ		95.0
2 σ		95.5	
	2.58 σ		99.0
3 σ		99.7	

When a value for σ has been determined from a large number of measurements, then for a single further measurement there is, for example, a probability of 95% that this value will be within ±1.96 σ of the population mean value. This is illustrated by the ratio of the unshaded area under the curve of Fig. 2 to the total area under the curve.

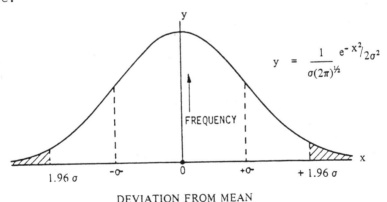

$$y = \frac{1}{\sigma(2\pi)^{\frac{1}{2}}} e^{-x^2/2\sigma^2}$$

DEVIATION FROM MEAN

Fig. 2   Normal frequency distribution curve.

When a number of measurements are made and their mean value is $\bar{x}$ then, based on the Central Limit Theory of statistics, confidence in that value can be expressed by assigning the limits,:

$$\bar{x} \pm \frac{K\sigma}{\sqrt{n}}$$

If K is again given a value of 1.96, then there is a probability of 95% that the population mean value lies between the limits so obtained.

Often in practice one has to accept that only an estimate of σ based on a relatively small number of measurements is available. To allow for the additional uncertainty that this will entail, reference can be made to Student's t-distribution (after W S Gosset) which provides a factor 't' for the chosen probability in place of the normal distribution factor 'K'. This takes account of the actual number of measurements made (see Appendix I). The confidence limits for the mean value are then:

$$\bar{x} \pm U_r$$

where the random uncertainty $U_r = \dfrac{t\ \sigma(\text{est})}{\sqrt{n}}$ ....(3)

From Appendix I and equation (3) the benefit of reduced uncertainty by increasing the number of measurements can be seen. This benefit gets progressively less as the number is increased and it is usually not necessary to make more than about ten measurements. However, for a measurement system and procedure that is going to be used frequently it is advantageous initially to make a large number of measurements on a component of high stability to obtain a value for σ(est). Then, provided there are no changes in the system or procedure, only a relatively small number of measurements need be made on future occasions. The significance of random uncertainty compared with other sources of uncertainty will need to be taken into account. It will be noted that Student's t factor becomes equal to the normal distribution factor K, when n equals infinity.

4. SYSTEMATIC UNCERTAINTY

Mention has been made of systematic uncertainty arising from the calibration certificate for an instrument. Although this is often the dominant contribution there are usually other contributions to systematic uncertainty in measurement that arise in the instrument user's own laboratory that must not be overlooked. The successful identification and evaluation of these contributions is very dependant on a detailed knowledge of the measurement process and the experience of the person making the measurements; a point that cannot be overstressed. By planned variation of the conditions of measurement and the measurement process and averaging results, errors can be revealed and corrected and thus prevented from contributing to the total uncertainty. Whenever possible a correction for an error should be made, but when the error is small it may be preferred to treat it as a contribution to systematic uncertainty equal to (±) the uncorrected error magnitude.

In determining systematic uncertainty it is necessary to consider in turn the measuring apparatus, the operational procedure, and the item under test. Although

the preparation of a truly comprehensive list of sources of
systematic uncertainty that should be considered is not
practical, some contributions are fairly common in electrical measurements and guidance on their identification and
assessment is given in Appendix II.

In dealing with systematic uncertainty it is first
necessary to group contributions in accordance with their
known or assumed probability distributions and then
determine the standard deviation for each such group. If
only the limits for systematic contributions to
uncertainty are known then, as already mentioned, it is
safe to assume that they have rectangular distributions.
The standard deviation for such a distribution when the
limits are ± a is

$$\sigma_s = \frac{a}{\sqrt{3}} \quad \dots\dots\dots\dots\dots(4)$$

If there are a number of such contributions a_1, a_2, etc
that are quite uncorrelated then the standard deviation for
their combination is

$$\sigma_s = \sqrt{\frac{a_1^2 + a_2^2 + \dots\dots}{3}} \quad \dots\dots\dots(5)$$

and, provided the values of the semi-ranges are similar in
magnitude, the resultant distribution of the combination
approximates to the normal form. If σ_s is multiplied by
the factor K for a normal distribution (Table 1) then it
can be shown that the probability of values falling within
the range ± $K\sigma_s$ will always be greater than for a truly
normal distribution of the same standard deviation,
provided K is greater than 1.8. This is illustrated in
Figure 3, which compares the result of combining three
rectangular distributions with a normal distribution
having the same standard deviation.

In so far as the limit of systematic uncertainty,
$U_s = K\sigma_s$, will not of course exceed the arithmetic sum
of the semi-range values a_1, a_2 etc (an estimated 100%
probability of values falling within such limits) it is
necessary to check that the calculated value of U_s does not
do so in order to avoid undue pessimism. This can occur
when one of the contributions is dominant over the others
in value. In these circumstances the dominant

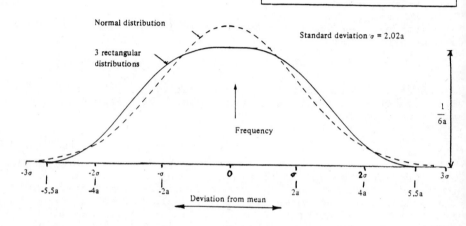

Fig. 3   Combination of three rectangular distributions with semi-ranges of 3a, 1.5a and
a with a normal distribution of equal
standard deviation superposed.

contributions, a_d should be separated out and the systematic uncertainty expressed as

$$U_s' = a_d + U_s \quad \dots\dots\dots\dots\dots(6)$$

where $U_s = K\sigma$ for the remaining contributions. The
confidence probability for association with U_s' will not be
less than for U_s.

When there are uncorrelated contributions to systematic uncertainty present with known forms of probability
distributions such as the theoretically predictable
U-shaped distribution for rf mismatch (Appendix II) or
perhaps an approximately normal distribution obtained
from the results of prior experimental work then the
standard deviation should be calculated for each
distribution. The result of the combination of the
distributions for a minimum confidence probability
corresponding to the chosen normal distribution factor K is

$$U_s = K \sqrt{\sigma_{s1}^2 + \sigma_{s2}^2 + \sigma_{s3}^2 + \quad \ldots} \quad (7)$$

where

σ_{s1} = standard deviation for 'rectangular' probability distributions

σ_{s2} = standard deviation for 'normal' probability distributions

σ_{s3} = standard deviation for 'U-shaped' probability distributions

etc for all forms of contributing distributions. Although the contributing distributions may be very different in form, the resultant distribution will approximate to the normal one, and it will be true to say that the probability of values falling within the range $\pm U_s$ will be greater than for a truly normal distribution of the same standard deviation for a value of K greater than 1.8. For example if K is 1.96 there is a probability of not less than 95% that the true value lies between the limits $\pm U_s$ (if it is assumed for the moment that random uncertainty can be neglected).

5.   STATEMENT OF TOTAL UNCERTAINTY

It will have been noted that the determination of both the random and systematic components of uncertainty has been based on the characteristics of probability distributions. The only significant difference is that the probability distribution for the random component can be determined from repeated measurements in the instrument user's own laboratory, whereas the probability distribution for the systematic component has been derived from a prior knowledge of contributing distributions either based on theory, or on the results of other experimental work, or on conservative assumptions. Provided both the main components of uncertainty have been derived for the same minimum confidence level in terms of probability, that is in choice of Student's t factor (random uncertainty) and normal distribution K factor (systematic uncertainty), the total uncertainty is given simply by

$$U = \sqrt{U_r^2 + U_s^2} \quad \ldots\ldots\ldots\ldots (8)$$

When there is a dominant contribution to systematic uncertainty present (equation 6), then the total uncertainty should be expressed as

$$U = a_d + \sqrt{U_r^2 + U_s^2} \quad \ldots\ldots\ldots (9)$$

Unless the systematic uncertainty, U_s, in equation 8 is insignificant, the confidence level for U will be greater than that for the normal distribution factor K chosen for U_s (provided K is greater than 1.8) and can be stated as an estimated minimum probability of,

say, 95%. To ascertain an actual confidence probability for
the stated limits of total uncertainty would require a
knowledge of the probability distributions and the variances
of the uncertainty contributions. Such knowledge is not
usually available.

When the random uncertainty, U_r, in equation 8
insignificant, which can often be the case in electrical
measurements, then from equation 7

$$U = U_s = K \sqrt{\sum_1^n \sigma_s^2} \quad \ldots\ldots (10)$$

Further, if it is only possible to assign realistic
limits, $\pm a_i$ where $i = 1, 2 \text{---} n$, to the contributions of
uncertainty, the total uncertainty is then calculated
simply from

$$U = \frac{K}{\sqrt{3}} \sqrt{\sum_1^n a_i^2} \quad \ldots\ldots\ldots (11)$$

For a general case when both random and systematic
uncertainty contributions have to be considered it is worth
appreciating that when the number of measurements made is
large, so that the value of Student's t factor approximates
to the normal distribution factor K, then

$$U = K \sqrt{\frac{\sigma^2}{n} + \frac{1}{3} \sum_1^n a_i^2} \quad \ldots\ldots (12)$$

That is the total uncertainty is given by a confidence
factor times the square root of the sum of the variance
of the mean value and the variances of the systematic
contributions to the total uncertainty.

6.   CHOICE OF CONFIDENCE LEVEL AND USE OF A SAFETY FACTOR

In an organized hierarchical national system of
calibration laboratories that provides assurance about
authentic traceability to national standards, it is not only
necessary that confidence levels associated with
uncertainties be stated but very desirable that the
confidence level for the measurement of a particular
quantity be the same for all laboratory levels or echelons
in the national system. The effect of the choice of
confidence level in determining uncertainties at the work
place in industry can be quite dramatic and is illustrated
in Fig. 4.

In this hypothetical example of the cumulative effect
of uncertainty through the echelons of measurement in the
traceability chain, it is assumed for simplicity that all
the uncertainty contributions are systematic with
rectangular distributions. Equation 11 has been used
to give the single values of total uncertainty shown in the
triangles on the diagram using values of K taken from
Table 1 for the selected confidence levels. It will be
noted that if the statistical approach to the treatment of

uncertainties is followed then the difference in uncertainties at the work place compared with the result of an
arithmetic summation of the contributions to provide the
utmost confidence is most marked. For a minimum confidence
probability of 99.7% the work place uncertainty in the
example is half that obtained by arithmetic summation and
well illustrates the price of including an estimated last
0.3% of the population within the uncertainty limits. The
result of reducing confidence probability to 50%, that is,
a 1 in 2 chance that the uncertainty for a corrected
instrument reading does not include the true value, has been
included in Figure 4, for that is what is meant when the
accuracy of an instrument is quoted in terms of probable
error. A clear need for the purchaser of an instrument to
read the manufacturer's specification carefully and understand its meaning.

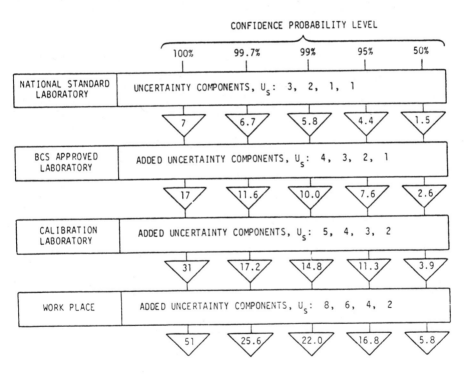

Fig. 4   Choice of confidence probability

To report limits of uncertainty beyond which it is
estimated that there is no chance that the true value of
a quantity will lie is bound to be a desirable objective.
This utmost confidence level can in practice be provided
by the arithmetic summation of the contributions to
uncertainty. However Fig. 4 has illustrated that at the

work place or at the point of sale of a product the total uncertainty may then be quite unrealistically large, and a manufacturer may lose sales to a competitor who offers assurance about his product in terms of a lower estimated confidence level based on a statistical rather than this arithmetic approach to assessing uncertainties. In deciding the confidence level for a particular field of measurement a balance has to be struck between uncertainty and confidence level on the one hand and on the other hand the cost of providing it (or the price paid if it is not provided). This is a matter for consideration by Government through the research establishment having the responsibility for the development and dissemination of national standards, and by industry itself. Although the numbers in Fig. 4 are hypothetical examples of uncertainties in a national system of calibration laboratories, they are sufficiently realistic to illustrate that the most benefit may come from devoting effort to reduce additional uncertainties as a measured quantity is transferred from one echelon to the next in the measurement system.

In the British Calibration Service, advice has been given by the BCS Technical Panels (appointed by the Advisory Council for Calibration and Measurement) that an uncertainty reported on a BCS Certificate of Calibration is to be for an estimated confidence probability of not less than 95% for all fields except dc and lf electrical measurements. In this field it has been traditional to report maximum limits of uncertainty based on the arithmetic summation of estimated contributions. However with increasing demands for smaller uncertainties by industry this practice may have to be reconsidered.

The statistical approach that has been described for arriving at a total uncertainty is considered to be most suitable in fields of measurement where confidence levels between 95% and 99% are considered adequate. When, as in dc and lf electrical measurements, it is desired to report an utmost level of confidence and there is measurable random uncertainty, then U_r should be calculated for a 99.7% confidence probability level and added arithmetically to the systematic uncertainties to provide the total uncertainty. It is to be appreciated that a considerable margin of confidence is inherent in the arithmetic summation approach for it must be unlikely that all contributions to systematic uncertainty will have values either all equal to their positive limits or to their negative limits. However such a margin of confidence does have the attraction that it provides some safety factor against underestimation of uncertainties or the possibility of mistakes and blunders.

To employ an additional safety factor may of course be very important to the ultimate user of a calibrated instrument. For example, if the success of a multi-million pound sterling project depends critically on a performance parameter that is verified by instruments, then the user's use of a safety factor is perfectly understandable. However

it is maintained that it should be left to the user to
employ safety factors if he wishes and not for the
calibration laboratory to incorporate such factors when
reporting uncertainties of measurement.

7.   EXPRESSION OF RESULTS

After the total uncertainty for a set of measurements
has been estimated for the appropriate confidence level,
the result should be reported as
$$x \pm U$$
with the accompanying statement:
'The uncertainty is for an estimated confidence
probability of not less than p%' and the recommendation
is that the confidence, p, in the uncertainty for reported
corrected mean value should be in the range 95% to 99%.
Except for dc and lf electrical measurements it has
already been mentioned that BCS has decided on the
confidence level of 95%.
Although it is maintained that a statistical approach
makes the best use of the information available to provide
an uncertainty qualified by a confidence probability, the
limitations in our knowledge of the underlying estimates
of contributions to uncertainty must not be forgotten. Thus
it is usually quite pointless to report uncertainties to
more than two significant figures and often only one figure
may be justified. Rounding down an uncertainty to two
figures is considered to be justifiable, but when only one
figure is reported the two or more figures of the calculated
uncertainty should be rounded up to the next higher single
figure.
Lastly and more generally regarding uncertainty and
confidence, it has to be remembered that an assurance in a
calibration laboratory depends heavily on the experience of
the person making the measurements. Only with experience
will there be success in identifying uncertainty contributions and then quantifying them. Also good work can be
spoilt by mistakes or blunders, for example in correcting
mean values. No allowance can be made for such human error
by any prior application of probability theory as there is
no basis for doing so. The aim must always be to apply
checks, including if necessary other methods of measurement,
to prevent human errors spoiling the assurance that a
reputation for authenticity of measurement provides.

8.   PRACTICAL APPLICATION

Step-by-step procedures for arriving at the total
uncertainty in a measurement are given in Appendix III.   In
Appendix IV three rather different examples of the
application of the code of practice are given.
Appendix IV(a) deals with the calibration of a power
meter. Two points are worth noting. The first is that the
total uncertainty for a minimum confidence probability of
95% is only half that which would be reported if the
contributions were added arithmetically. The second is that

although there was a spread of readings, this random uncertainty makes an insignificant contribution to the total uncertainty at the 95% confidence level.

In the example of uncertainty in vswr measurements given in Appendix IV(b), attention is drawn particularly to the need to identify and establish the effects of sources of uncertainty. Some errors may unavoidably degrade the voltage reflection coefficient and consequently the value of the standing wave ratio. On the other hand the error of the detector system may be avoidable if it became significant.

In the calibration of the fixed attenuator (Appendix IV(c)) allowance for variation in results is made by assigning realistic limits (as also in the vswr examples). The calculation of the total uncertainty from equation 11 is then a simple matter. The importance of the mismatch uncertainty contribution should not be overlooked in making this type of measurement as Warner et al have pointed out. (8)

9. REFERENCES

1. Dietrich, C.F., 1973, Uncertainty, Calibration, and Probability, Adam Hilger, Chapter 4 discusses the combination of rectangular probability distributions and compares the results with normal distributions of equal standard deviations.

2. Harris, I.A., 1972 Guide to the Expression of Uncertainty of Measurement in Relation to the Accuracy of Instruments and Specified Test Limits, MOD/PE Joint Service Specification REMC/31/FR.

3. Harris, I.A., Hinton L.J.T., 1977 The Expression of Uncertainty in Electrical Measurements. BCS Guidance Publication 3003 - Issue 1.

4. BS 5233, 1975 'Glossary of Terms Used in Metrology'.

5. Müller J.W., 1981 The Assignment of Uncertainties to the Results of Experimental Measurements. Paper presented at the Second International Conference on precision Measurement and Fundamental Constants at Gaithersburg.

6. Harris, I.A., Warner F.L., 1981 Re-examination of mis-match uncertainty when measuring microwave power and attenuation. IEE Proc. Vol. 128. Pt H No 1.

7. The Expression of Voltage Standard Wave Ratio. BCS Guidance Publication 4301.

8. Warner, F.L., 1977 Microwave Attenuation Measurement. Peter Pereginus.

10. SYMBOLS

The meaning of symbols of measurement have been given in the main text where they occur but are repeated here for convenience of reference:

x_i : a value among n measurements where i = 1,2...n

$\bar{x}$: arithmetic mean value of a sample of n measurements

s : standard deviation of a sample of n measurements

σ : standard deviation for the distribution of a random variable

σ(est): an estimate of the standard deviation of the distribution of a whole population (infinite number) of measurements based on a sample of n measurements:

$$\sigma \text{ (est)} = s \sqrt{\frac{n}{n - 1}}$$

K : factor for a normal distribution with associated probability

t : Student's t factor for the difference between sample mean, $\bar{x}$, and population mean, μ, for a specified confidence probability

$$|t| = \frac{|\bar{x} - \mu|}{\sigma \text{(est)}} \sqrt{n}$$

U_r : value of uncertainty in a measurement result due to random (and measurable) effects for a specified confidence probability

σ_s : standard deviation for a systematic contribution to uncertainty in measurement having an assumed or given distribution of possible values

a : semi-range of a systematic contribution to uncertainty

U_s : value of uncertainty in a measurement result due to systematic effects for a specified confidence probability

U : single value of uncertainty in a measurement result obtained from combining U_r and U_s for a specified confidence probability

APPENDIX I - STUDENT'S t DISTRIBUTION

Values of t for specified confidence probability P as
a function of the number of measurements, n.

n \ P	0.500	0.683	0.950	0.955	0.990	0.997
2	1.000	1.84	12.7	14.0	–	–
3	0.817	1.32	4.30	4.53	9.92	–
4	0.765	1.20	3.18	3.31	5.84	9.22
5	0.741	1.14	2.78	2.87	4.60	6.62
6	0.727	1.11	2.57	2.65	4.03	5.51
7	0.718	1.09	2.45	2.52	3.71	4.90
8	0.711	1.08	2.36	2.43	3.50	4.53
9	0.706	1.07	2.31	2.37	3.36	4.28
10	0.703	1.06	2.26	2.32	3.25	4.09
11	0.700	1.05	2.23	2.28	3.17	3.96
12	0.697	1.05	2.20	2.25	3.11	3.85
13	0.695	1.04	2.18	2.23	3.05	3.76
14	0.694	1.04	2.16	2.21	3.01	3.69
15	0.692	1.04	2.14	2.20	2.98	3.64
16	0.691	1.03	2.13	2.18	2.95	3.59
17	0.690	1.03	2.12	2.17	2.92	3.54
18	0.689	1.03	2.11	2.16	2.90	3.51
19	0.688	1.03	2.10	2.15	2.88	3.48
20	0.688	1.03	2.09	2.14	2.86	3.45
∞	0.675	1.00	1.96	2.00	2.58	3.00

Values obtained from New expressions and some precise values
for Student's t factor by I.A. Harris and F.L. Warner.
Proc IEE Vol 125 p902-904 September 1978.

APPENDIX II - SOME SOURCES OF SYSTEMATIC UNCERTAINTY
IN ELECTRICAL MEASUREMENT

1.   Instrument Calibration

The uncertainties assigned to the values on a BCS
Certificate for the calibration of an instrument are
certified as being correct at the time and under the
conditions of calibration. The values are used to correct
the instrument indications and/or recorded values.

2.   Secular Stability

The performance of all instruments, whether measuring
equipment or reference standards, must be expected to drift
slowly with time. Thus the instrument user in seeking the
most probable values at the time of his measurements has to

assess the drift since the last calibration. This assessment will need to be based on the results of previous
calibrations. It is helpful to display the accumulated data
in graphical form for this purpose. The magnitude of the
drift in values compared with other sources of systematic
uncertainty will of course determine the required periodicity
of calibration.

3.   Value of a Quantity and Electromagnetic Frequency of Measurement

In considering a calibration for an instrument there
are always practical and economic factors which limit the
number of calibration points that can be provided. Thus
it becomes probable that the quantity to be measured and its
frequency, if it is an rf measurement, will be different
from the values for which a calibration has been provided.
When the quantity is between two calibration values, then
consideration needs to be given to any systematic uncertainty due, for example, to scale non-linearity, and range
switching. If the measurement frequency falls between two
calibration frequencies, it will also be necessary to
assess the additional interpolation uncertainty that this
can introduce. At high rf and microwave frequencies, one
can only proceed with a degree of confidence if

a)   a theory of the instrument is available to predict a
     characteristic for the interpolation between the
     calibration frequencies, or there is accumulated
     evidence of the performance of many other models of the
     same instrument; and

b)   the performance of the actual instrument being used has
     been explored with a swept frequency measurement system
     to verify the absence of effects due to manufacturing
     imperfections.

4.   Interconnection of Apparatus

The manner in which an instrument is connected to other
items of apparatus and the characteristics of these items
can significantly affect the performance of an instrument.
Thus, unless the method of use of an instrument is identical
to that used during calibration, additional contributions
to systematic uncertainty in the results obtained are likely.
The possible causes of such uncertainty are legion and their
successful identification and evaluation can prove difficult.
The classic approach under such circumstances is a planned
variation of the measurement conditions to assist detection.
In power and attenuation measurements at rf and microwave frequencies, mismatch uncertainty at the terminal
planes of an instrument that is being calibrated can make
a very important contribution to the total uncertainty in
the calibration. This is well known but to evaluate
mismatch uncertainty fully requires measurements of both the
magnitude and phase of the reflection coefficient.

Unfortunately many measurement systems only provide the modulus of a reflection coefficient and Harris and Warner have re-examined the nature of mismatch uncertainty (see Reference 6) taking this into account. Reference should be made to the work of Harris and Warner for a full treatment of this topic. In this appendix only the mismatch uncertainty for a simple power measurement, when phase angles are not measured, will be briefly considered.

The true power P_O measured by an absorption wattmeter is related to the **actual power** absorbed P_L by the equation

$$P_O = P_L \ \frac{|1 - \Gamma_G \Gamma_L|^2}{1 - |\Gamma_L|^2} \quad \dots\dots\dots\dots (13)$$

where P_O is the power that would be absorbed by a load equal to the transmission line characteristic impedance and Γ_G, Γ_L are the voltage reflection coefficients of the generator and the wattmeter.

The multiple reflection factor in the numerator of the above expression depends on the phase angles of the reflection coefficients. When these angles are not known the upper and lower limits of the true power are

$$P_{O_{\text{limits}}} = P_L \ \frac{\left[1 + |\Gamma_G|^2 |\Gamma_L|^2 \pm 2|\Gamma_G||\Gamma_L|\right]}{1 - |\Gamma_L|^2} \quad \dots (14)$$

and when Γ_G and Γ_L are small this becomes

$$P_{O_{\text{limits}}} = \frac{|1 \pm 2 \ |\Gamma_G||\Gamma_L||}{1 - |\Gamma_L|^2} \quad \dots\dots\dots\dots (15)$$

Although the actual phase angles of the reflection coefficients are not known, the multiple reflection factor $|1 - \Gamma_G \Gamma_L|^2$ can be rewritten as $[1 - 2|\Gamma_G||\Gamma_L|]$ cosϕ when Γ_G and Γ_L are small and ϕ is the relative phase of generator and wattmeter reflection coefficients. If any value of ϕ is equally probable, that is an assumption of a rectangular probability distribution, Harris and Warner show that the distribution for the mismatch uncertainty will be U-shaped and that the standard deviation will be a $/ \sqrt{2}$ where a $= 2|\Gamma_G||\Gamma_L|$.

5. Measurement (or Service) Conditions

If the laboratory measurement environment is different from that for a calibration, then due allowance has to be made for any influence conditions such as temperature, relative humidity, presence of objects in the measurement electromagnetic field. Also it is necessary to be aware of the possible effect of operating electrical conditions, such as power dissipation and applied voltage, differing from those for a calibration. Examples of the importance of such contributions to systematic uncertainty, although more common in dc and if measurements, do also occur in rf measurements.

APPENDIX III - STEP BY STEP PROCEDURES FOR THE
DETERMINATION OF MEASUREMENT UNCERTAINTY

a)   Estimated Confidence Level in Range 95% to 99%

1. Identify and list all corrections that have to be
    applied to measurements of a quantity.

2. Assign estimated semi-range limits, for which there
    is the utmost confidence, to the residual uncertainties
    in these corrections.

3. Assign a distribution form to each of the residual
    uncertainties and calculate the standard deviations
    (assume rectangular distributions if only limits can be
    estimated and refer to equation 5.

4. Choose the confidence level for the statement of
    measurement uncertainty and calculate the systematic
    uncertainty component U_s from equation 7.

5. Assign a good estimate for the standard deviation based
    on the results of previous repeated measurements or
    if this is not available make a minimum of 4 measure-
    ments to obtain an estimate from equation 2.

6. Repeat measurements to obtain a mean value for the
    quantity and apply all corrections (step 1). A mean
    value is, of course, available from step 5 when the
    standard deviation of the measurement process has not
    been previously determined.

7. For the confidence level of step 4, calculate the random
    uncertainty component U_r from equation 3. The value
    of t is selected from Appendix I for the number of
    measurements leading to the estimate of the standard
    deviation. The actual number of measurements that
    should be made will be determined by the relative
    importance of U_r to U_s in equations 8 or 9. A
    measurement should be repeated at least once however,
    to avoid operator mistake even when U_r is insignificant
    compared with U_s, as it can be in certain electrical
    measurements.

8. Calculate a single value for total uncertainty from
    equations 8 or 9.

9. Report the result of the measurement process as the
    corrected mean value that has an uncertainty which is
    qualified with an estimate of confidence level as in
    Section 7.

b)   Estimated Utmost Confidence Level

1.   As for steps 1, 2, 6, in a) above.

2.   Calculate the random uncertainty U_r as in step 7 in
a) above for a confidence probability of 99.7%, or
otherwise determine that it can be neglected.

3.   Calculate a single value for total uncertainty as U_r
plus the arithmetic sum of the semi-range limits
of the contributions to systematic uncertainty.

4.   Report the result of the measurement process as the
corrected mean value that has an uncertainty which is
qualified by the following statement of confidence.
'The uncertainty is based on an arithmetic summation of
the contributions'.

APPENDIX IV - EXAMPLES OF MEASUREMENTS AND INSTRUMENT
CALIBRATIONS

a)   Calibration of a Coaxial Power Meter at 10 mW and
9.4 GHz by Direct Substitution for a Working Standard Power
Meter

1.   Contributions to systematic uncertainty

	Semi range limits (%)
Working Standard Power Meter	
Calibration	1.0
Stability (between calibrations)	0.5
Frequency interpolation	0.2
Mismatch	
Generator vswr 1.06 (Γ = 0.031) Meter Head Unit vswr 1.10 (Γ = 0.048) 200 $\lvert\Gamma_G\rvert$ $\lvert\Gamma_L\rvert$	0.3
Unknown Power Meter	
Mismatch Generator vswr 1.06 (Γ = 0.031) Meter Head Unit vswr 1.34 (Γ = 0.145) 200 $\lvert\Gamma_G\rvert$ $\lvert\Gamma_L\rvert$	0.9
Arithmetic sum	2.9

2.   Systematic uncertainty. Assume rectangular
distributions for the calibration, stability and frequency
interpolation uncertainty contributions of standard. From
equation 24.5.

$$\sigma_{s_1} = 0.66\%$$

For the mismatch contributions, the probability distributions are U-shaped with a standard deviation equal to a/ $\sqrt{2}$

$$\sigma_{s_2} = 0.67\%$$

From equation 7 with K equal to 1.96

$$U_s \text{ (est CL: 95\% min)} = 1.84\%$$

3.   Random uncertainty. Repeatability of readings in mW of standard power meter (all corrections applied) for a reading of the unknown power meter of 10.00 mW. Measurement results:

    10.04; 10.07; 10.03; 10.06

Mean value = 10.05 mW

From equation 2 the estimated standard deviation = 0.018 mW. The random uncertainty, U_r is calculated from equation 3 for values of Student's t for the required confidence levels.

$$U_r \text{ (est CL: 99.7\%)} = 0.9\% \text{ (0.09 mW)}$$
$$U_r \text{ (est CL: 95.0\%)} = 0.3\% \text{ (0.03 mW)}$$

4.   Total uncertainty

    U(est CL: 95% min) = 1.87% (0.187 mW)

By arithmetic summation

    U = 3.8% (0.38 mW)

5.   Recommended expression of results. The incident rf power meter reading of 10.00 mW is

    10.05 mW ± 0.19 mW

The uncertainty is for an estimated confidence probability of not less than 95%.

b)   Measurement of VSWR of Nominal Values 1.0 and 4.0 at 3 GHz

    In this example it is assumed that only realistic semi-range limits can be assigned to all contributions to uncertainty. Included as limits is the random uncertainty revealed by repeating measurements

1.   Contributions to uncertainty

                                             VSWR = 1.0   VSWR = 4.0

a)   Errors in voltage reflection
      coefficient (vrc)

      Residual vrc of slotted line   0.03   0.03
      Line coupling variation   0.003   0.003
      Probe penetration loading   0.002   0.004
      Effects of signal source
        harmonics and slotted line
        attenuation established as
        negligible   -   -
      Repeatability of readings   0.005   0.005
                                          ————————   ————————

                    Arithmetic sums   0.040   0.042

b)   Errors in vswr indication

      Detector system deviation from
        square law   -   0.01

2.   Uncertainty in vrc. From equations 5 and 11.

      U(est CL: 95% min) = 0.035 for vswr values of
      1.0 and 4.0

3.   Total uncertainty in vswr. From the relation

$$VSWR = \frac{|VRC| + 1}{|VRC| - 1}$$

the uncertainty in vswr for an uncertainty in vrc of
0.035 is given in the table of BCS publication 4301
(reference 7) as

VSWR = 1.0:   U(est CL: 95% min) = 0.073
VSWR = 4.0:   U(est CL: 95% min) = 0.48

For comparison, from an arithmetic summation of
contributions to vrc

VSWR = 1.0:   U = 0.083
VSWR = 4.0:   U = 0.59

To obtain the total uncertainties for vswr of 4.0, a
detector system contribution of 0.01 must be added to the
above values.

4.   Recommended expression of results. The
measured values of vswr are

$$1.0 \begin{cases} +0.08 \\ -0.00 \end{cases}$$
$$4.0 \quad \pm 0.5$$

The uncertainties are for an estimated confidence probability of not less than 95%.

Note   The uncertainty for vswr equal to 1.0 is not expressed as a bilateral limits (±) as the ratio as defined cannot have a value less than unity. BCS publication 4301 shows that for values greater than 1.0 the limits are never in fact equal but for simplicity of expression and usage the larger limits value from the table is reported with suitable rounding.

c)   Measurement of a Fixed Attenuator of a Nominal Value of 30 dB at 10 GHz using a Power Ratio Method

In this example the signal source is modulated at, say, 1 kHz and the ratio of detected rf powers is determined by reference to a 1 kHz calibrated attenuator. As in b) above it is assumed that only realistic semi-range limits can be assigned to all contributions to uncertainty.

1.   Contributions to uncertainty, dB

Uncertainty and/or uncorrected errors in
reference attenuator   0.01
Resolution and noise of detector   0.01
Estimated maximum deviation of detector
system from a square law characteristic   0.01
Mismatch uncertainty (see Note below)
    vswr of source and load 1.05
    vswr of both ports of attenuator 1.08   0.02
                                                                        _____
                              Arithmetic sum   0.05

2.   Total uncertainty in attenuation, dB. From equations 5 and 11.

U(est CL: 95% min) = 0.03

3.   Recommended expression of result. The measured value of attenuation is

30 dB ± 0.03 dB

The uncertainty is for an estimated confidence probability of not less than 95%.

Note   The mismatch uncertainty limits for a two port fixed attenuator when phase relationships are not known is given in reference 9. Equation 2.71 on page 18 of this reference gives this mismatch uncertainty as

$$20 \log_{10} \frac{1 \pm \left[|\Gamma_G S_{11}| + |\Gamma_L S_{22}| + |\Gamma_G \Gamma_L S_{11} S_{22}| + |\Gamma_G \Gamma_L S_{12} S_{21}| \right]}{1 \mp \Gamma_G \Gamma_L}$$

where Γ_G is the source reflection coefficient

Γ_L is the load reflection coefficient

$S_{11}, S_{22}, S_{12}, S_{21}$ are the scattering parameters for the two port item under test.

APPENDIX V - GUIDE TO REPEATABILITY OF CONNECTOR PAIR INSERTION LOSS

     For some rf and microwave measurements the repeatability of the insertion loss of rf connectors can be a major contributor to measurement uncertainty. The assessment of this contribution involves many measurements with connector pairs and can readily be beyond the resources of a measurements laboratory and consequently reliable information is scarce. An American source of information on the performance of precision (sexless) and semi-precision coaxial connectors describes the results of repeated measurements on GPC-14, GPC-7, type N, and SMA connector pairs over the frequency range 2 to 18 GHz. It has been possible to supplement this information with the measurement experience of NPL RSRE and EQD on both coaxial and waveguide connectors and arrive at the limits for insertion loss repeatability shown below. The values given are considered to be safe limits provided it is assured that the connector pairs are in good mechanical condition and clean, and in use are not subjected to stresses and strains due to misalignment or transverse loads. The marked improvement in repeatability of using shims with WG 16 bolted flanges will be noted. It can be expected that the effect on performance of using shims for the larger waveguide sizes will be somewhat less.

a)   Coaxial Connectors

GR 900 - 14mm	DC - 0.5GHz 0.001 dB	0.5GHz - 4GHz 0.001 dB	4GHz - 8.5GHz 0.002 dB
APC 7 - 7mm[+] (Confidence level 95%)	2GHz 0.0002 dB	12GHz 0.007 dB	18.0GHz 0.015 dB
TYPE N - 7mm	DC - 1GHz 0.001 dB	1 GHz - 12GHz 0.002 dB	12GHz - 18GHz 0.004 dB
SMA APC 3.5 - 3.5mm	DC - 1GHz 0.001 dB	1 GHz - 12GHz 0.003 dB	

b)   Waveguide Connectors (Brass)

WG 11A   (dowelled flanges.   0.0005 dB
          metal shim gaskets)   (0.001%)

WG 15   (dowelled flanges,   0.0005 dB
          metal shim gaskets)   (0.01%)

WG 16   (square, fitted bolts,   0.001 dB
          metal shim gaskets)   (0.02%)

WG 16   (square, fitted bolts,   0.01 dB
          no gaskets)   (0.2%)

*Insertion Loss Repeatability Versus Life of Some Coaxial
Connectors by Dietrich Bergfield and Helmut Fischer.
IEE Trans. Instrumentation and Measurement - November 1970.

†Report on Repeatability of APC 7 Connectors by
J D Bradshaw and E.J. Griffin BCS Panel 2.2 Paper,
BCS/2.2/83/197 - September 1983.

Chapter 22

Who makes measurements and why?

C. H. Dix

PROLOGUE

It would be useful and desirable to people making measurements to be able to provide the same sort of cost/benefit
information that can be obtained from many other activities,
such as the introduction of new capital plant or making design
changes. For reasons that are well-known, and will be reinforced
by this paper, this is a difficult task, and would require a
great deal of study to do with any accuracy. What follows is an
attempt to find at least a lower limit for the cost, and to make
an estimate of the financial benefit at least for industry and
commerce. Only electrical measurement is considered here,
although the same approach could well be used for both wider or
more restricted fields of measurement.

1. WHO MAKES MEASUREMENTS, AND WHY?

The following is a list, doubtless incomplete, of some
measurement environments and purposes:

1.1 Electrical industries

In industry, which is taken as including both manufacturing
and the provision of services such as telecommunications,
transport, broadcasting etc., measurements are made of the
product, and of materials, components, sub-assemblies etc., and
of the performance of systems to ensure that the final product,
whether it is a precision instrument, an aircraft, a telephone
call or an air journey meets its specification or lives up to the
claims made in advertising, both of which are usually themselves
based on measurements made during the development. It also often
happens that potential customers are greatly influenced by
producer demonstrating that the process is carried out with
adequate traceability and uncertainty, and in the case of
tendering to many large organisations, who will have to take
responsibility for subsequent performance, purchases often will
not be considered unless this aspect of quality assurance is
provided. Various supervisory organisations such as NATLAS,
BSI, Weights and Measures Service, Civil Air Authority, etc.
exist to ensure that standards are maintained. In the present
context this situation applies principally to the supply of
products with some electrical content.

1.2 Other manufacturing industries

Over a much larger fraction of industry, the principal variables used in process control depend on measurement of flow and temperature and the behaviour of electrical or electronic control systems. In most such industries, e.g. oil-refining, chemical manufacture, food processing and ceramics, there is typically a large throughput of material, with relatively few staff. The ratio of output/staff may be much larger than in the case of technical products, but because of the size of these industries there are actually many thousands of people engaged on regular measurement to maintain the efficiency of the process control, and particularly in the case of temperature, this is usually electrical measurement.

1.3 Research & development establishments

In both research and development there is usually a much higher level of measurement activity than in production to ensure for example that environments or variables are accurately described, to establish dependence of quantities on each other and/or for many other widely varied reasons which can perhaps be summarised as the need to know. Although R & D is only a small fraction of national activity, it is important in this context because of this relatively large measurement content. Reference to Table 1 shows that the total measurement activity is largely made up of a quite small content in very large industries plus a relatively large content in smaller industries.

1.4 Armed services

A very large amount of electrical measurement is carried out in the armed services and MOD organisations, both in controlling the quality of equipment supplied by manufacturers, and to maintain the efficiency of the extremely wide range of equipment.

1.5 Regulating bodies

For the protection of the public in trading there are two laws, The Electricity Meters Act and the Weights and Measures Act. The Electricity Meters Act governs the measurement of the supply of electricity, and all the measurement activity which takes place in order to comply with it is carried out by the National Physical Laboratory, the CEGB and the Electricity Area Boards, who are supervised in this respect by the Department of Energy. The costs of all these are therefore included under these headings. Electrical measurement is also involved, however, in the activities of the Weights and Measures Service, which has to approve the design of a very large range of weighing and measuring apparatus, much of which is now electrical or electronic. Some other regulating bodies whose work involves electrical measurement are:

National Measurement Accreditation Service
Air Registration Board
Dept. of Environment
Health & Safety Executive
National Radiation Protection Board

1.6 National Physical Laboratory

Lastly, and probably smallest in total cost and numbers of
staff, there is the national measurement and standards laboratory
in UK, the National Physical Laboratory. A major part of the job
of NPL is developing, maintaining and improving the standards of
measurement of the United Kingdom, and ensuring that they conform
to the internationally agreed standards (principally formulated in
the Systeme International, or SI) and of participating in the
relevant international discussions and agreements. The costs of
the measurement activity at NPL can be quite accurately
established; the benefits are much more difficult to quantify.

A large fraction is undoubtedly reflected in the savings in
costs or improvement in the quality of measurement throughout the
whole of industry. Another important factor is the acceptability
of UK technical products throughout the world because of the
awareness of the measurement infra-structure which NPL provides.

In the absence of such quantified benefits, it is suggested
that a reasonable basis for the justification of this effort at
NPL (although this is not the purpose of this paper) is to compare
the expenditure on electrical measurement at NPL with that
throughout the whole of UK, and perhaps further to compare the
fraction of GNP that this expenditure represents with that of
other technologically successful countries.

2.   COSTS AND PROFITS OF MEASUREMENT IN INDUSTRY

2.1 Basis of estimates

In this treatment it is necessary in order to proceed to make
some broad-brush assumptions. It would be possible by very much
more work to replace these by better-based estimates, and hence to
improve the accuracy, however it is the author's belief that the
figures used lead to a lower limit, which may be below the
"true value" by a factor of several times, but which is unlikely
to be as much as ten times.
There does not seem to be any reason for treating the cost
of measurement any differently from any other component of
production costs. It can be presumed that no producer puts more
measurement (or any other expensive ingredient) into his
production process than he believes to be necessary, and so if
there is an overall profit, it is reasonable to regard the
appropriate fraction of this as being the profit arising from the
measurement. This is the approach which has been taken, and the
first broad-brush assumption that has been made is that industry
operates at a gross profit margin of 15%.
For each industry an estimate has been made of the effective
fraction of staff engaged on electrical measurement. This figure
has been arrived at from observation, discussion, and experience,
supplemented in some cases by correspondence. This is one aspect
whose accuracy could undoubtedly be improved by further study and
observation, however it must be appreciated (as became clear from
the correspondence) that the correct appropriate figure is
frequently unknown even by the relevant production manager.
The next "broad-brush" assumption is that the effective cost of
this effort is given by using an overheaded figure of £30,000 per

man year where the overheads arise from management and office support, measurement premises and instruments, and generally anything that would not be there if the measurements were not being made. The accuracy of this also could be improved by discussion and observation; the figure given is (intentionally) believed to be low.

2.2 Manufacturing industry

The figures of employment in the first column of Table 1 were obtained by selecting those industries in which electrical measurement has a significant role from the information in the Annual Abstract of Statistics (1984) produced by the Central Statistical Office, published by HMSO.

TABLE 1. Electrical measurement effort and costs in manufacturing industries

Industry	Employment (thousands)	Measurement fraction(%)	Measurement employment	Cost (M£)
Oil processing	24	0.3	72	2.2
Aluminium & Al.alloys	27	0.1	27	.8
Copper & Cu alloys	23	0.1	23	.7
Glass & Glassware	50	0.1	50	1.5
Ceramics	50	0.2	100	3.0
Chemical Industry	331	0.5	1655	49.7
Man-made fibres	15	0.2	30	.9
Refrig.machy., heating & ventilation eqpmt.	42	0.3	126	3.8
Insulated wires & cables	38	1	380	11.4
Batteries	92	2	1840	55.2
Basic elec.eqpmt.	118	2	2360	70.8
Telegraph & telephone eqpmt.	56	3	1680	50.4
Telecommunication eqpmt	200	3	6000	180.0
Radio & electronic capital goods	88	5	4400	132.0
Passive components	31	10	3100	93.0
Other elec. eqpmt	126	10	12600	378.0
Domestic appliances	43	0.1	43	1.3
Lighting eqpmt. and installation	25	0.2	50	1.5
Aerospace eqpmt	163	5	8150	244.5
Measuring instruments	103	10	10300	324.0
Food processing & manf., drink & tobacco	2109	1	21090	632.7
Totals	3759	(1.98)	74576	2237.3

The conclusions from Table 1 are that of roughly three and three-quarter million people working in manufacturing industries in which electrical measurement is significant, 74.6 thousand, or nearly 2% are effectively engaged full-time in

measurement, at a cost of M£ 2237.   This is approximately 2% of
the value of the products of these industries, and 1.6% of GNP.
From the first assumption made above, the gross profit which can
be ascribed to this activity is M£ 335.

As Table 1 is rather unsystematic, since its entries are in
the order in which they appear in the CSO Abstract, it is
summarised in Table 2, which perhaps enables one to make a better
assessment of the validity of the assumptions.

TABLE 2.   Summary of Table 1

Industry	Employment (thousands)	Measurement fraction(%)	Measurement effort	Cost (M£)
Electrical & Electronic products inc. components & sub-assemblies	157	10	15700	471.0
Measuring instruments	108	10	10800	324.0
Aerospace eqpmt.	163	5	8150	244.5
Telecommunication eqpmt.	256	3	7680	230.4
Radio & electronic capital goods	88	5	4400	132.0
Insulated wires & cables, Batteries, Basic electrical eqpmt, Domestic appliances, Lighting equipment	316	1.5	4673	140.2
Non-electrical products	2671	1.0	23173	695.2
Totals	3759		74576	2237.3

2.3   Non-manufacturing industries

The non-manufacturing industries which are considered here
are
1. Electricity Industry
2. Air Transport
3. British Telecom
4. Broadcasting Services
5. Other service industries

2.3.1 Electricity

a) Generation. The CEGB Report and Accounts 1984 gives the
following:
Total generated energy   212 728 GWh
No. of generating stations   90
Tech. & scientific staff   13063
Total staff   52250
Total staff cost   M£ 736

It seems reasonable to take a fraction of 10% of the technical
and scientific staff, i.e. 1306, as the effective effort on
measurement. Taking a pro rata fraction of the staff cost, this
gives M£ 18.4   Overheads are taken to raise this to at least M£30.

b) Distribution and Sales. The Electricity Council provided the following figures:

No.of meter testing stations   22
Staff at meter testing stations 1500
No.of meter readers   5000 (aaarox)

Neglecting any other effort on measurement, this gives a total of 6500 men, whose overheaded cost is taken as being M£100 .

2.3.2 Air Transport. The CSO Abstract gives total employment (1984) as 35,000, and total number of flights on UK registered aircraft as 327,000. This is taken to imply about 500 aircraft, and a figure of effectively 50 man years per year is taken as representing the appropriate electrical measurement effort, at a cost of M£ 1.5

2.3.3 British Telecom. The Post Office Report & Accounts (1980/1) gives:

	M£	
Maintenance costs	621	
Operating costs	237	totalling M£1059 out of
Management & support costs	201	total costs of M£3835

These are the activities in which measurement would be significant (other than R & D which is included in 2.) and a fraction of 5% is taken for electrical measurement, which is therefore M£53.

2.3.4 Broadcasting Services. The BBC Annual Report & Handbook 1984 lists approximately 730 television transmitters, 143 VHF radio transmitters and 126 Medium and Long-wave transmitters. Most television sites are shared, so a similar number of television transmitters is assumed for IBA. The BBC maintenance costs are given as M£9.1 for TV, and M£3.6 for radio, and M£2.4 for external services. It is assumed that half these costs arise from measurement, giving a total of M£7.5 for the BBC. IBA has far fewer radio transmitters, so a figure of M£5 is taken, giving a total of M£12.5 for this industry.

2.3.5 Other service industries. These include road and rail transport and maintenance services, police and fire services, water supply, radio and television maintenance and repair etc. There is undoubtedly a large amount of electrical measurement which is not included under any other heading. No figure is included for it; it is just one more omission that supports the notion that this is a lower-limit estimate.

3. RESEARCH AND DEVELOPMENT

Total figures for 1981 are given in Table 3, below, amounting to M£ 5903. If it is assumed that half of this involves electrical measurement, at a fraction of 10% of the time, the cost is approx. M£ 300.

TABLE 3.   Research & Development (1981)

Government	M£	Non.Government	M£
Defence	558	Universities	670
Research Councils	250	Public corporations	365
Other	527	Research assns.	88
Local Govt.	5	Private industry	3320
		Other	160
Total	1340	Total	4563

4.   ARMED SERVICES

The most extensive information readily available is in the Statement on the Defence Estimates Vol.1., published by HMSO 1985. If we exclude R & D (covered in section 3), the principal civilian expenditure is M£6338 on production of equipment. Bearing in mind the nature of military equipment and the rigid Q.A. requirements involved, a figure of at least M£50 for measurement appears reasonable.

The Statement also gives the number of direct civilian employees as 174,700. This includes the effort to maintain equipment, and if we assume that 1% of this is engaged on electrical measurement, at an overheaded cost of k£20 per man, we get a further M£350.

The total of Service personnel (1983/4) was 327 thousand, and clearly many of these in the technical arms of all three services carry out a vast amount of electrical measurement. However no published information was readily available, and no figure is included in the total which is therefore M£400.

5.   REGULATORY BODIES

These include bodies such as:

British Calibration Service
Weights and Measures Service
Civil Aviation Authority
Home Office Radio Establishment
British Standards Institute
British Electrical Approvals Board
Dept. of Energy Meters Inspectorate
National Radiological Protection Board
MOD Procurement Executive Q.A. Inspectorate

Most of these employ relatively few staff, whose job is principally to ensure that others make valid traceable measurements, rather than to make measurements themselves. As an example, BCS has three professional engineers in the electrical field, who are responsible for the operation of over 35 laboratories employing a staff of over 150. However the activity of the laboratory staff should have already been included in the industrial category (Section 2), and so only the 3 are properly included here.

An exception to the above comment is the case of MOD., where the numbers are very much greater, but it is assumed that these are included in the Defence contribution (Section 4).

A figure of 50 is taken for this category, with an overheaded rate of k£30, as in general they have a relatively high level of office support; the total cost is therefore put at M£1.5.

6. THE NATIONAL PHYSICAL LABORATORY

One of the principal jobs of NPL, which occupies more of the laboratory's total effort than any other, is the establishment, maintenance, dissemination and improvement of measurement standards, and all that this entails.

The Dept. of Trade and Industry Science and Technology Report 1983/4 indicates that of the M£322 spent by the Department on Science and Technology, M£19.5 went to NPL, and approximately one-third of this (M£6.2) to Electrical Metrology. The accuracy of all the measurement activity considered above is derived from and depends on this work, which is appropriately regarded as the bed-rock of the infra-structure which measurements provide to a technological society. The justification of this expenditure on a direct quantified cost/benefit basis is as difficult as (and in some ways similar to) that of justifying education. Observation shows that those countries which support adequate national metrological laboratories (e.g. Germany, U.S.A., Japan) enjoy greater success in technical fields than those which do not. Perhaps in this context it is most appropriate to regard the expenditure at NPL as a necessary overhead on the cost of all the measurement activity which is derived from it. On this basis, it represents an overhead of 0.2%.

7. SUMMARY AND CONCLUSIONS

Summarizing the previous Sections, we have in Table 4 a rough estimate of the total cost of electrical measurement in UK:

TABLE 4. Total costs of electrical measurement in UK.

	M£
Manufacturing industry	2237
Non-manufacturing industry	197
Research and development	300
Armed services	400
Regulatory bodies	2
National Physical Laboratory	6
Total	M£ 3142

The gross profit which can be ascribed to the first two entries in Table 4 is of the order of M£350. It may be noted that this will generate a tax income to the Treasure of the order of M£100, (which might be regarded as its return on the outlay of M£6 on NPL).

The expenditure on measurement in R & D should be regarded in the same way as measurement in industry, i.e. that it is included only when necessary, and profitable and justifiable to the same extent as the expenditure on any other aspect of the work.

The estimate of expenditure in the Armed Services, which is suspected to be unrealistically low is, like the Services themselves, necessary to preserve the defence of the realm. This, like continued existence, is not usually regarded as necessary to justify on a cost/benefit basis.

The overall conclusion that clearly emerges is that provided measurement is limited to what is necessary, it is unquestionably profitable, and the answer to the question "Does electrical measurement pay?" must be an unequivocal "Yes, it does!"

The more difficult question of how much is necessary, or rather how much is desirable to maximise either the profitability or the profit can only be determined by experience, observation, and analysis. It would be interesting to compare the output and measurement fraction (preferably using more accurate estimates) of the industries listed in Section 2 with the corresponding quantities in other countries. Unfortunately few reliable figures are available.

Signal flow graphs

F. L. Warner

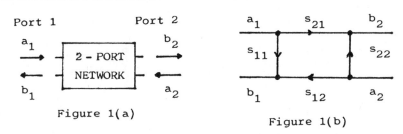

Figure 1(a)

Figure 1(b)

The characteristics of a two-port microwave network are now usually specified using s parameters. In fig. 1(a), a_1 and a_2 represent complex entering wave amplitudes while b_1 and b_2 represent complex outgoing wave amplitudes. When s parameters are employed, the relationships between these wave amplitudes are as follows:

$$b_1 = s_{11}a_1 + s_{12}a_2 \qquad (1)$$

$$b_2 = s_{21}a_1 + s_{22}a_2 \qquad (2)$$

When a matched load is connected to port 2, $a_2 = 0$ and when a matched load is connected to port 1, $a_1 = 0$.

Thus it follows from equations (1) and (2) that:

$s_{11} = b_1/a_1$ when port 2 is perfectly matched,

$s_{12} = b_1/a_2$ when port 1 is perfectly matched,

$s_{21} = b_2/a_1$ when port 2 is perfectly matched, and

$s_{22} = b_2/a_2$ when port 1 is perfectly matched.

In a signal flow graph, complex wave amplitudes such as a_1, a_2, b_1 and b_2 are represented by points or nodes and the s parameters are represented by directed lines. Fig. 1(b) shows the signal flow graph for the two-port network of fig. 1(a). The arrow directions are from the independent variables (a_1 or a_2) to the dependent variables (b_1 or b_2). The value of a node is the sum of all signals entering it, each signal being the value of the node from which it comes multiplied by the path coefficient; for example, in fig.1(b) $b_1 = s_{11}a_1 + s_{12}a_2$, in accordance with equation (1).

Figure 2 shows the signal flow graphs and the associated equations for:

(a) an imperfectly matched generator with a reflection coefficient Γ_g,

(b) a termination with a reflection coefficient Γ_L, and

(c) a detector with a reflection coefficient Γ_d. In this case, K is a factor relating the incoming wave amplitude, a, to a meter reading M. The value of K will depend to some extent upon Γ_d.

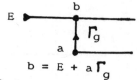

$$b = E + a\,\Gamma_g$$

(a) Generator

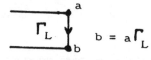

$$b = a\,\Gamma_L$$

(b) Termination

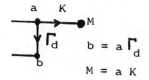

$$b = a\,\Gamma_d$$
$$M = a\,K$$

(c) Detector

Fig. 2:
Signal flow graphs.

When several microwave networks are cascaded, the signal flow graphs of the individual networks can be joined together and the ratio between any two complex wave amplitudes can then be written down straight away from the overall flow graph using Mason's non-touching loop rule, which will be discussed next.

In a signal flow graph, the following rules apply:

RULE 1.   The value of a path is the product of all path coefficients encountered en route.   A path must always follow the directions of the arrows and no node may be passed more than once.   In figure 3, the value of one path from a_1 to b_1 is equal to s_{11} and the value of the second path from a_1 to b_1 is equal to $s_{21}\Gamma_L s_{12}$.

RULE 2.   A first order loop is a closed path which can be followed, always in the direction of the arrows, without passing any node more than once. Its value is the product of all path coefficients encountered en route.   In figure 3, there are three first order loops. Their values are seen to be:

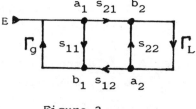

Figure 3

$$\Gamma_g s_{11},\ s_{22}\Gamma_L \text{ and } \Gamma_g s_{21}\Gamma_L s_{12}.$$

RULE 3.   A second order loop is the product of the values of any two first order loops which do not touch at any point. In figure 3, there is one second order loop whose value is:

$$\Gamma_g s_{11} s_{22}\Gamma_L.$$

RULE 4.   A third order loop is the product of any three non-touching first order loops, etc.

RULE 5.   The ratio T of the complex wave amplitude at point Y to that at an independent point X is given by:

$$T = \frac{P_1(1 - \Sigma^1 L_1 + \Sigma^1 L_2 - \cdots) + P_2(1 - \Sigma^2 L_1 + \Sigma^2 L_2 \cdots) + \cdots}{(1 - \Sigma L_1 + \Sigma L_2 - \cdots)}$$

(3)

where P_1 is one path from X to Y,

P_2 is a different path from X to Y, etc,

ΣL_1 is the sum of all first order loops,

ΣL_2 is the sum of all second order loops, etc,

$\Sigma^1 L_1$ is the sum of all 1st order loops not touching P_1

$\Sigma^1 L_2$ is the sum of all 2nd order loops not touching P_1

$\Sigma^2 L_1$ is the sum of all 1st order loops not touching P_2

$\Sigma^2 L_2$ is the sum of all 2nd order loops not touching P_2

etc.

Equation (3) is Mason's non-touching loop rule.   A proof has been given by Lorens (1).

Various rules exist for simplifying signal flow graphs and an example will now be given.

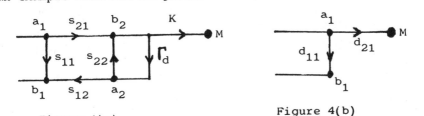

Figure 4(a)

Figure 4(b)

The signal flow graph for a two-port network followed by a detector is shown in fig. 4(a).   In this diagram there is one first order loop, $s_{22}\Gamma_d$, and no higher order loops. Using equation (3), we find:

$$\frac{M}{a_1} = \frac{s_{21}K}{1 - s_{22}\Gamma_d}$$

(4)

$$\frac{b_1}{a_1} = s_{11} + \frac{s_{21}\Gamma_d s_{12}}{1 - s_{22}\Gamma_d}$$

(5)

Thus the signal flow graph shown in figure 4(a) can be replaced by the much simpler one in figure 4(b), where

$$d_{21} = s_{21}K/(1 - s_{22}\Gamma_d)$$

(6)

and $$d_{11} = s_{11} + s_{21}\Gamma_d s_{12}/(1 - s_{22}\Gamma_d)$$

(7)

Reference: 1. Lorens, C.S.: MIT Research Laboratory of Electronics, Q. Prog. Rept, 1956, pp 97 - 102.

Index